Galactic Nebulae and Interstellar Matter

JEAN DUFAY

Late Director of the Lyon and Haute-Provence Observatories

Translated by

A. J. POMERANS

DOVER PUBLICATIONS, INC., NEW YORK

Published in Canada by General Publishing Company, Ltd., 30 Lesmill Road, Don Mills, Toronto, Ontario.

Published in the United Kingdom by Constable and Company, Ltd., 10 Orange Street, London WC 2.

This Dover edition, first published in 1968, is an unabridged republication of A. J. Pomerans' translation of *Nébuleuses Galactiques et Matière Interstellaire,* originally published by Hutchinson & Co. in 1957, to which has been added a brief note by the author's son, Professor Maurice Dufay. New photographs, supplied by the author, have been used for plates I, II, IV, XVIII, XIX and XXIII. This edition is published by special arrangement with Hutchinson & Co.

Library of Congress Catalog Card Number: 68-23803

Manufactured in the United States of America
Dover Publications, Inc.
180 Varick Street
New York, N. Y. 10014

CONTENTS

Part Three

FROM ATOMS TO GRAINS AND FROM GRAINS TO STARS

Part Four

DIFFUSE MATTER OUTSIDE THE MILKY WAY

ILLUSTRATIONS

NOTE TO THE DOVER EDITION

JEAN DUFAY died on November 6, 1967. Before his death he authorized Dover Publications to reprint the English translation of *Galactic Nebulae and Interstellar Matter*, though he felt that certain chapters needed to be brought up to date. He had intended to mention in a new preface the most important works published in the field since the appearance of the first French edition in 1954, but died before he was able to write it.

Readers should refer to the latest volumes of the *Transactions of the International Astronomical Union* (Academic Press, London and New York) for abstracts and bibliographical information on recent books and articles on galactic nebulae and interstellar matter. They can also consult the draft reports of Commissions 33 and 34.

Maurice Dufay

Villeurbanne,
January 25, 1968

TRANSLATOR'S NOTE

M. Dufay has here assembled a very large amount of essential information on recent research into the problems associated with the existence of interstellar matter. To do so it has been necessary to describe many different techniques of observation ranging from spectroscopy to electron optics, and to use many concepts from widely different branches of physics, such as the quantum theory of radiation, the optics of small solid particles, etc.

It has accordingly not always been possible to develop a particular topic in historical order, nor yet to treat a particular question in comparative isolation. I have therefore endeavoured to give in the following preface a short account of the scheme of the book. In it I have borrowed freely from M. Dufay's own preface to the French edition, but I should like to emphasize that none of its deficiencies should be attributed to him. For the present English edition, M. Dufay has supplied revisions of certain sections which are now (1955) up to date.

Conventional astronomical and physical notations have been followed except in the cases of parsec and kiloparsec which, for brevity, have usually been retained as pc. and kpc. respectively. The footnotes in the original French edition are often very long, and have therefore been placed at the ends of the chapters for the reader's convenience.

Finally, I should like to express my indebtedness to Dr. A. Klug for his invaluable assistance, and to Dr. M. Davidson, F.R.A.S., for his most helpful suggestions with the manuscript. I should also like to thank Miss S. Senior for her untiring efforts in preparing the typescript, and to express my acknowledgement to Mr. W. H. Johnson of Hutchinson's for his encouragement and co-operation.

A. J. P.

PREFACE

It is now known that between the stars of our Galaxy there are present enormous quantities of diffuse matter, both in the form of highly rarefied gases and of fine dust. In fact, this interstellar matter constitutes as much as half the total mass of the Galaxy, but it reveals its presence only by its effects on the light of the stars. It is likely that stars are continually being formed out of it by a process of condensation, first into grains and then into larger masses. It also appears that interstellar matter is present in the spiral arms of the external galaxies, with which our Galaxy is comparable.

The study of interstellar matter has thus become of great importance in astrophysics; this book attempts to give an account of the many different observational techniques and theoretical methods involved in such a study and to trace the history of how the evidence of its existence has been built up, mainly in the last thirty years.

By comparison with what we know today, twenty-five years ago our knowledge of gaseous nebulae was very meagre, and we hardly ever took into consideration the existence of interstellar matter. It is really only since 1927 that progress in theoretical and experimental spectroscopy has led to the identification of the principal bright lines of the gaseous nebulae, and it was not till 1930 that the work of Schalén and Trumpler finally established, beyond all possible doubt, the existence of a general absorption of light in the neighbourhood of the plane of the Galaxy. Since then, research on nebulae and diffuse matter in space has developed with astonishing rapidity.

The development of radio-astronomy, and the discovery that decimetre waves are emitted by the hydrogen atoms of space, have resulted in a considerable modification of our ideas on the structure of the universe. The implications of both this research and the new study of the dynamics of interstellar space, are here adequately expounded.

This book is intended not only for the professional astronomer, but also for those generally interested in this subject, fascinating for its own sake, and also as a field of active research. In the first part the types of spectra produced by the atoms and molecules in space are classified and explained, and there is an interpretation of the "forbidden" lines, which are the most intense and most characteristic lines of nebular spectra.

Apart from the brightest gaseous nebulae, certain regions of the Milky Way contain ions of hydrogen, oxygen and sulphur, which may also give rise to emission lines in the visible and ultra-violet; the relevant observations and an appropriate theory are described in Chapter IV. Here is also described the radio-frequency emission arising from neutral hydrogen atoms, which now enables us to study the rotation of the internal parts of the Milky Way.

The second part of the book is devoted to the solid particles of space. Here we study the dark nebulae and review the statistical methods based on star counts, from which the total absorption and distance of the clouds may be found. The notions of differential absorption and colour excess are dealt with particularly in Chapter IX.

The general absorption outside the great clouds is considered in Chapter VII, which also discusses Schalén's and Trumpler's demonstration of an absorbing layer in the Galactic plane. The relation of differential absorption to total absorption is covered in Chapter VIII, and an account of the curve of absorption against wave-length, and in particular, its extension into the infra-red is given in Chapter IX. The general scattering of light in the Milky Way is dealt with in Chapter XI, but the study of its polarization is deferred until Chapter XV.

In the third part of the book there are put forward the modern ideas concerning the co-existence and equilibrium of atoms and of dust within the same clouds, and the physical conditions prevailing in interstellar space are discussed, but the density of interstellar matter is more fully treated in Chapter XVII. The formation of supergiant stars by the condensation of clouds is discussed in Chapter XVI, and this is followed by a theory of stellar dynamics.

In the fourth part the behaviour of diffuse matter in the external galaxies is examined, and finally the problem of intergalactic absorption is discussed, and there is an account of Stebbins and Whitford's recent observations on the excess reddening in distant galaxies.

M. Dufay concludes his own preface with the following remarks:

> "I hope that, in spite of its elementary character, and although it omits some recent points of view, this work may render some service to both the astronomer and the student. It is for this reason that I have added a bibliography which, though incomplete, is sufficient to lead the reader to a closer study of those questions which may interest him.
>
> "I wish to express my appreciation to Mr. Carl Schalén whose advice has been most useful to me in all that concerns interstellar absorption, and to Mr. Vladimir Kourganoff, to whom I am indebted for many useful suggestions.

"All the photographs reproduced in this book have been obtained at the Observatories of Haute-Provence and of Lyon, and also by M. de Kerolyr, at the Astrophysic Department of the Paris Observatory, which was at Forcalquier previous to the foundation of the Haute-Provence Observatory. I thank M. Danjon, Director of the Paris Observatory, for permitting me to reproduce their photographs, which are largely the work of Messrs. Mevolhon, Senouque, Gueriau and Texereau."

Saint-Michel Observatory,
March 1953.

INTRODUCTION

SOME FUNDAMENTAL IDEAS OF ASTROPHYSICS

WE characterize the radiation received from a star by its *stellar luminosity*; i.e. the illumination which it produces on a plane normal to the rays of light, otherwise called the flux which falls on unit surface. In the case of stars whose apparent diameters are too small to be measured even by the largest instruments, this luminosity is the only magnitude directly accessible to photometric measurements.

Stellar luminosities and magnitudes

Astronomers evaluate stellar luminosities on a logarithmic scale. By definition, the *difference of magnitude*, $m_2 - m_1$, of two stars of luminosity l_1 and l_2 is given by

$$m_2 - m_1 = 2{\cdot}5 \log \left(\frac{l_1}{l_2}\right) \qquad (0.1)$$

Thus, a given difference of magnitude corresponds to a certain luminosity *ratio*. This ratio is 2·512, 100 or 10,000 when the difference of magnitude is -1, -5 or -10 respectively. The scale of magnitudes was established arbitrarily with a view to retaining as nearly as possible the differences in magnitude of the ancient observers who, long before any photometric measurement, divided those stars visible to the naked eye into six classes.

Let us first suppose that the measurements are made by isolating radiation of determined wave-length. For monochromatic radiation the difference in magnitude has a well-defined meaning, altogether independent of the detector of radiation used. Naturally, if the energy is not identically distributed in the spectra of the two stars, the difference in magnitude will vary with wave-length.

Most often, however, measurements are made over an extended spectral interval, and if the distribution of energy differs in the spectra of the two stars compared, the sensitivity curve of the detector must be taken into account, at least if we use a thermal detector, equally sensitive to all radiation (bolometer, radiometer or thermo-electric couple). The difference of magnitude found is not the same in the case of visual measurements, photographic measurements or photo-electric measurements.

Thus, to each type of selective detector there corresponds a particular system of magnitudes. So far we have made use of *visual* magnitudes (m_V), *photographic* magnitudes (m_{pg}) and *photovisual* magnitudes (m_{pv}) but, more and more frequently, photoelectric cells are being used which, combined with colour filters, permit the construction of detectors of varied sensitivities. Photographic magnitudes are determined on so-called "ordinary" plates, practically insensitive to radiation of wave-lengths greater than 0·5 micron. The measurements are thus essentially made on the blue and violet regions of the spectrum and at the beginning of the ultra-violet. Photovisual magnitudes are determined on so-called "orthochromatic" plates whose sensitivity extends up to almost $0{\cdot}59\mu$, and in front of which is placed a yellow filter, eliminating radiation of wavelengths less than $0{\cdot}5\mu$ approximately. The sensitivity curve of the plate and filter combination reaches a maximum towards $0{\cdot}55\mu$, and is thus similar to that of the eye, although its maximum is sharper.

To relate each stellar luminosity with a given magnitude, it is necessary to fix the magnitude of a certain star arbitrarily taken as standard. The convention of the Harvard visual magnitude system assigns the magnitude +6.55 to the star λ Ursae Minoris, near the N. celestial pole and which, in consequence, can be observed throughout the year at almost the same height above the horizon. On the other hand, to fix the zero of the scale of photographic magnitudes, it was decided that those hydrogen stars of the type A0 (see page 26, Table I) of the sixth magnitude ought to be assigned the same visual and photographic magnitudes. In practice, the international system of photographic magnitudes follows the *polar sequence*: this is a sequence of stars near the north pole, whose magnitudes range between the second and the twentieth and which have been very carefully compared among themselves. In the same way, the international system of photovisual magnitudes is graduated up to the seventeenth magnitude, according to the photovisual magnitudes of the stars of the polar sequence.

The principal difficulties encountered in astronomical photometry are due to atmospheric absorption. Ozone completely absorbs ultra-violet radiation of wave-lengths less than $0{\cdot}3\mu$; of the other gases, oxygen and water vapour in particular, also have troublesome absorption bands in the visible region and even more so, in the infra-red. Scattering by gas molecules and atmospheric dust, however, weakens radiation in quite different proportions. For a given wave-length, absorption varies with the zenith distance of the stars, the altitude of the place of observation, and the meteorological conditions; it sometimes undergoes notable changes within a short interval of time. Also, it demands special precautions of astronomical photometry and often, specialized methods of measurements. Without insisting on these, we shall assume from here on, that all the magnitudes we speak of have been corrected for atmospheric absorption.

Stars with a measurable apparent diameter

Let us recall that we call the *brightness* of a source its luminous intensity per unit apparent surface. Let J be the luminous intensity of the source whose surface has a projection S on a plane normal to the light rays (S=apparent surface). The average brightness B of the source, in the direction considered, is thus defined by the relation

$$J=BS$$

Further, we know that the illumination produced on a plane normal to the rays, at a distance r from the source, is

$$l=\frac{BS}{r^2} \tag{0.2}$$

This equation simply expresses the conservation of luminous flux; it is thus not valid except in the absence of all absorption. But the quotient $\frac{S}{r^2}$ measures the solid *angle* subtended by the source at a distance r, so that equation (0.2) can be written:

$$l=B\omega \tag{0.3}$$

From this results a new definition of brightness, equivalent to the first: namely *the stellar luminosity per unit solid angle*. It is particularly useful in astrophysics, for it presents the advantage of requiring neither dimensions nor the distance of the source, which are often uncertain. It also permits of an unambiguous definition of the brightness of a fictitious source, e.g. the celestial vault.

In particular, if we consider a circular star, of (linear) diameter D and of apparent (angular) diameter α, we have

$$\alpha=\frac{D}{r} \quad \text{and} \quad \omega=\frac{\pi D^2}{4r^2}=\frac{\pi\alpha^2}{4} \tag{0.4}$$

whence:

$$l=\frac{\pi B\alpha^2}{4} \tag{0.5}$$

The equations (0.3) and (0.5) can be applied to any source whatever, but we rarely make use of them in the case of stars whose extremely small apparent diameter is generally unknown. They are, on the other hand, very useful in the case of nebulae. In their case, the stellar luminosity l is then no longer the only quantity accessible to photometric measurements: these can equally be made on the brightness B.

Let the total magnitude m correspond to the stellar luminosity l, the magnitude μ to the brightness B. The definition of magnitudes (0.1), gives

$$m-\mu=-2{\cdot}5 \log \left(\frac{l}{B}\right)=-2{\cdot}5 \log \omega. \tag{0.6}$$

The usual solid angle of the physicists—the *steradian*—is much too large for most astrophysical purposes (the entire celestial sphere corresponds to only four steradians), rather do we employ as the astrophysical unit the *square degree*: the solid angle contained at the interior of a pyramid of square base, each of whose faces makes an angle of 1°. One steradian contains $\left(\frac{180}{\pi}\right)^2=3282{\cdot}79$ square degrees and there are approximately 41253 square degrees in the celestial sphere. We also frequently make use of the *square minute* (1/3600 square degree) and of the *square second* (1/3600 square minute), defined in an analagous manner.

In the case of a circular star of apparent diameter α, the relation (0.6) becomes

$$m=\mu-5 \log \alpha-\log \left(\frac{\pi}{4}\right)=\mu-5 \log \alpha+0{\cdot}26 \tag{0.7}$$

The magnitude μ here corresponds to one steradian, one square degree, one square minute, or one square second, according to whether α is expressed in radians, in degrees, in minutes or in seconds of arc.

Distances, luminous intensities and absolute magnitudes

The measurement of the distance of stars is based on that of their *annual parallax*. Let p_m be the maximum value of the angle p which the radius R of the terrestrial orbit would subtend, if seen from the star in question. (This orbit may be considered circular, without affecting the calculations.) As a result of the orbital movement of the Earth around the Sun, the angle p varies in the course of the year, and is at a maximum value p_m twice yearly. The parallax expressed in radians, is

$$p_m=\frac{R}{r}$$

where r denotes the distance of the star.

The radius R is taken as the astronomical unit (A.U.) of length for the solar system. It is $1{\cdot}496.10^{13}$ cm. (almost 150 million km.). But this unit is much too small to be convenient in the evaluation of the distance of the stars. We therefore, take as unit the distance of a star whose annual parallax p_m would be exactly one second of arc; this is called the parsec (pc). Between the distances r, expressed in parsecs, and the parallax p_m, expressed in seconds of arc, we have the relation

$$r=\frac{1}{p_m}$$

Since 1 radian is $\frac{180\times 60\times 60}{\pi}=2{\cdot}063.10^5$ seconds of arc, 1 pc$=2{\cdot}063\times 10^5$ A.U.$=2{\cdot}063.10^5\times 1{\cdot}496.10^{13}=3{\cdot}086\times 10^{18}$ cm. (more than 30,000 milliards km.).[1]

The study of the small apparent displacements from their average positions which certain stars show annually permits the determination of their annual parallax p_{m}, and thus of their distance r. As we also know their stellar luminosity l, we can evaluate their luminous intensity J by means of the fundamental relation

$$l=\frac{J}{r^2} \tag{0.8}$$

which accounts for the conservation of luminous flux. If, for example, l were expressed in *lux* (lumen per square meter), and r in metres, we would get the luminous intensities J in candles. But the distances of the stars in metres and their luminous intensities in candles are represented by extremely unwieldy numbers, and astronomers justly have a horror of so-called "astronomical" figures. On the other hand, the evaluation of stellar luminosities in lux requires the translation of visual magnitudes into the usual photometric units and introduces a supplementary uncertainty. It is thus preferred to compare the luminous intensities of the stars directly. If these were all situated at the same distance from the observer, their stellar luminosities would evidently be proportional to their intensities. We take, as the measure of intensity, the stellar luminosities or the magnitudes which would be observed if all the stars were at the same fixed distance. The distance of 10 pc. was arbitrarily chosen (there is no star at a distance of 1 pc. from the Sun, but there are several at less than 10 pc.). The magnitude converted into the fixed distance of ten pc is called the *absolute magnitude* (M). Conversely, magnitudes measured directly are called *apparent magnitudes* (m).

It is quite easy to establish a relation connecting the absolute magnitude, the apparent magnitude and the distance. The stellar luminosities l_{r} and l_{10} measured at distances of r and of ten parsecs respectively are, according to (0.8), inversely proportional to the square of the distances

$$\frac{l_{\mathrm{r}}}{l_{10}}=\frac{10^2}{r^2}$$

By taking logarithms of both sides, this relation becomes

$$\log\left(\frac{l_{\mathrm{r}}}{l_{10}}\right)=2-2\log r$$

and, according to (0.1), the difference of magnitude $m-M$ (called the *modulus of distance*) is expressed

$$m-M=-2{\cdot}5\log\left(\frac{l_{\mathrm{r}}}{l_{10}}\right)=5\log r-5 \tag{0.9}$$

Naturally, this expression, which assumes the conservation of luminous flux, would not be exact if there existed an absorbing medium between the star and the observer.

By the so-called trigonometrical method, based on the study of the small annual displacements of the stars, the parallaxes of some thousands of stars situated at least 100 pc. from the Sun, have been determined and indirect methods have led to the evaluation of the distances of even more distant stars. Equation (0.9) permits the calculation of their absolute magnitudes.

To the apparent visual, photographic and photovisual magnitudes there naturally correspond absolute visual (M_v), absolute photographic (M_{pg}) and absolute photovisual (M_{pv}) magnitudes. A particular interest is attached to the so-called *bolometric* absolute magnitudes (M_{bol}) obtained by measurements made with a thermal detector, and corrected for absorption by the terrestrial atmosphere. Bearing in mind the sensitivity curve of the eye and the transparency of the atmosphere, correction tables were drawn up which permit us to translate visual magnitudes into bolometric magnitudes (Eddington, Kuiper).

Classification of the stellar spectra

Just like the Sun, almost all the stars show continuous spectra crossed by absorption lines. Some also show emission lines on the continuous background, with or without dark lines. But the nature and the intensity of the lines are extremely variable.

With some slight modifications made from time to time, the actual classification of stellar spectra is essentially that of the Harvard Observatory. It is mainly based on the relative intensity of the lines.

The stars are placed into a dozen classes, each of which is quite arbitrarily denoted by a capital letter. The first ten classes, in decreasing order of excitation and temperature, are written:

```
           C   S
           /   /
W—O—B—A—F—G—K—M
```

The bifurcations from the classes G and K signify that the C stars seem to be directly attached to the G stars and that both the M and S stars can be attached to the class K.

Two other classes Q and P are not included. The first is reserved for temporary stars or *novae*, whose characteristics are very specialized, the second to gaseous nebulae which we shall study in the following chapters.

In principle, each spectral class is, in its turn, sub-divided into ten groups called the *spectral types* and each one of them is designated by the numbers 0, 1, 2, ... 9, placed after the capital letter of the class (but all these sub-divisions are not currently used). Generally speaking, we were forced to employ an even gradation between successive types: a spectrum of type B9, for instance, is almost half-way between those of types B8 and A0.

The most characteristic lines of the various spectral classes are mentioned in the second column of Table I, on which we shall now comment very briefly. First, it must be stressed that the lines considered as characteristic of a given class can also be met in other classes, but generally with less intensity. Thus hydrogen lines, preponderant in the A class, are found as absorption lines in all the classes from O onwards, and as emission lines already in the class W. The lines of neutral helium characterize the B stars, but they also exist as emission lines in the W stars, as absorption lines in the O stars side by side with ionized helium lines, which disappear from the first B types onwards. The CN bands, intense in the C class, are not shown from the F class onwards and are already very strong with certain G stars. The bands of titanium oxide are dominant in the class M, but they accompany those of zirconium oxide in the S class, etc.

The Wolf-Rayet stars, which constitute the class W, have a very peculiar spectrum without absorption lines, but with strong and very often large, emission lines. Although intermediary cases exist, they can be grouped roughly into two parallel series: in the one (WC) side by side with the He^+ lines multiple-ionized carbon and oxygen lines dominate, and in the other (WN), those of multiple-ionized nitrogen.

The O and B stars characterized respectively by absorption lines of the He^+ ion and of the neutral He atom, also show hydrogen absorption and sometimes hydrogen emission lines. In them we also find much weaker absorption lines belonging to the once- or multiple-ionized metaloids. The first lines of metallic ions appear in the B class; the line K 3933·7 A of Ca^+ in particular, is clearly visible from the type B5 onwards. They are reinforced in the A class, whereas the helium rays disappear after those of O^+, N^+, etc.

Whereas the B and A stars are very white, those of the classes F, G, K, appear to us as more and more yellow. The spectra of F stars strongly resemble those of the A stars, but with hydrogen lines weaker and a greater number of metallic lines. In the class G, to which our Sun belongs, the abundance of metallic lines, mainly in the neutral state, renders the spectrum more and more complex; the H (3968·5 A) and K (3933·7 A) lines of ionized calcium are by far the most intense. The yellow colour is accentuated in the K class, where the continuous spectrum is rapidly weakened towards the violet, while the metallic lines become more and more numerous. H and K are very large and very strong, but the resonance line of neutral calcium 4226·7 A becomes prominent.

Finally, in the following classes, molecular bands are very apparent and play a preponderant role in the classification: bands of TiO and ZrO, shaded off towards the red in the classes M and S, bands of C_2 and CN shaded off towards the violet in the class C. All these stars are more or less red and many among them present variations which are of long period, or else irregular.

The great variety of spectra does not imply any difference of chemical composition in the stellar atmosphere (save perhaps in some exceptional cases). The absence of helium lines, for instance, starting from the A class does not mean that the atmosphere no longer contains any helium, but that the physical conditions necessary for the production of these lines are not realized (insufficient excitation energy).

Generally speaking, the greater the energy necessary for the excitation of the line of an element, the sooner is the maximum intensity of its absorption lines observed in the spectral series. Where the same element can appear in the neutral and in the ionized states, we first meet lines of the ionized atom, then those of the neutral atom, as we run through the spectral series from the class O to the class M (He^+ before He, Ca^+ before Ca, etc.).

In the case of the Wolf-Rayet stars, the excitation there is still larger than in the absorption-line O stars. Emission of rays is in effect produced by recombination of ions and electrons. The presence of absorption lines of ionized helium only shows that in the atmosphere of the O stars many He^+ ions are capable of being excited. The presence of the same emission lines in the atmosphere of the W stars, however, requires the presence of a great number of twice ionized helium atoms (He^{++}).

Thus, from the class W to the class M the excitation diminishes gradually all along the spectral series.

TABLE I

Spectral Classification

Classes	*Characteristic Lines*	*Types*	*Colour Temperatures* c_T *(K°)*	*Colour Indices C.I.*
W { WC	Emission He^+, He, H C^{+++}, C^{++}, O^{++++}, O^{++}			
W { WN	. . . . He^+, He, H N^{++++}, N^{+++}, N^{++}			
O	Absorption He^+			
B	. . . . . He	B0	28,000	−0.29 magn.
		B5	18,000	−0.17
A	. . . . . H	A0	16,500	0.00
F	. . . . . H and metallic lines	F0	8,900	+0.33
G	. . . . . metallic lines	G0	6,000	+0.70
K	. . . . . metallic lines	K0	4,500	+1.12
M	. . . . . bands of TiO	M0	3,500	+1.66
S	. . . . . bands of ZrO			
C	. . . . . bands of C_2 and CN			

The colour-temperatures of stars

The distribution of energy in the continuous spectra of stars varies much with their spectral type. To study it, we must compare wave-length by wave-length the radiation coming from a star with that of a terrestrial standard source. We generally employ photographic or photoelectric spectrophotometry. Atmospheric absorption makes measurement a difficult matter since it affects the radiation of the star and not that of the terrestrial source. Since it varies with wave-length, it deforms the energy curve of the stars, and as it changes from one night to the next, it must be carefully studied in the course of each night, to correct the observations.

Physicists usually compare spectral energy curves with those of a *black body*, that is to say, of a body that would absorb all radiation perfectly. The energy curves of the black body can easily be constructed for all values of temperature. The energy dW_λ radiated by unit surface of the black body in all directions, in a small interval of wave-length $d\lambda$, is proportional to the length of the interval $d\lambda$ and can be written

$$dW_\lambda = \mathbf{R}_\lambda d\lambda.$$

$\mathbf{R}_\lambda$ is, by definition, the *spectral radiance* of the black body for the wave-length λ at the temperature considered. It is given, as a function of the absolute temperature T (temperature Centigrade $+273°$), by Planck's now classical formula

$$\mathbf{R}_\lambda = 2\pi hc^2\lambda^{-5}\left(e^{hc/k\lambda T} - 1\right)^{-1} \tag{0.10}$$

where e is the modulus of the Napierian logarithms ($e=2{\cdot}718$), c is the velocity of light in vacuo; h and k represent Planck's and Boltzmann's constants respectively. In the C.G.S. system we have:

$$c=2{\cdot}9977.10^{10} \text{ cm. sec}^{-1},\ h=6{\cdot}62.10^{-27},\ k=1{\cdot}380.10^{-16}.$$

λ is expressed in cms, $\mathbf{R}_\lambda$ in ergs per cm^2 per sec.

For practical purposes, this is simplified to:

$$\mathbf{R}_\lambda = C_1\lambda^{-5}\left(e^{C_2/\lambda T} - 1\right)^{-1} \tag{0.11}$$

by putting

$$C_1 = 2\pi hc^2 = 3{\cdot}70.10^{-5} \text{ ergs cm}^{-2}$$

$$C_2 = \frac{hc}{k} = 1{\cdot}4323 \text{ cm. degree } ^{(2)}.$$

The curves $\mathbf{R}_\lambda$, plotted by means of formula (0.11) all have the same general appearance. They are at a maximum for the wave-length λ_m such that

$$\lambda_m T = 0{\cdot}2884 \text{ cm-degree (Wien's law)} \tag{0.15}$$

or, if we express λ in microns (10^{-4} cm.)

$$\lambda_m T = 2884 \text{ micron-degree} \tag{0.15'}$$

The maximum is thus shifted towards the small wave-lengths as the temperature is increased. It corresponds, for instance

to $\lambda_m = 1\mu$ for $T = 2{,}884°K$

and to $\lambda_m = 0{\cdot}2884\mu$ for $T = 10{,}000°K$

The curves are not symmetrical about the maximum and fall more rapidly on the side of small wave-length.[3]

It may be further proved that the *brightness* of a black body for any wave length, is the same in all directions. From this results that the spectral brightness $\mathbf{B}_\lambda$ can be obtained by dividing the spectral radiance $\mathbf{R}_\lambda$ by π.

Let us construct energy curves of the black body for a sufficiently large number of temperatures, and let us superimpose a curve representing the distribution of energy in the spectrum of any source, *on an arbitrary scale* (all the ordinates can be multiplied by an appropriate constant factor). Often we find that, in a more or less extended spectral interval, the energy curve of the studied source will nearly correspond with that of a black body at a certain temperature T. In the visible region this is notably the case with tungsten filament lamps. *We then call the colour temperature of the source, in the spectral interval considered, that temperature of the black body whose energy distribution most nearly resembles that of the source, in that interval.* It is very convenient thus to characterize the distribution of energy in a spectrum by means of a single parameter.

Now, experience has shown that the energy curves of most of the stars sufficiently resemble that of a black body, for the definition of colour temperature to have a precise meaning. This, for instance, is true for an interval of about 1000 to 1500 A, around 4500 A. We have indicated the approximate values of the average colour temperature relative to whatever spectral types in the third column of Table I. Their determination is much more difficult for stars of the first type, where we only observe the descending branch of the energy curve on the red side of the maximum ($\lambda m > 0{\cdot}3\mu$), while terrestrial standard sources have their maximum in the near infra-red. The numbers given for the stars of the classes B and F are the result of a very careful and very extended research by Chalonge, Barbier and their collaborators.

Examination of the Table immediately leads to a very important conclusion: *the colour temperature gradually diminishes as does excitation, all along the spectral series.* The temperature is, in effect, one of the essential factors which govern the excitation and the ionization in the atmosphere of the stars.

For the practical evaluation of the colour temperature T_2 of a star, we generally proceed as follows:

For a certain number of wave-lengths the differences in monochromatic magnitudes $(m_2 - m_1)$ between the star and the standard source, whose temperature T_1 we know (for instance a tungsten filament lamp), are measured. Then a graph is plotted of m_2-m_1 against the wave number $\tilde{\nu}$. If the energy curve of the star sufficiently resembles that of a black body, the points should lie along a straight line whose slope gives us the temperature T_2.

According to Planck's formula (0.11), the ratio of spectral brightness for the same wave number $\tilde{\nu}$ of two sources is:

$$\frac{B_2}{B_1}=\frac{\left(e^{C_2\tilde{\nu}/T_1}-1\right)}{\left(e^{C_2\tilde{\nu}/T_2}-1\right)}$$

and the difference of monochromatic magnitude is expressed as

$$m_2-m_1=2\cdot 5\left[\log\left(e^{C_2\tilde{\nu}/T_1}-1\right)-\log\left(e^{C_2\tilde{\nu}/T_2}-1\right)\right]+K,$$

where K denotes an unknown constant, independent of the wave-lengths (but dependent on the apparent diameters of the two sources). The derivative of m_2-m_1 with respect to $\tilde{\nu}$ can be written as

$$\frac{d\,(m_2-m_1)}{d\tilde{\nu}}=2\cdot 5\log_{10}e\,[\phi\,(T_2)-\phi(T_1)] \tag{0.17}$$

by making

$$\phi(T)=\frac{C_2}{T}\left(1-e^{C_2\tilde{\nu}/T}\right)^{-1} \tag{0.18}$$

The function $\phi(T)$, often called the *absolute gradient* at the temperature T, only varies slightly with the wave number $\tilde{\nu}$ and can be considered as independent of $\tilde{\nu}$ in an extended spectral interval. From this it follows that $\frac{d\,(m_2-m_1)}{d\tilde{\nu}}$ is practically independent of $\tilde{\nu}$ and, in consequence, that m_2-m_1 is a linear function of $\tilde{\nu}$.

We call the relative gradient of the source two compared to the source one, the quantity

$$G_{2\cdot 1}=\frac{d\,(m_2-m_1)}{d\tilde{\nu}}\cdot\frac{1}{2\cdot 5\log_{10}e}=0\cdot 921\,\frac{d\,(m_2-m_1)}{d\tilde{\nu}} \tag{0.19}$$

which is expressed as

$$G_{2\cdot 1}=\phi(T_2)-\phi(T_1) \tag{0.20}$$

The slope of the straight line gives us the relative gradient $G_{2\cdot 1}$, which, added to $\phi(T_1)$ gives $\phi(T_2)$. Whence, finally the colour temperature T_2 of the star is obtained.

If the points do not fall along a straight line, we must conclude from this that the energy distribution in the spectrum of the star cannot be regarded as that of a black body. The colour temperature is then no longer defined.

Colour indices

In the absence of an energy curve which has only been determined for a small number of sufficiently bright stars, certain *colour indices* permit the very rough characterization of the "colour" of a great number of faint stars.

On spectrograms of small dispersion one need only measure the darkening for two chosen darkening wave-lengths. Let us suppose, for simplicity, that the same curve applies to the two wave-lengths. Without difficulty we thus obtain the ratio of photographic luminosities l and l' corresponding to wave-lengths λ and λ'. The difference in magnitude

$$C = m - m' = -2{\cdot}5 \log\left(\frac{l'}{l}\right)$$

depends both on the ratio of luminous energies and on the spectral sensitivity of the plate, which is not the same for the two wave-lengths. If, and only if, we always use the same plates, C can be considered as a characteristic *colour index* of the star.

More generally, we can take as colour index the difference of magnitudes measured with two selective detectors of different spectral sensitivities, or yet with the same detector used together with two colour filters. The colour index most frequently used is the difference between the photographic magnitude and the visual magnitude

$$C.I. = m_{\text{pg}} - m_{\text{v}} \tag{0.21}$$

or yet the difference between the photographic magnitude and the photovisual magnitude

$$C.I'. = m_{\text{pg}} - m_{\text{pv}} \tag{0.22}$$

The indices $C.I.$ and $C.I'.$ differ little from each other. We know that, by convention, the stars A0 of the sixth magnitude have, on the average, the same photographic and visual (or photovisual) magnitudes. Their colour index is thus, on the average, nil. The hottest stars, and thus the "bluest", have negative colour indices, the coldest stars, and thus the "reddest", positive colour indices. We have placed the average colour indices $C.I.$ of various spectral types in the fourth column of Table I, according to the old measurements of King. They agree well with the order of colour temperatures.

Many other combinations are possible. It is naturally of interest to separate two measured spectral regions from each other so as to increase the variations of the colour index. The difference between magnitudes measured in blue and in red light have been considered at Harvard for the last fifteen years (panchromatic plates and appropriate filter). But more and more frequently use has been made of *photoelectric colour indices* measured by using successively two colour filters with the same cell. They have the advantage of being determined rapidly and with greater

PLATE I.—The Great Nebula in Orion, M 42 (NGC 1976). 193 cm. telescope of the Haute-Provence Observatory.

PLATE II.—The Great Nebula in Andromeda, M 31 (NGC 224) and its two companions. Schmidt telescope of the Haute-Provence Observatory.

PLATE III.—Central region of the Great Nebula in Andromeda (M 31). 120 cm. telescope of the Haute-Provence Observatory (R. Mévolhon).

precision. We shall see that their use has been particularly fruitful in the study of interstellar absorption.

For a first approximation we often proceed as if the measurements had been made in monochromatic light (as is the case with spectrograms of weak dispersion). But in that case the *effective* wave-lengths which we must consider will depend not only on the spectral sensitivity of the detector (often used together with a filter) but also on the colour temperature of the star. Thus, in the case of the international photographic magnitudes, Seares and Miss Joyner [268] have shown that if the colour temperature decreases from 30,000 to 3,000°K, the effective wave-length increases from 4163 to 4559 A. It varies much less in the case of photovisual magnitudes and only changes from 5415 to 5478 A.

Absolute spectroscopic magnitudes and the Hertzsprung-Russell Diagram

For a long time it was impossible to evaluate the absolute magnitudes of the stars if their distances (or, which comes to the same, their parallaxes) were not known. Considering the problem in reverse, the spectroscopic method has permitted the direct determination of the absolute magnitude and it is, this time, the distances which are deduced from the equation (0.9), the apparent magnitude being known.

The method is based on the relative intensities of certain spectral lines. By carefully comparing the spectra of stars of the same type, known either to have great or very feeble luminous intensities, Adams and Kohlschütter (1914) discovered that in the former, certain lines belonging to ionized atoms, such as Sr^+ 4077 and 4215 A, Fe^+ 4233 A, Ti^+ 4161 and 4399 A, were particularly strong. On the other hand, lines belonging to neutral atoms such as Cr 4324 A, Ca 4435 and 4454 A, Ti 4535 A were particularly weak in these stars of great luminosity. The theoretical interpretation of the facts, on which we cannot insist here, were furnished later by Saha. Ionization in stellar spectra is essentially governed by temperature and by pressure: it grows when the temperature is increased, or when the pressure diminishes. If certain lines of ionized atoms are stronger in the atmosphere of the stars of great luminous intensity, it is because the pressure there is much reduced.

For a certain number of stars of the same spectral type, we already know the distances (evaluated either by the direct trigonometrical method, or by some indirect method), and, in consequence, the absolute magnitudes. We measure, in their spectra, the ratio of intensities of two neighbouring lines, of which one is "sensitive" to the absolute magnitude, and we plot a graph of the ratio of intensities of the two lines against the absolute magnitudes. Each star furnishes a point and all the points fall along a regular curve which we shall be able to use for determining the absolute magnitude of stars of the same spectral type. We ignore their distances: it suffices to measure the ratio of intensities of the two lines.

By this entirely empirical procedure, Adams, Joy, Humason and Brayton rapidly determined the absolute magnitudes of more than 4000 stars of the classes F, G, K and M (1931). Other criteria have also been used: one of the most interesting is furnished by the bands of CN (3590, 3883, 4216 A), much stronger in the stars of large luminous intensity of the classes G and K than in the stars of faint luminosity (Lindblad). Finally, after much trial and error, analagous methods could be applied to stars of the classes B and A, where the metallic lines used by Adams are absent or very weak.

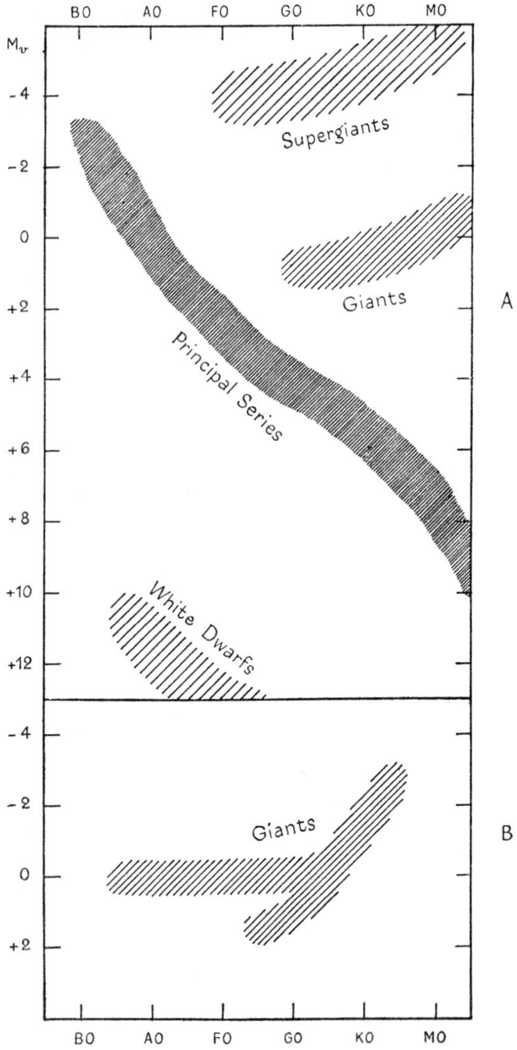

Fig. 1.—Russell Diagram. **Absolute** visual magnitude m_v against spectral type. A=Population I. B=Population II, following Baade (giant branch).

The distances deduced from the absolute spectroscopic magnitudes are never determined with more than 25 per cent accuracy. But the relative error is independent of the distance and this is the enormous advantage over the trigonometrical method where the relative error is proportional to the distance, and may become of the order of 100 per cent, and thus exceeds some hundreds of parsecs.

The knowledge of a large number of absolute magnitudes has permitted the completion of the diagram which Hertzsprung and Russell had already constructed with the data at their disposal before the discovery of spectroscopic methods (Fig. 1 A). Taking spectral types as abscissae (with equal intervals between B0 and B5, B5 and A0, etc. . . .), we plot a graph with absolute magnitude as ordinates. Most of the points fall in the neighbourhood of two regular curves, thus constituting two families of stars. In the first, called the *principal series*, the luminous intensity diminishes gradually from the class O to the class M; in the second, called the *giant branch*, it remains large in all the spectral classes, from F to M. Naturally, if we include in the diagram all the stars of known absolute magnitude, we observe a cluster of points around each of the above-mentioned curves and the two clusters are not completely separated in any but the K and M classes. In the latter we only have very luminous stars, the giants (around $M_V=0$), and very weak luminous stars, the *dwarfs* (around $M_V=+10$). The ratio of their intensities is of the order of 10,000. This separation of the two groups is less clear in the classes F and G; finally, the classes O and B do not contain stars of other than very great luminosity.

The relative proportion of giants and dwarfs, of which the spectra have been observed, is completely thrown out of balance by a selection effect: giants can be photographed from further away, by reason of their great luminous intensity. In reality, their number is very small compared to the stars of the principal series.

Well above the branch of the giants, we still find, in all the spectral classes, from F to M, some rare stars called *supergiants*, whose luminous intensities are at least 100 times larger than those of the giant types ($M_V \simeq -5$ to -6 and perhaps -7). Very much below the principal series we also know a certain number of *white dwarfs* (classes A and F), whose intensity is extremely small. In reality, their number ought to be very large for the selection effect tends to make us under-estimate them. Finally, between the white dwarfs and the principal series, there are *subdwarfs*, which seem to constitute a group apart, still relatively unknown.

Fig. 1 A is of rather bright stars, relatively near the Sun. The diagram of stars contained in the *globular cluster* presents quite a different aspect (Fig. 1 B).[4] There the most luminous stars are the giants K whose absolute photographic magnitude is $M_{pg}=-1{\cdot}1$ and whose absolute photovisual magnitude $M_{pv}=-2{\cdot}4$ approximately ($C.I'.=+1{\cdot}3$). The luminous intensity diminishes up to the first types G ($M_{pv} \simeq 0{\cdot}0$) where the branch of giants is divided into two. Following the one which extends up to the F5 type, we find that the luminous intensity continues to decrease. The other traverses the diagram horizontally up to the class B ($M_{pv} \simeq 0{\cdot}0$). We have no direct information about the dwarfs, which are probably very numerous, but which cannot be observed in these distant formations.

Thus, the globular clusters do not contain very luminous O and B stars nor any more super-giants of the classes F and M. The lacuna in the classical diagram, on either side of the F class, between the principal sequence and the branch of giants, is, however, packed in the diagram of globular clusters. Amongst the stars which fill up the gap *Cepheids of short period*—also called the *variable clusters*—figure notably. Their brightness varies with a period smaller than one day, and they play a very important role in the determination of the distances of very distant stellar systems.[5]

It would thus seem that there exist two types of stellar population: those of stellar regions, in the Milky Way, which Baade called the "Population I" and those of globular clusters which Baade called "Population II" [12]. We shall see that these two types of populations are also found in external galaxies.

Radiation temperature and its applications

Let us now consider the total energy radiated by any source for all wave-lengths. This is the product of the surface of the source and its *total radiant energy*. By definition, the *radiation temperature* T_r of the source is the temperature of a black body having the same total radiant energy as the source considered. It will only coincide with the colour temperature T_c if the source effectively radiates like a black body. In general, the two definitions lead to different values; thus we find for the Sun, $T_r = 5{,}750°K$, $T_c = 6{,}500°K$.

We also know that the total radiant energy of a black body is proportional to the fourth power of the absolute temperature. This is Stefan's law which is written

$$\mathbf{R} = \sigma T^4 \tag{0.23}$$

where $$\sigma = \frac{2\pi k^4}{15c^2h^3} = 5{\cdot}68.10^{-5} \ \text{(erg. sec}^{-1}\text{.degree}^{-4}\text{)}.$$

The total energy radiated by a star of radius R is thus

$$W = 4\pi R^2 \mathbf{R} = 4\pi R^2 \sigma T_r^4$$

where T_r is its radiation temperature.

Let us compare the energy radiated by the Sun, of radius R_s, to the temperature T_{rs}.

$$W_s = 4\pi R_s^2 \, \sigma T_{rs}^4$$

We obtain by division,

$$\frac{W}{W_s} = \left(\frac{R}{R_s}\right)^2 \times \left(\frac{T_r}{T_{rs}}\right)^4$$

Using logarithms, and then absolute bolometric magnitudes, we have:

$$M_b - M_{bs} = -2{\cdot}5 \log\left(\frac{W}{W_s}\right) = -5 \log\left(\frac{R}{R_s}\right) - 10 \log\left(\frac{T_r}{T_{rs}}\right)$$

With the numerical values $M_{bs} = +4{\cdot}62$, $T_{rs} = 5750°K$, this relation becomes

$$\log\left(\frac{R}{R_s}\right) = 0{\cdot}970 - 0{\cdot}2\, M_b - 2 \log\left(\frac{T_r}{5750}\right) \qquad (0.24)$$

It is often utilized for calculating the order of the magnitude of the radius of the stars. We start with the absolute visual magnitude M_v, which we correct by means of the Kuiper Tables to pass over to the absolute bolometric magnitude M_b. The choice of radiation temperature T_r inaccessible to direct measurement in the case of point sources, carries with it much uncertainty. It is generally lower than the colour temperature.

Let us, as an example, try to determine the radii of two stars, both of spectral type M0 of temperature $T_r = 3{,}270°K$, of which one is a giant ($M_v = 0$) and the other a dwarf of 10,000 times weaker intensity ($M_v = +10$) (although in reality the giants always have a somewhat lower colour temperature than dwarfs of the same type). The Kuiper Tables give bolometric corrections which differ little for the cases of giant and dwarf of the M0 type ($-1{\cdot}55, -1{\cdot}43$ magnitudes, respectively). Let us take for the two stars the mean value $M_b - M_v = -1{\cdot}49$. Thus, we obtain for the giant:

$$\frac{R}{R_s} = 51{\cdot}5 \ (2R—72.10^6 \text{ km.}).$$

and for the dwarf:

$$\frac{R}{R_s} = 0{\cdot}515 \ (2R = 0{\cdot}72.10^6 \text{ km.}).$$

The radius of the second is naturally 100 times smaller than that of the first and the terms "giant" and "dwarf", which until now have not denoted anything but stars of large or small luminous intensity, now correspond to a geometric reality.

Let us further calculate the radius of an average white star of absolute visual magnitude $M_v = +1$, and whose temperature of radiation is $T_v = 9{,}700°K$. The bolometric correction this time is $M_b - M_v = -0{\cdot}51$ and we have

$$\frac{R}{R_s} = 2{\cdot}35 \ (2R = 3{\cdot}27.10^6 \text{ km.}).$$

For a white dwarf of the same temperature, but of absolute magnitude $M_v = +11$, we naturally have a radius 100 times smaller

$$\frac{R}{R_s} = 0{\cdot}0235 \ (2R = 3{\cdot}27.10^4 \text{ km.}).$$

The diameter of the white dwarf is only two and a half times that of the earth ($1{\cdot}27.10^4$ km.).

NOTES TO INTRODUCTION

[1]In many popular works the light-year is taken as unit. This is the distance traversed by light in one year ($3{\cdot}156.10^7$ secs.).

Since the speed of light, $c=2{\cdot}998.10^{10}$ cm./sec. (almost 300,000 km./sec.)—

$$\text{One light year}=2{\cdot}998.10^{11}\times 3{\cdot}156.10^{7}=9{\cdot}462.10^{17}\text{ cm. and}$$

$$\text{One parsec}=\frac{3{\cdot}086.10^{18}}{9{\cdot}462.10^{17}}=3{\cdot}26\text{ light years.}$$

[2]It is often useful to take as a variable in place of the wave-length λ either the number of waves per cm. (called the wave number)

$$\bar{\nu}=\frac{1}{\lambda}\,(\text{cm}^{-1})$$

or the frequency

$$\nu=\frac{c}{\lambda}\,(\text{sec.}^{-1})$$

If we represent the spectral radiances of the intervals $d\bar{\nu}$ and $d\nu$ by $\mathbf{R}_{\bar{\nu}}$ and $\mathbf{R}_{\nu}$ respectively we have

$$\mathbf{R}_{\nu}=2\pi hc^2\bar{\nu}^3\left(e^{hc\bar{\nu}/kT}-1\right)^{-1} \tag{0.12}$$

$$\mathbf{R}_{\nu}=2\pi hc^{-2}\nu^3\left(e^{h\nu/kt}-1\right)^{-1} \tag{0.13}$$

The energy is emitted in light quanta or *photons*, each of which possesses an energy $h\nu$ and it will sometimes be necessary to evaluate the number $\mathbf{N}_{\nu}$ of photons emitted per second in the band of frequency $d\nu$. By dividing both sides of (0.13) by $h\nu$ we obtain

$$\mathbf{N}_{\nu}=2\pi c^{-2}\nu^2\left(e^{h\nu/kt}-1\right)^{-1} \tag{0.14}$$

Fig. 9 (p. 68) shows the curve of $\mathbf{N}_{\nu}$ when $T=30{,}000$°K.

[3]Left of the maximum ($\lambda<\lambda_m$), we obtain an exact representation (at least within nearly 1 per cent) neglecting unity in comparison with the exponential in Planck's formula and simply write it as

$$\mathbf{R}_{\bar{\nu}}=C_1\,\lambda^{-5}e^{-C_2/\lambda T} \tag{0.16}$$

This is Wien's approximation, which permits the simplification of some calculations.

[4]Stars of the same cluster are all equally distant from the Sun and have an equal modulus of distance $m-M$.

[5]The period of the variation is a function of the absolute magnitude, and permits its determination (Miss Leavitt's " period-luminosity" relation).

Part One

ATOMS AND MOLECULES IN SPACE

CHAPTER I

ATOMS IN THE VICINITY OF HOT STARS: NEBULAE WITH BRIGHT-LINE SPECTRA

1. General

HERE and there, the exploration of the sky reveals pale cloudy patches which the first observers called *nebulae*. These are easily distinguished from stars which always appear as more or less bright points both to the telescope and to the naked eye. The reader is referred to Plate 1 showing the Great Nebula in Orion, discovered by Fabri de Peiresc in 1610, and to Plates II and III of the Great Nebula in Andromeda, found two years later by Simon Marius.

With the increase in telescopic resolving power we have seen that many of the stars first catalogued as nebulae are, in reality, nothing but concentrated stellar clusters. As distinct from these, the name *nebula* is currently applied to two entirely different categories of celestial bodies.

Galactic nebulae and external galaxies

The first, like the Great Nebula in Orion, retain their diffuse character even on photographs taken with the largest telescopes. These are clouds of gas or of fine solid particles which, as we shall see, are almost always associated with stars. They belong to our stellar system, and are *Galactic nebulae* whose distances from the Sun are counted in hundreds or, at most, thousands of parsecs.

The others owe their diffuse appearances only to their extreme remoteness. Our great modern telescopes are capable of resolving the stars in some of the nearest of them, such as the Great Nebula of Andromeda. They are stellar systems comparable to our Milky Way both in structure and dimensions: *external galaxies* whose distances are not less than a million parsecs, sometimes hundreds of millions.

In general, it is easy enough to distinguish these two categories, even on small-scale photographs. Whereas Galactic nebulae present very diverse forms, to which we shall refer later, many nebulae of the external galaxies are characterized by their spiral form. Among them only the elliptic nebulae appear at first sight to resemble certain types of Galactic nebulae—i.e. planetary nebulae (see below)—but the brightness of the latter is far greater.

Plates IV, V, VI, IX, X, XII, XVI, XX, XXI and XXII show the appearances of some galactic nebulae on photographs. Plates II, III, XXIII and XXIV represent spiral nebulae seen "face on" or "in profile".

Usually nebulae (and stellar clusters) are denoted by the number of their order, either in the first catalogue compiled in Paris by Messier and published in 1787 (103 nebulae and clusters of greater declination than 35°S.), more often in the *New General Catalogue of Nebulae and Clusters of Stars* (N.G.C.) of Dreyer (1888), which is much more comprehensive, in the *Index Catalogue* of the same author (1895), or finally, in more recent specialized catalogues. Thus the Great Nebula in Orion is denoted by M 42 or NGC 1976, the Great Nebula in Andromeda by M 31 or NGC 224.

Planetary nebulae and diffuse nebulae

We shall concern ourselves here with Galactic nebulae only. Their appearance permits their immediate classification into two quite distinct groups:

1—*Planetary nebulae* presenting a more or less spheroidal or annular symmetry with well-defined edges. Looked at through the telescope they vaguely resemble planets, hence their name. Photographs often show an apparently stellar *nucleus* in their centre as, for example, in the nebulae NGC 40, Dumb Bell, and the famous Ring-Nebula of Lyra (Plate IV).

2—*Irregular or diffuse nebulae* having no well-defined geometric form. Their contours can either be very blurred or sharply serrated. Some are drawn out in long filaments like the Lace-work Nebula in Cygnus (Plate IX). The Great Nebula in Orion belongs to this category, as do the nebulae IC 405, NGC I 59, shown in Plates V and VI.

Nebulae with line spectra and nebulae with continuous spectra

The above classification must needs be completed by the examination of spectra which alone are capable of explaining the nature of the phenomena that cause the luminosity of Galactic nebulae. All planetary nebulae and certain diffuse nebulae have an absolutely characteristic bright line spectrum (Plates VII and VIII), while other irregular nebulae give a continuous spectrum containing dark lines like that of a star. We shall, therefore, distinguish between *line spectrum nebulae*, also called *gaseous nebulae*, because such a line spectrum usually indicates emission by a rarefied gas, and *continuous spectrum nebulae*. The latter are also referred to as *reflection or diffuse nebulae*, because it is known that they diffuse or scatter in all directions the light received from a star.

The mere appearance of an irregular nebula does not allow us to predict accurately what its spectrum will be like. In general, those nebulae that are serrated or filamented give a line spectrum, and those that have vague and cloudy contours, a continuous spectrum. Great brightness is often associated with a line spectrum, great extension with a continuous spectrum. These indications, however, which we owe to Hubble [146], cannot be regarded as absolute rules.

Furthermore, there are many intermediate cases between these two main types. A more or less apparent continuous spectrum nearly always accompanies a bright-line spectrum (not necessarily due to the scattering of the light of a star). Conversely, continuous spectrum nebulae frequently show some emission lines.

Galactic distribution

Less than 400 planetary nebulae are known. Their apparent diameter is rarely as large as 10′ (the greatest NGC 7293, "Helix", measures between $15' \times 12'$) and is, in general, smaller than 1′. The largest planetary nebulae are almost uniformly distributed over the entire sky, but the smaller ones are fairly strongly concentrated in the plane of the Galaxy and especially near the centre of the Milky Way.

Irregular nebulae, which are more numerous, show a different distribution. They describe two circles around the celestial sphere, one of which coincides with the plane of the Galaxy, whereas the other is inclined at 20° to it. The second more or less coincides with the mean plane of the B stars, which would seem to imply a relationship with these stars. The distribution in longitude is far from being uniform: diffuse nebulae show a tendency to group themselves in clusters. This fact is particularly clear in the second circle which is almost discontinuous and contains the majority of continuous spectrum nebulae, whereas the bright-line spectrum nebulae are more abundant in the plane of our Galaxy.

The distribution of external galaxies is entirely different and, in a way, complementary. Very numerous in the vicinity of the poles of the Milky Way, they become progressively rarer towards the lower latitudes, and near the Galactic plane there is an irregular zone—the " zone of avoidance "—where we find neither spiral nor elliptical nebulae.

Spatial distribution and linear dimensions

The distances of planetary nebulae are only roughly known as yet [359], but well enough, nevertheless, to give us an idea of the actual distribution of planetary nebulae in space. We find that their percentage above and below the Galactic plane is relatively small. The most recent statistics, due to Minkowski [208], and based on the observation of 371 planetary nebulae, indicate that the majority of these are located in the central nucleus of the Milky Way. By various methods, based mainly on the associated stars, Cederblad [54] tried to determine the distances of diffuse nebulae also, and to study their spatial distribution.

The measurement of the apparent diameter of a nebula of known distance permits the evaluation of its linear diameter. In the case of planetary nebulae we usually find figures of tenths to some hundredths of parsecs (Vorontzov-Velyaminov [360], Berman [36], Parenago [234]). Minkowski [208] has studied the distribution of linear diameters on the assumption that the small planetary nebulae, observed near the centre of the Milky Way, are located 9,000 pc. from the Sun. His results, which correspond well enough with those of Berman and Parenago, show that the most frequent diameter is approximately 0·15 pc. (about 30,000 times the distance between the Earth and the Sun).

In the case of diffuse nebulae we shall only consider their greatest dimension D. The values obtained by Cederblad range between 0·02 pc. (Nebula round Merope, in the Pleiades) and 38·3 pc. (Nebula η Carinæ). In approximately half the cases (93 out of 181), D is found to be between 1 and 10 pc. Its mean value $D=3{\cdot}6$ pc. is, therefore, considerably greater than that of typical planetary nebulae. These figures are obviously affected by the fact that the sample is limited, since the probability of finding nebulae of small dimensions diminishes with increase of distance. It is, nevertheless, instructive to compare the dimensions D observed in the Galaxy with those in external formations. In the Smaller Magellanic Cloud, Shapley and Miss H. H. Wilson [281] have found that the dimensions D of diffuse nebulae range from 0·8 to 13·8 pc., with a mean value $D=5{\cdot}0$ pc. Now, if in Cederblad's count we eliminate objects of dimensions less than 0·8 pc., which cannot be observed in the Magellanic Clouds because of their too small apparent diameter, we obtain precisely the same mean $D=5{\cdot}0$ pc. (1,000,000 times the distance between the earth and the sun) for the Galactic nebulae. The Great Cloud contains, in addition, a nebula of 40 pc. in length (30 Doradus), very close to that of η Carinæ. The diffuse nebulae of the Magellanic Clouds seem identical to those of the Milky Way.

2. Bright line Spectra in Gaseous Nebulae

The spectral study of gaseous nebulae was begun quite a long time ago. In 1918 Wright [385], in a brilliant thesis, published a detailed and very accurate description of the spectra of numerous planetary nebulae. At the same time, Campbell and Moore studied their radial velocities [52]. Since then, numerous and usually very weak lines have been discovered by Bowen and Wyse [50], and by Wyse [386]. They have also collaborated in measuring the intensity of the lines, as have H. H. Plaskett [238], Berman [35], T. L. Page [226] and Aller [3].

Permitted lines

Plate VII represents the characteristic emission spectra of some planetary nebulae, Plate VIII that of the diffuse nebula in Orion (from 3700 to 5100 A). The Balmer lines of the hydrogen atom are always prominent. At first spaced very widely in the visible region (Hα at 6563, Hβ at 4861, Hγ at 4340, $H\delta$ at 4102 A, etc.), they become more and more close towards the ultra-violet and tend towards a limit in the vicinity of 3650 A. From there on, the series gives way to a continuous spectrum, still emitted by the hydrogen atom, and particularly intense in the planetary nebulae.

The lines of the neutral atoms of helium are also constantly present (for example He I 6678, 5876, 5016, 4922, 4713, 4472, 4388, 4144, 4121, 4026 A, etc.) and quite frequently, those of ionized helium (He II, Fowler series 4686 and 3203 A; Pickering series 5412, 4542, 4200 A, etc.).[1] Among the weak lines of the nebulae we can also frequently identify the most characteristic ones of the once, twice or three times ionized carbon atom (C II 4267 A, C III 4650 A, C IV 4658 and 5812 A), and sometimes of O^+ and N^+. In the infra-red region (up to 8700 A), Swings and José [331] observed some hydrogen lines belonging to the Paschen series, the line 8237 A of He II and, in NGC 6572, a line of neutral oxygen (8446 A).

The behaviour of the lines O^{++} and N^{++} merits further examination. Certain lines only of O^{++} ion, all situated in the ultra-violet, appear with a fairly great intensity (O III 3760, 3444, 3429, 3341, 3312, 3299 A, etc.). Similarly, among the lines of the N^{++} ion, some are quite intense at times (N III 4641, 4634, 4103, 4097 A), while others, usually as strong or even stronger in the laboratory, may be totally absent (N III 4379 A, for example). Consequently, in the nebulae there is often a curious selection from the lines of the O^{++} and N^{++} ions which we have been unable to explain.

The forbidden lines of the nebulae

We have omitted a fairly large number of lines, often the most intense, and always the most characteristic of nebular spectra (5007, 4959, 4363, 3729 and 3726 A for example). Their wave-lengths have often been measured very accurately. Thus, in 1914, Fabry, Buisson and Bourget determined those of the doublet 3728·9–3726·2 A by means of an interferometer with semi-silvered plates. The values they found, however, did not correspond with the known lines of any element. The identification of the principal nebular lines, therefore, presented a problem to the spectroscopists which, for a long time, remained unsolved.

In the same way as helium was discovered in the Sun by means of its spectral lines before it was isolated on Earth, one was tempted to attribute these nebular lines to an element unknown on Earth—nebulium. Sometimes it was even thought that two unknown elements existed. But physico-chemical progress soon showed that there was no room in the periodic tables of the elements for a new light atom, and this hypothesis had to be abandoned. Croze and Mihul [62], and Russell and Fowler did important work on this problem which was finally solved by Bowen [36] in 1928. They are, in fact, lines belonging to atoms which have strong common ions like N^+, O^+ or O^{++}, but lines which are very difficult to produce in the laboratory (the majority have not yet been re-produced) because their emission is forbidden by the selection rules of spectroscopy.

Energy levels of atoms

Let us recall in the first place that an atom emits a *photon* when it passes from an energy level W_2 to an inferior level W_1. The energy of the photon is the difference between the energies W_2 and W_1 and the frequency ν of the corresponding line is given by the fundamental relation

$$h\nu = W_2 - W_1 \qquad \text{(I.1)}$$

where h is Planck's constant.

Since an atom can only occupy discrete energy levels, which are well defined, the frequencies of the lines which it is capable of emitting are, therefore, perfectly defined also. Conversely, the atom can absorb these lines in passing from an energy level W_1 to a higher level W_2.

It is easy to connect the frequency ν with the wave-length (in vacuo) or, which amounts to the same thing, $\tilde{\upsilon} = \frac{1}{\lambda}$, which represents the *number of waves per centimetre*, if λ is expressed in centimetres. In effect, if we denote the speed of light in vacuo by c ($c = 2{\cdot}998.10^{10}$ cm.-sec.$^{-1}$).

$$\lambda = \frac{c}{\nu}; \; \tilde{\upsilon} = \frac{1}{\lambda} = \frac{\nu}{c} \qquad \text{(I.1)}$$

By dividing the two sides of the equation (I.1) by hc, we thus obtain

$$\tilde{\upsilon} = \frac{W_2}{hc} - \frac{W_1}{hc}$$

Each energy level of an atom can be represented by a horizontal line whose vertical distance from the ground level of the atom (where the energy is considered to be zero) is proportional to $\frac{W}{hc}$. In this way,

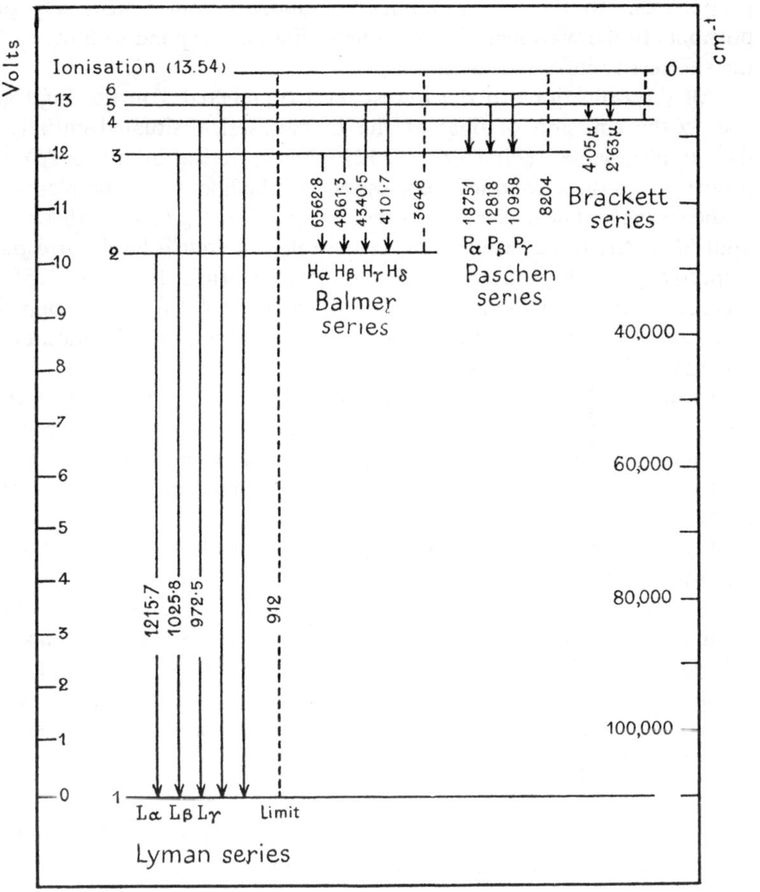

Fig. 2.—Diagram of the energy levels of the hydrogen atom.

Fig. 2 represents the energy diagram of the hydrogen atom. The vertical axis on the right-hand side of the figure is graduated in cm.$^{-1}$, that on the left-hand side, in volts. Instead of expressing the energy in ergs, for example, one can, in effect, look for the difference in potential V which it would be necessary to apply to an electron to give it this energy. The energy can thus be expressed in *electron volts* (eV) and the levels $\frac{W}{hc}$ can be counted in volts, one volt corresponding to 8106 cm.$^{-1}$. (2)

The arrows of Fig. 2 represent the various lines which the hydrogen atom can emit in passing from one energy level to a lower one. Their length is proportional to their wave number $\tilde{\nu}$, and, of course, inversely proportional to the wave-length. The longest lines correspond to ultra-violet, the shortest to infra-red radiations.

All the transitions to the lowest level (numbered 1 in the Fig.) give rise to the emission of lines of the Lyman series, situated entirely in the far ultra-violet. (First line L_α 1216 A, theoretical limit 912 A.) The transitions to the second level correspond to the lines of the Balmer series in the visible and near ultra-violet regions (first line H_α 6563 A, theoretical limit 3646 A); those terminating in the third or fourth level correspond respectively to the infra-red series of Paschen (first line P_α 18751 A, theoretical limit 8204 A) and of Brackett (first line 4·05μ). In the opposite direction to the arrows the transitions correspond to the absorption of the same radiations.

In Bohr's elementary theory, the various levels of the hydrogen atom correspond to the stable, elliptic orbits which the planetary electron can describe around the nucleus. The lines are emitted when an electron jumps from an exterior orbit to one which is nearer the nucleus and are absorbed in the converse case.

Fig. 2 is quite simple because the hydrogen atom has only one planetary electron. If we consider heavier atoms which possess an increasing number of peripheral electrons, the diagrams become correspondingly more complicated. There are more and more energy levels and the transition between these different levels can give rise to the emission of a considerable number of lines. But among all the lines which we could predict by combining the different levels in all possible ways, some are not observed—they are excluded by the *selection rules.*

*Selection rules**

It is impossible to enter into a discussion of selection rules without referring to the concepts of quantum mechanics which, to all intents and purposes, are outside the scope of this work. We shall only recall that every electron of the atom is characterized by three *quantum numbers*—meaning that they cannot change in anything but a discontinuous manner, and only take integral or half integral values. We have: the *principal quantum number n*, analagous to the energy state of the Bohr hydrogen atom, the azimuthal (angular) momentum *l*, which is an integral multiple of $\frac{h}{2\pi}$ and the spin of the electron whose value is always $\frac{1}{2}$ of $\frac{h}{2\pi}$.

The atom itself, in a fixed energy level, is characterized by three quantum vectors: a vector $\vec{L}$ which is the resultant of the angular momenta

*The reading of this paragraph is not absolutely essential for the understanding of what is to follow.

l of all the electrons; a vector $\vec{S}$ which is the resultant of all the spins s of the electrons, and a vector $\vec{J}$, for the *total angular momentum*, which is the resultant of the vectors $\vec{L}$ and $\vec{S}$ (we are dealing here with a case where the vectors l of the different electrons, as also the spins s, are strongly coupled; this is the Russell-Saunders coupling).

The addition of the vectors obeys a quantum rule: L is necessarily an integral multiple of $\frac{h}{2\pi}$ and S an integral multiple of $\frac{1}{2}\times\frac{h}{2\pi}$, in such a manner that the different vectors $\vec{l}$ and $\vec{s}$ can only take well-defined orientation with respect to each other. The vectors $\vec{L}$ and $\vec{S}$ are added in accordance with the same rules. Their resultant $\vec{J}$ is an integral multiple of $\frac{h}{2\pi}$ if the atom contains an even number of electrons, and an integral multiple plus half of $\frac{h}{2\pi}$ when the total number of electrons is odd.

Each energy level is represented by a symbol which consists of:

1—A group of numbers and of small letters which denote the *electronic configuration* (the distribution of electrons in successive shells, values corresponding to l);

2—A capital letter corresponding to a spectral term, which indicates the value of L (taking $\frac{h}{2\pi}$ as unity) in accordance with the following:

S means $L=0$	P means $L=1$
D ,, $L=2$	F ,, $L=3$
G ,, $L=4$	etc.

The letter S is preceded by a superscript which shows the *multiplicity* of the level equal to $2S+1$. The superscript 2 meaning $S=\frac{1}{2}$ corresponds to a doublet term, the superscript 3 meaning $S=1$ to a triplet term, etc. Finally, the letter is followed by a subscript which gives the value of J.

It is unnecessary, for what follows, to enter into the details of notation corresponding to the electronic configuration. For this, we must stress the importance of knowing whether the arithmetic sum Σl of the azimuthal quantum numbers is an even or an odd multiple of $\frac{h}{2\pi}$. In the first case we say it is an even term, in the second case, that it is odd. Since we do not write down the details of electronic configuration, we distinguish the odd terms by placing the superscript 0 in the upper right-hand corner after the capital letter.

In this manner, the normal level of the neutral oxygen atom is written as

$$^3P_2.$$

This means that we are dealing with a triplet level (superscript 3) characterized by $L=1$ (P) and $J=2$ (subscript 2). The two other adjoining levels of the triplet will have 1 and 0 as subscripts. Since a subscript is an integral number, the atom has an even number of electrons (it has 8). Finally, since the superscript 0 does not figure in this symbol, the term is even (Σl=an integral multiple of $\frac{h}{2\pi}$).

The normal level of the nitrogen atom is written as

$$^4S^{\circ}_{\frac{3}{2}}$$

It is, therefore, a quadruplet (superscript 4), characterized by $L=0$ (S) and $J=\frac{3}{2}$ (subscript). This subscript being equal to a whole number plus $\frac{1}{2}$, the atom has an odd number of electrons (7). The superscript 0 finally indicates an odd term (Σl=an odd multiple of $\frac{h}{2\pi}$).

We can now clearly state the selection rules:

1—The total angular momentum J of the atom cannot change except by $\Delta J=0, \pm 1$, with the restriction that $J=0$ cannot be combined with $J=0$.

2—Even terms cannot combine with odd terms and vice versa (Laporte's rule).

3—L can only change by $\Delta L=0, \pm 1$, and at the same time for the electron which is undergoing the transition we must have $\Delta l=\pm 1$.

4—S may not change ($\Delta S=0$).

The last two rules apply strictly to the Russell-Saunder's coupling only, to which all atoms do not conform. They are, furthermore, much less strict than the first two and are often broken, particularly in the case of heavy atoms. In accordance with rule 3 for example, the term S could only combine with another term S, or a term P; a term P only with terms S, P or D. The last condition prohibits the so-called intercombination lines corresponding to transitions between levels of different multiplicities, since the multiplicity is $2S+1$.

Electric dipole, electric quadrupole, and magnetic dipole radiation

How can we explain the fact that the strictest selection rules, confirmed by wave mechanics, are so often broken in the nebulae? To understand this, it is necessary to distinguish between electric dipole, electric quadrupole and magnetic dipole radiations.

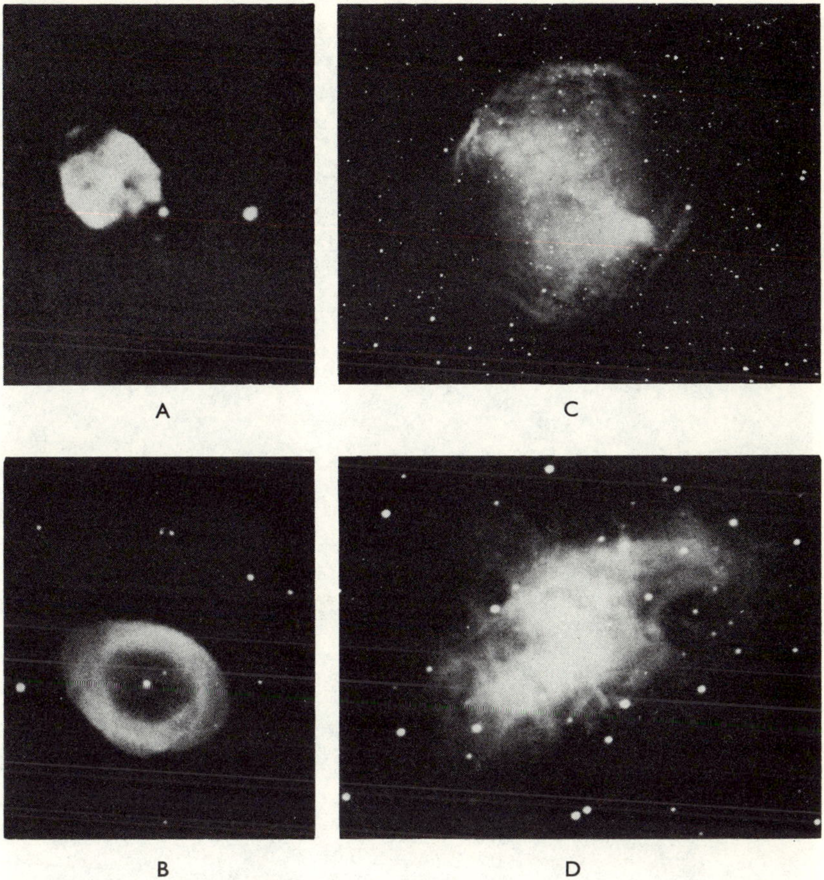

PLATE IV.—Planetary Nebulae and the Crab Nebula. (A) NGC 40 in Cepheus much enlarged. 120 cm. telescope (J. Dufay). (B) NGC 6720, Ring Nebula in Lyra. 193 cm. telescope. (C) NGC 6853 (M 87), Dumb-Bell Nebula in Vulpecula. 193 cm. telescope. (D) Crab Nebula (M I or NGC 1951), in Taurus. 80 cm. telescope (M. de Kérolyr).

PLATE V.—The Nebula IC 405 in Auriga. 80 cm. telescope (M. de Kérolyr).

The simplest model which we can imagine as a source of electromagnetic radiation is the electric dipole, analagous to the Hertzian oscillator and represented by Fig. 3A. Oscillations of frequency ν of the charges $+e$ and $-e$, about their equilibrium positions, produce a variation in the electric movement of the dipole and create, in space, a variable electric field of frequency ν. An electric quadrupole, which might correspond to Fig. 3B, is also capable of radiation but its field diminishes much more rapidly with increased distance. A simple calculation will show that the potential, due to an isolated charge, is inversely proportional to the distance, that of a dipole inversely proportional to the square of the distance, and that of the quadrupole, inversely proportional to the cube of the distance. The radiation of a quadrupole is, therefore, always weaker than that of a dipole.

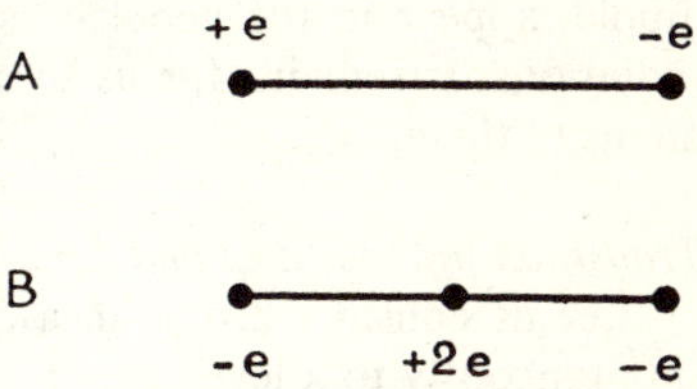

Fig. 3.—A, electric dipole; B, electric quadrupole.

The study of the splitting of spectral lines, when the source is placed in a magnetic field (Zeeman effect) allows us to distinguish between the lines due to electric quadrupole and electric dipole radiations. It also brings into evidence *magnetic dipole* radiation which can exist in the absence of an electric dipole. The moment of a magnetic dipole is also capable of creating a variable field and of radiating feebly, rather like a coil through which an alternating current is passed.

Furthermore, wave mechanics shows that the above selection rules can only be applied to electric dipole radiation. For lines due to electric quadrupole or magnetic dipole radiations the selection rules are different, or even opposite. Thus, in both their radiations, even terms can only combine with even terms, and odd terms with odd terms. This is the exact opposite of Laporte's rule. The lines which are forbidden by this rule for electric dipole radiation are, therefore, the only ones which are allowed for the case of the electric quadrupole and the magnetic dipole.

The selection rule for the total angular momentum is the same in the case of the magnetic dipole as for the electric dipole, but in the case of the electric quadrupole, it becomes $\Delta J=0$, ± 1 or ± 2, with this restriction that the sum of the values of J for the two combined levels must be greater than 2.

In all other cases, intercombination lines remain forbidden, no matter for what radiation, and the rule concerning the vector $\vec{L}$, unchanged for the magnetic dipole, becomes $\Delta L=0$, ± 1, ± 2, for the electric quadrupole, with this addition, that $L=0$ cannot be combined with $L=0$. Once again, these two rules relating solely to the Russell-Saunder's coupling, are only followed very approximately. Henceforth it will be understood that the lines, strictly forbidden by Laporte's rule, relating to the electric dipole, appear in the nebulae as electric quadrupole and electric dipole radiations. It remains for us to explain how they can achieve such great intensity there.

Transition probabilities and duration of excited states

Let us consider atoms in a certain energy level W_n. They can drop spontaneously to a level W_m of lower energy, emitting a line of frequency such that $h\nu = W_n - W_m$. Since, however, there are, in general, a number of levels with an energy lower than W_n, our atoms can emit a number of distinct lines. The fact that they exhibit very different intensities shows that the atoms drop to certain levels rather than to others. In other words, the various transitions are not equally probable.

The probability of a spontaneous transition from a level *n* to a level *m* is characterized by the number $A_{n,\,m}$ of transitions produced per atom per second. We obtain the intensity of a line by multiplying the energy $h\nu$ of the photon by the probability of transition $A_{n,\,m}$ and by the number of atoms present on the higher level *n*.

The transition probability $A_{n,\,m}$ depends mainly on the characteristics of the levels *n* and *m* and, by methods which are often very laborious, they have been calculated in a number of cases. It is much smaller for those lines which are only permitted for magnetic dipole or electric quadrupole radiation, than for those lines that are permitted for electric dipole radiation. To abbreviate, we shall in future, call these lines "forbidden" and "permitted", it being clearly understood that we are referring to the electric dipole.

The transition probability is the reciprocal of time and is expressed in seconds, as is the sum $\Sigma A_{n,\,m}$ of the probabilities of all the possible transitions from the same level. The reciprocal of this sum, which is expressed in seconds, is the mean life-time of the level *n*, and the law regulating the depopulation of the level *n* by radiation is identical with that of the rate of radio-active transmutations.

The *ground* level of an atom (minimum energy level) has an indefinite life-time, since no transition can lead to a lower energy level. It is, therefore, perfectly stable. The *excited* levels, on the other hand, which can give rise to permitted rays have a very short life-time, of the order of 10^{-8} secs. But, when all those lines which are capable of being emitted from a certain excited level are forbidden, the sum of the transition probabilities is very small and the mean duration of the level can be very long. We call this a *metastable state*. The mean life-time of metastable states is frequently of the order of quarter second; often it amounts to minutes or even hours.

From this we can clearly understand why it is usually so difficult to excite forbidden lines in the laboratory. Even at the lowest pressures produced in discharge tubes, a metastable atom undergoes a great number of collisions before spontaneously losing its energy by the emission of a forbidden line. There are thus many chances of its losing this energy to one of the particles colliding with it. It is also capable of absorbing a photon which will carry it to a higher energy level, from which it will drop again almost immediately, emitting a permitted line.

One of the processes which have enabled us to reproduce certain forbidden rays in the laboratory consists of diluting the atoms in the experiment in a large excess of monatomic gas, since the collisions of these gases are usually elastic and do not make them lose their energy to the metastable atoms. We are thus able to excite forbidden rays even under relatively high pressures. The presence of a large excess of monatomic gas, in reducing the mean free path of the atoms, diminishes the frequency of collisions against the enclosure of the recipient, and contributes largely to the de-excitation of metastable states.

In this manner, diluting oxygen with helium, argon or neon, McLennan and Shrum (1924) managed to excite (much more easily than in pure oxygen) the green line 5577·35 A which is emitted nightly in the high terrestrial atmosphere and which is of great intensity during polar auroras. These notable experiments led to our attributing the green line to the forbidden transition $^1D-{}^1S$ of the neutral oxygen atom.[3] At the same time, a weak ultra-violet line of 2972 A appeared, which corresponds to the transition $^3P_1-{}^1S$ (see Fig. 14). The mean life-time of the level ^1S is 0·46 secs. Similarly, by diluting oxygen with xenon, we can excite the ultra-violet line 2972 A with far greater intensity than in pure oxygen, and also the two red lines of the auroras and of the nocturnal sky 6300·3 and 6363·8 A, which represent the transitions $^3P_2-{}^1D$ and $^3P_{21}-{}^1D$ (R. and L. Herman). Since the level D has a much longer life-time than the level S (95 secs.), in the nebulae the red lines are generally much more intense than the green.

We shall soon see that the density of the nebulae is about 10^{-20} gm./cm.^{3}. This represents a vacuum which is far greater than the best vacuum which we can produce by means of laboratory pumps. In order to obtain a density of that order by evacuating the lightest gas, hydrogen, at ordinary temperature, we would have to reach a pressure below 10^{-13} mm. of mercury. The collisions are, therefore, extremely rare; the time between successive collisions is about 10 to 10,000 seconds (at a temperature of 10,000°K). Further, far enough from a star, the radiation which strikes the atoms is weak (it is called "diluted") and the probability that metastable atoms will absorb a photon becomes very small. Very low pressures and diluted radiation are conditions extremely favourable to the emission of forbidden lines. It will be understood that, in these conditions, those atoms which are situated on a level with a long life-time will all finally drop spontaneously to lower levels, emitting forbidden lines. If ever these levels happen to be densely populated, forbidden lines are capable of assuming an intensity comparable to that of the permitted lines.

Identification of forbidden lines in the nebulae

When the principal nebular lines were identified, not only were the transition probabilities unknown, but even the position of many metastable levels was ignored. Some examples will show how, nevertheless, we have succeeded in identifying forbidden lines.

[O III]. Fig. 4 shows arrangements of the lowest energy level for the ion O^{++}. Above the fundamental triplet level ^3P, two metastable singlet levels ^1D and ^1S are found. These three levels, having as they do, exactly the same electronic configuration, all transitions between them are strictly forbidden by Laporte's rule (for the electric dipole). Further, for whatever manner of radiation, the probability of transition between the triplet ^3P and the two singlet levels (intercombination lines) is weak.

Two permitted ultra-violet lines (between 500 and 800 A) of which one arises by transition to the level ^1D and the other to the level ^1S were known. The difference between their wave numbers measures directly the wave number of the transition ^1D to ^1S. The wave-length thus calculated was (in air) $\lambda=4363{\cdot}54$ A. Now, in the nebulae, we can observe the strong line 4363·21 A. The difference 0·33 A (or in wave numbers 3 cm.$^{-1}$) would correspond to an error of 0·01 A in the wave-lengths of the ultra-violet lines. There is, therefore, no doubt as to the identification (Bowen [47]).

Further, the position of the level ^1D below ^3P was unknown. But ultra-violet lines bordering on $^3P_2$ and $^3P_1$ respectively, were known. The difference in their wave numbers $\Delta\tilde{\nu}=192$ cm.$^{-1}$ measures the separation of their two levels. This is almost exactly the difference between the wave numbers of the two strong nebular lines 5006·8 and 4958·9 A (193 cm.$^{-1}$). It is, therefore, quite probable that these two rays represent the transition $^3P_2-{}^1D$ and $^3P_1-{}^1D$ (Bowen).

[O II]. The diagram of the lower levels of the ion O^+ is also represented by Fig. 4. The two metastable doublets ^2D and ^2P are found above the ground level ^4S. There can be no question of electric dipole transition between them as all three have the same electronic configuration. Furthermore, the transitions between metastable states and the basic level are intercombination rays.

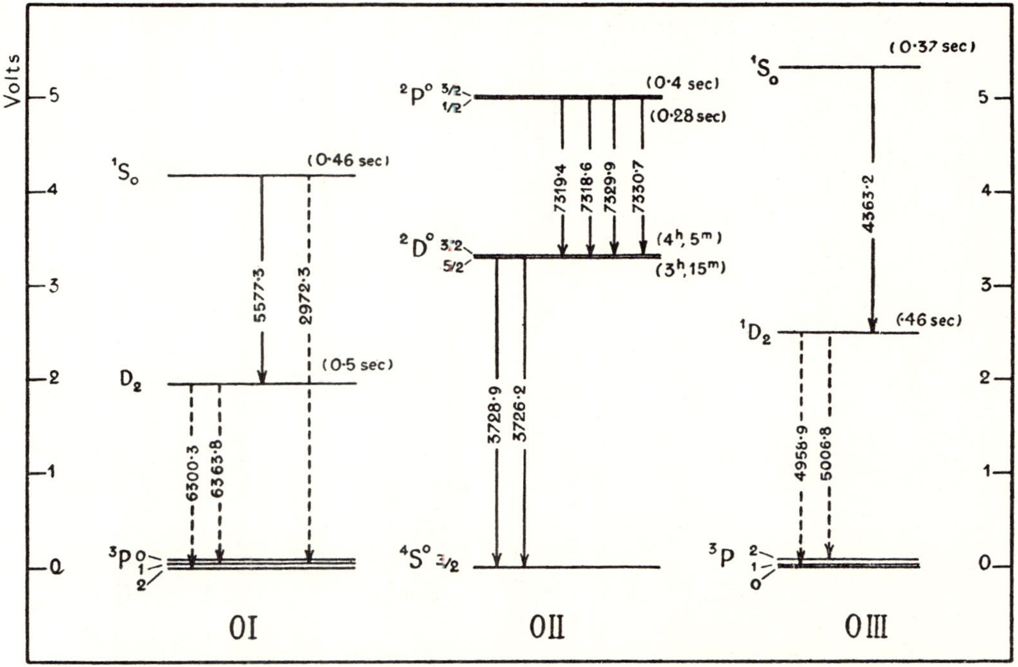

Fig. 4.—The lowest energy levels of O I, O II and O III. The separations in the multiple levels have been exaggerated. The straight lines show transitions due to electric quadrupole radiation, the dotted lines those due to magnetic dipole radiation. The lifetimes of the metastable levels are given in parentheses and follow from the transition probabilities calculated by Pasternack.

The difference in the wave numbers of the ultra-violet rays to the levels ^2D and ^2P respectively, also provides us with the wave number of the transition $^2D-^2P$. The wave-length is found to be 7326·2 A and since, in the nebulae, we can observe, with weak dispersion, the line 7325 A, the latter is thus identified. In reality, since each level is double, a number of components exist which are known today.

The identification of the rays $^4S^\circ_{\frac{3}{2}} - {}^2D^\circ_{\frac{5}{2}}$ and $^4S^\circ_{\frac{3}{2}} - {}^2D^\circ_{\frac{3}{2}}$ has increased the difficult Boies.wen calculated their *approximate* wave-lengths by considering analagous levels of other atoms of the same electronic configuration. He thus obtained 3681·25 and 3678·81 A. The values are fairly far removed from the wave-length of the two strong nebular rays 3728·91 and 3726·16 A, but the difference in the calculated wave-lengths (2·34 A) is much more definite than the wave-lengths themselves, and is sufficiently near the difference of observed wave-lengths to allow identification.

These examples illustrate the three types of reasoning most frequently utilized in these researches:

1—The difference in the respective wave numbers of the two ultra-violet lines converging to the two levels considered, immediately gives the wave numbers of the forbidden line ([O III] $^1D-^1S$, [O II] $^2D-^2P$). This is the most favourable case.

2—The difference in the wave numbers of the ultra-violet rays to the two components of the same multiple level, provides the separation of these levels ([O III] $^3P-^1D$).

3—The evaluation of the unknown separation of the two levels by means of the known separation of the analagous levels of atoms with the same electronic configuration ([O II] $^4S-^2D$). In fact, in this iso-electronic series, the separation of the levels is a function of the atomic number. This is obviously the most doubtful case.

The progress which has been made in the last twenty years in the study of ultra-violet spectra has led to the identification of a great number of forbidden lines in the nebulae. Table II gives a list of the principal ones among them in the region 3700–6800 A, which is the best-known one. For [Fe II], [Fe III], [Fe VI] and [Fe VII], only the most intense lines of the most characteristic multiplets have been retained. In this Table, other lines belonging to [Fe V], [K V], [K VI], etc., are not shown. The excitation energy *W* of the superior level is given in electron volts.

Beyond 6800 A, forbidden lines are still found belonging particularly to [O II] (7319, 7330 A), [A III] (7136, 7751 A), [A IV] (7169, 7236, 7263 A) and [AV] (7006 A).

TABLE II

Principal Forbidden Lines identified in the Nebulae between 3700 and 6750

Identification			*(eV)*	*Identification*			*(eV)*
*3726·16	O II	$^4S^0_{\frac{3}{2}}-{}^2D^0_{\frac{3}{2}}$	3·31	5158·3	Fe VII	$^3F_3-{}^3P_1$	2·52
				5177·0	Fe VI	$^4F_{\frac{9}{2}}-{}^2G_{\frac{9}{2}}$	2·63
*3728·91	O II	$^4S^0_{\frac{3}{2}}-{}^2D^0_{\frac{5}{2}}$	3·31				
*3868·74	Ne III	$^3P_2-{}^1D_2$	3·19	5198·5	N I	$^4S^0_{\frac{3}{2}}-{}^2D^0_{\frac{3}{2}}$	2·37
*3967·51	NE III	$^3P_1-{}^1D^2$	3·19				
*4068·62	S II	$^4S^0_{\frac{3}{2}}-{}^2P^0_{\frac{3}{2}}$	3·03	5200·7	N I	$^4S^0_{\frac{3}{2}}-{}^2D^0_{\frac{5}{2}}$	2·37
*4076·22	S II	$^4S^0_{\frac{3}{2}}-{}^2P^0_{\frac{1}{2}}$	3·03	5276·1	Fe VII	$^3F_4-{}^3P_2$	2·63
4243·98	Fe II	$^4F_{\frac{9}{2}}-{}^4G_{\frac{11}{2}}$	3·14	5308·9	Ca V	$^3P_2-{}^1D_2$	2·32
				5322·2	Cl IV	$^1D_2-{}^1S_0$	4·02
4276·83	Fe II	$^4F_{\frac{7}{2}}-{}^4G_{\frac{9}{2}}$	3·19	*5517·2	Cl III	$^4S^0_{\frac{3}{2}}-{}^2D^0_{\frac{5}{2}}$	2·24
4287·40	Fe II	$^6D_{\frac{9}{2}}-{}^6S_{\frac{5}{2}}$	2·88	*5537·7	Cl III	$^4S^0_{\frac{3}{2}}-{}^2D^0_{\frac{3}{2}}$	2·23
4359·34	Fe II	$^6D_{\frac{7}{2}}-{}^6S_{\frac{5}{2}}$	2·88	5577·35	O I	$^1D_2-{}^1S_0$	4·18
*4363·21	O III	$^1D_2-{}^1S_0$	5·33	5631·6	Fe VI	$^4F_{\frac{7}{2}}-{}^4P_{\frac{3}{2}}$	2·34
4416·27	Fe II	$^6D_{\frac{9}{2}}-{}^4F_{\frac{9}{2}}$	2·79	5678·0	Fe VI	$^4F_{\frac{9}{2}}-{}^4P_{\frac{5}{2}}$	2·42
4658·1	Fe III	$^5D_4-{}^3F_4$	2·65	*5720·9	Fe VII	$^3F_2-{}^1D_2$	2·16
4701·5	Fe III	$^5D_3-{}^3F_3$	2·68	*5754·8	N II	$^1D_2-{}^1S_0$	4·04
*4711·4	A IV	$^4S^0_{\frac{3}{2}}-{}^2D^0_{\frac{5}{2}}$	2·62				
				*6085·5	Fe VII	$^3F_4-{}^1D_2$	2·16
*4740·3	A IV	$^4S^{0\,2}_{\frac{3}{2}}-D^0_{\frac{3}{2}}$	2·60	6085·9	Ca V	$^3P_1-{}^1D_2$	2·32
4930·5	Fe III	$^5D_1-{}^3P_0$	2·62	*6300·30	O I	$^3P_2-{}^1D_2$	1·96
4931·8	O III	$^3P_0-{}^1D_2$	2·50	*6363·8	O I	$^3P_1-{}^1D_2$	1·96
*4958·91	O III	$^3P_1-{}^1D_2$	2·50	6434·9	A V	$^3P_2-{}^1D_2$	2·01
4985·9	Fe III	$^5D_4-{}^3H_6$	2·48	*6548·1	N II	$^3P_1-{}^1D_2$	1·89
*5006·84	O III	$^3P_2-{}^1D_2$	2·50	*6583·6	N II	$^3P_2-{}^1D_2$	1·89
5032·7	Fe III	$^5D_3-{}^3H_5$	2·51	6717·0	S II	$^4S^0_{\frac{3}{2}}-{}^1D^0_{\frac{5}{2}}$	1·84
5146·8	Fe VI	$^4F_{\frac{7}{2}}-{}^2G_{\frac{7}{2}}$	2·54	6731·3	S II	$^4S^0_{\frac{3}{2}}-{}^2D^0_{\frac{3}{2}}$	1·83

*Denotes most intense wave-lengths.

NOTES TO CHAPTER I

[1]The spectrum of a neutral atom is denoted by the symbol of the element followed by the Roman numeral I (first atomic spectrum), that of the ionized atom by the symbol followed by the Roman II, etc. Thus H I represents the spectrum of the neutral helium atom, H II that of H+. Similarly, Fe VI, Fe VII would represent the spectra of iron five or six times ionized.

[2]We express the fact that the energy is the product of the difference of potential V and the charge e of the electron $(4{\cdot}803 \times 10^{-10}$ e.s.u.) by

$$h\nu = hc\bar{\nu} = eV \text{ (erg.)},$$

which holds in the electrostatic C.G.S. system. V is thus expressed in Volts (1 volt $= \frac{1}{300}$ e.s.u.) by

$$V = \frac{300\, hc\bar{\nu}}{e}$$

With the values given for h, c and e, we obtain

$$V = \frac{\bar{\nu}}{8085} \ (\bar{\nu} \text{ in cm}^{-1}).$$

[3]We symbolize electronic transitions by first writing the symbol of the lower level, followed by that of the higher level.

CHAPTER II

THE EXCITATION OF BRIGHT LINES IN THE NEBULAE

1. Association of Nebulae and Stars

MOST planetary nebulae are organized around a central star. Although quite apparent in the *ring* nebulae,* this is often less visible when the brightness decreases from the centre outwards. In those cases where it is not clearly distinguishable, we are probably correct in thinking that it is masked by the great luminosity of the central nebular region.

Stras associated with planetary and diffuse nebulae

The existence of one or more stars associated with diffuse nebulae is far from being evident, a priori. But after thorough investigations, Hubble found evidence for them in almost all those regular nebulae which he studied.[1] He also discovered an important relation between the nature of the nebula and the spectrum of the associated star [147].

The nuclei of planetary nebulae are always stars of high temperature and rich in ultra-violet radiation. Their spectra often resemble those of a Wolf-Rayet star, but they have less broad emission lines. In certain cases the lines of carbon and nitrogen in them are of comparable intensities. The nuclei of other nebulae show both emission and absorption lines; others again only show absorption lines characteristic of O stars. Finally, we meet planetary nebulae whose nuclei are apparently very hot stars with entirely continuous spectra and showing neither emission nor absorption lines.

The stars associated with irregular nebulae of the line spectrum type are less hot. They belong either to the last type of the class O, or to the type B0. Those which accompany continuous-spectrum nebulae are at lower temperatures still: their spectral type is more "advanced" than B0. We can, therefore, observe a decrease in the temperatures of the associated stars, as we proceed from planetary nebulae to those with continuous spectra.

*At least on photographs, for the central star of the nebula can rarely be seen through the eye-piece of the telescope. By reason of its very high temperature the central star radiates mainly in the violet and in the ultra-violet.

This relation would suggest that *the luminosity of all the Galactic nebulae could, in a definite and unique manner, be derived from the illumination of the associated stars.* That of continuous-spectrum nebulae is due to the scattering of light from the star by atoms of gases or by the solid particles which constitute the nebula. Nebulae with bright-line spectra are the result of more complex phenomena: fluorescence and ionization of gases under the action of ultra-violet radiation of the star.

Hubble has endeavoured to be more precise by measuring the maximum angular distance α of the exciting star with which the nebulae can be photographed under standard conditions. He found that if we plotted logarithms of α against the magnitude m of the exciting stars, the respective co-ordinates of the eighty-two diffuse nebulae examined lay nearly in a straight line of slope -5. Hubble saw in this result the proof that the brightness of a nebula is proportional to the illumination produced by the star. Furthermore, the straight-line graph resulting from these observations would seem to indicate that the nebulae reflects or re-emits towards the observer, all the light which it receives from the star. This result is the more surprising in that Hubble utilized both line- and continuous-spectrum nebulae. Since then, Cederblad (1946) who considered the two classes of nebulae separately, found slightly different straight lines for the two cases [54].

A similar work, on fifty-six planetary nebulae, led to a similar graph (slope of approximately -5). But in this case the brightness of the nebulae exceeded by 4 to 4·5 magnitudes, that resulting from the complete re-emission of stellar light (Hubble) [147]. It is, therefore, no longer the illumination of stars observed on photographs which causes the luminescence of planetary nebulae. It is natural to suppose that the actual illumination is situated in a region of the far ultra-violet and cannot penetrate the terrestrial atmosphere. It may be extremely intense because of the high temperatures of the W and O stars.

In spite of the somewhat rough approximation and haphazard methods of reasoning involved in them, Hubble's statistical researches have been of very great importance and have served, as we shall see, as the basis for the work of Zanstra.

Here, then, is a rough reconstruction of Hubble's reasoning. In Fig. 5 let S be a star, M a point in the nebula at a distance SM$=p$, and O the observer, at a distance SO$=r$. The illumination produced at O by the star of luminous intensity J is $l=\frac{J}{r^2}$; the illumination produced at M is $\frac{J}{p^2}$. If the brightness of the nebula at M is proportional to the illumination

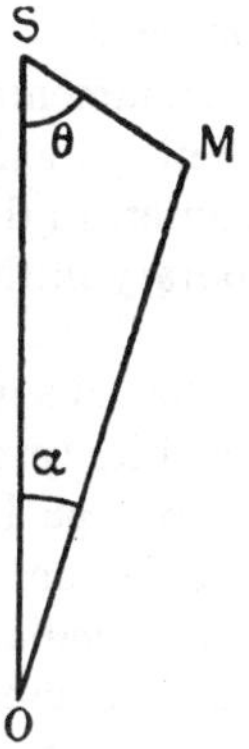

Fig. 5.

received we can write $B=\frac{CJ}{P^2}=\frac{Clr^2}{P^2}$ where C is a constant. Further, the angular distance $\widehat{SOM}=\alpha$ has a value $\frac{p \sin\theta}{r}$ and we can then write for the brightness

$$B=\frac{Cl \sin^2\theta}{\alpha^2}.$$

In considering a great number of nebulae the lines SM may be considered as orientated at random. We, therefore, take the mean value of $\sin^2\theta$ between O and π, i.e. $\frac{1}{2}$ (it is of course quite possible that this value is too small, for since the nebula entirely surrounds a star, θ should be taken as $\frac{\pi}{2}$. Hubble takes the mean value $\frac{2}{\pi}$ of $\sin\theta$ which leads to a somewhat less satisfactory numerical value). We therefore, write:

$$B=\frac{Cl}{2\alpha^2}.$$

By multiplying the logarithms of both sides by $-2{\cdot}5$, we obtain

$$-2{\cdot}5 \log B = -2{\cdot}5 \log C - 2{\cdot}5 \log l + 2{\cdot}5 \log 2 + 5 \log \alpha.$$

$-2{\cdot}5 \log l$ and $-2{\cdot}5 \log B$ are measures of (with almost the same arbitrary constant) the apparent magnitude m of a star and the magnitude μ of the nebula per unit solid angle. We then have

$$m=\mu+2{\cdot}5 \log C - 0{\cdot}75 - 5 \log \alpha.$$

Since, under standard conditions, the maximum angular distance α corresponds to a fixed brightness, μ is a constant and the variation of m with $\log \alpha$ is represented by a straight line of slope -5.

Further, the value of μ is known from Seares's observations ($\mu=11{\cdot}61$ magnitudes per square minute of arc). The experimental value of the origin ($+11{\cdot}02$ magnitudes) gives us $\log C=0{\cdot}064$, whence $C=1{\cdot}16$. In view of the lack of precision of the data, it is reasonable to conclude that C is almost unity.

All this assumes that the nebula *reflects* all the light received from the star towards the observer, and that it behaves like a perfect mirror and not like a perfect diffuser, for which C could not exceed the value of $\frac{1}{\pi}$.

The reasoning is obviously based on an over-simplification. It considers a nebula as a surface although it is in fact a volume. Light received by the observer does not come from a point but from the entire depth of the nebula. We must remember, nevertheless, that the light re-emitted by the nebula is of the same order of magnitude as that which it receives from the star.

Is the association of stars with nebulae permanent or only temporary?

It is of the utmost importance to know whether there is a fundamental physical connexion between the nebulae and the stars illuminating them, or if their association is simply the result of a fortuitous encounter. In the first case, the connexion would be permanent, but only temporary in the second.

In trying to settle this question it is natural to compare the velocity of the nebula with that of the star. For this we take radial velocities (projected on the line of vision) easily measured with the spectroscope by application of the Doppler-Fizeau effect. The displacement of emission or absorption lines towards the violet or the red permits the immediate evaluation of the speed of approach or separation of any star whatever. The fine bright lines of nebulae lend themselves particularly well to a determination of radial velocities, whereas measurement can become more difficult in the case of associated stars which do not always show fine lines.

This is one of the reasons why we know so little about the velocities of the nuclei of planetary nebulae, even though, in 1918, Campbell and Moore [52] had already measured the radial velocities of 125 planetary nebulae. Their observations led to the discovery of complex phenomena. The spectrum of a certain number of planetary nebulae photographed at high dispersion, shows lines of an irregular form interspersed here and there with regions of very close lines coming from the central regions of the nebulae. The radial velocity, therefore, varies from one point of the nebular image to the next, revealing the existence of turbulent internal movements, in the gaseous masses. A general inclination of the lines towards a spectrum juxtaposed for the purposes of comparison, may be explained by a very long period movement of rotation of the entire mass (thousands or ten thousands of years). This interpretation, however, is false (Zanstra). A most significant fact is the splitting of the lines, interpreted by Zanstra [389] as being due to the *continuous expansion* of planetary nebulae. The violet component of the line is emitted by the anterior face of the nebula which in receding from the nucleus, approaches the observer, and the red component is emitted from the posterior face which is receding from the observer. This expansion has been confirmed by all subsequent research. Its existence is well established in at least a dozen planetary nebulae with speed ranging between 4 and 53 km./sec., and it is suspected in twenty-five others (Stoy [312a]). In the case of NGC 3242, the large-scale spectra obtained by O. Wilson [380] have shown that the velocity of expansion varies with the atoms considered. It is 41 km./sec. for hydrogen and 34 km./sec. for Ne^+ and He^+, which are at a high ionization potential.

Thus, the problem which we have posed is solved for the case of planetary nebulae, not by a comparison of the radial velocities of the gas and the stellar nucleus, but by the expansion. This proves that there does exist, between the nebula and the star, a fundamental and permanent physical bond which the very appearance of the nebula originally led us to suspect. It also suggests that gases must be ejected by the star.

We are much less well informed on the radial velocities of diffuse nebulae. In a few cases which we have been able to measure, the value of the radial velocity of the entire nebula differs markedly from that of the star illuminating it. Thus, for the Great Nebula in Orion v_r was found to be $+17\cdot5$ km./sec. (separation velocity) and for the star θ_1 Orionis it was found to be approximately $+30$ km./sec. The Trifid Nebula gives $v_r = -11$ km./sec. and the star which illuminates it $+7\cdot5$ km./sec. This seems to favour the theory of fortuitous encounter. But published works are rare, and what is more, in most diffuse nebulae, turbulent movements must exist, making the determination of the radial velocity of the whole even more uncertain. These have been clearly shown to be present in the Great Nebula of Orion where their amplitude attains some 10 km./sec. [97], as shown by Fabry, Buisson and Bourget's interference measurements (1914).

We shall see further on (Chapter XI) that this problem has been resolved by statistical methods for the case of continuous-spectrum nebulae. The association of the nebula with the star may, therefore, be considered as fortuitous and transitory. It is most probable that this is also the case for irregular line-spectrum nebulae. This impression will be reinforced by the study of Galactic emission areas (Chapter IV). Thus, there seems to be a fundamental difference between planetary nebulae on the one hand, and bright-line, or continuous-spectrum diffuse nebulae on the other.

2. The Mechanisms of Line Excitation (Bowen)

We are now going to investigate according to what mechanisms the bright lines of the nebulae may be excited, assuming that the total energy is definitely derived from ultra-violet radiation of the associated star [48].

First Mechanism: photoexcitation, photoionization and recombination

The atoms of the nebulae are almost all at ground level, and are capable of absorbing certain well-defined wave-lengths from the spectrum of the star. They are thus raised to excited levels, from which they drop, in a very short time, emitting permitted lines.

Thus a normal hydrogen atom (level 1 of Fig. 2) can be raised to the level 2, absorbing a photon of wave-length 1215 A (line $L\alpha$), whence it can only redescend to the level 1 by emitting the line Lα. But, if our atom, originally on level 1, should absorb a photon of wave-length 1025 A (line Lβ) which raises it on to level 3, there are two paths by which it can return to the normal level: it can either return directly to the level 1, with an emission of the Lβ line, or it can first descend to the level 2, emitting the Hα line, then to the level 1, emitting the Lα line (the reader is again referred to Fig. 2). The absorption of radiation on a still higher level in the Lyman Series, naturally leads to even more varied possibilities.

This *direct photoexcitation*, however, only plays an accessory role, for it can be brought about only by radiation of well-determined frequency, which, to all intents and purposes, is only a small proportion of the energy radiated by the star. The ionization of hydrogen atoms of the nebula by the continuous spectrum of the star is certainly a much more important phenomenon, since all photons of frequency ν, greater than the limit ν_0 of the Lyman Series (wave-length 912 A) can be absorbed by the hydrogen atoms. The energy necessary for the ionization of an atom on the ground level (level 1, electron in an orbit of rank 1 in Bohr's scheme) is in fact $h\nu_0$. A photon of frequency higher than ν_0 thus ionizes the atom and communicates to the expelled electrons, of mass m, a certain speed v, and a kinetic energy $m\frac{\text{v}^2}{2}$ given by the relation

$$h\nu = h\nu_0 + \tfrac{1}{2}m\text{v}^2 \qquad \text{(II.1)}$$

which expresses the conservation of energy.

Conversely, the recombination of a proton and an electron liberates a certain quantity of energy, which appears in the form of radiation. If the electron of speed v settles in an orbit of rank 1 to give us a normal atom, the emitted frequency still satisfies the relation (II.1). The capture of electrons on this orbit thus leads to the emission of the continuous spectrum associated with the Lyman Series ($\nu > \nu_0$), situated in the unobservable region of the ultra-violet. The electrons, however, may just as well settle on an orbit of rank 2, thus producing emission of the continuous spectrum associated with the Balmer Series (wave-length $<$3646 A), which is in fact observed in the spectra of the nebulae (Wright). In redescending to the orbit of rank 1 (normal level of the atom), these electrons will consequently emit the ray Lα. The capture of electrons into orbits of higher rank explain, in the same way, the emission of all the hydrogen rays, particularly those of the Balmer Series.

This mechanism of *photoionization* followed by *recombination*, can of course be applied to all the other atoms which are contained in the nebula. The only difference is that, from one atom to another, the minimum frequency capable of producing the ionization varies. Here then are the ionization energies E_i of the most abundant atoms and ions, together with the limits of their corresponding wave-lengths:

	H	C^+	He	N^+	O^+	N^{++}	C^{++}	He^+	O^{++}	C^{+++}	
E_i	13·54	24·28	24·48	29·49	35·00	47·24	47·67	54·17	54·71	64·22	eV
λ	912	510	506	420	354	262	260	229	226	193	A

These *recombination spectra* are *complex* spectra, containing all permitted lines of the atoms. The mechanism thus explains the excitation of the lines of H I, C II, He I, N II, O II, He II, in the nebulae, but is incapable of accounting for the selection of lines which are often observed in the spectra of O III and N III (page 43).

The return of atoms to the normal level is always accompanied by ultra-violet radiation which is capable of ionizing or exciting other atoms in their turn. This is also so in the case of continuous ultra-violet spectra, i.e. those which lie beyond the Lyman Series. This produces secondary photoexcitation and photoionization. Those ultra-violet radiations which are capable of being absorbed in the nebula, as are the Lyman Series and its continuous spectrum, have therefore, little chance of leaving it. Radiation of greater wave-length, like the Balmer lines and the corresponding continuous spectrum, can leave it more easily, since they are no longer capable of ionizing other atoms.

Second Mechanism: Excitation by electron collisions

Instead of recombining with positive ions, the electrons, torn from the atoms by ultra-violet radiation of the star, can excite atoms at ground level by collisions, and so imparting to them all or part of their kinetic energy. The atom, raised to an excited level, descends in turn to the ground level and emits its characteristic lines.

This mechanism is not very effective in exciting permitted lines, since the energy of photo-electrons is rarely sufficient for raising the atoms to those levels corresponding to the emission of resonance lines. The latter are situated mainly in the far ultra-violet. On the other hand, electron collisions account for the excitation of forbidden lines. The examination of Table II shows that electrons of kinetic energies between two and five electron volts (velocities of the order of 840 to 1330 km./sec.) suffice to raise the major part of the atoms and of the ions to metastable levels. They descend from these levels, in a somewhat long time, emitting forbidden lines.

Third Mechanism: Fluorescence by chance coincidence

In those nebulae whose spectrum shows ionized helium lines, the return of the He^+ ion to the ground level is necessarily effected with emission of the resonance line (303·780 A). Now this is known to coincide almost exactly with the resonance line of O III (303·799 A). The difference between these two wave-lengths is negligible in comparison with the wave-lengths themselves, and the O^{++} ion can thus absorb the helium line and be carried from the ground level to a certain excited level $^3P_0$. It descends either by emitting the line 303·799 A or in a *series of jumps*, whereby it successively emits a certain number of permitted lines.

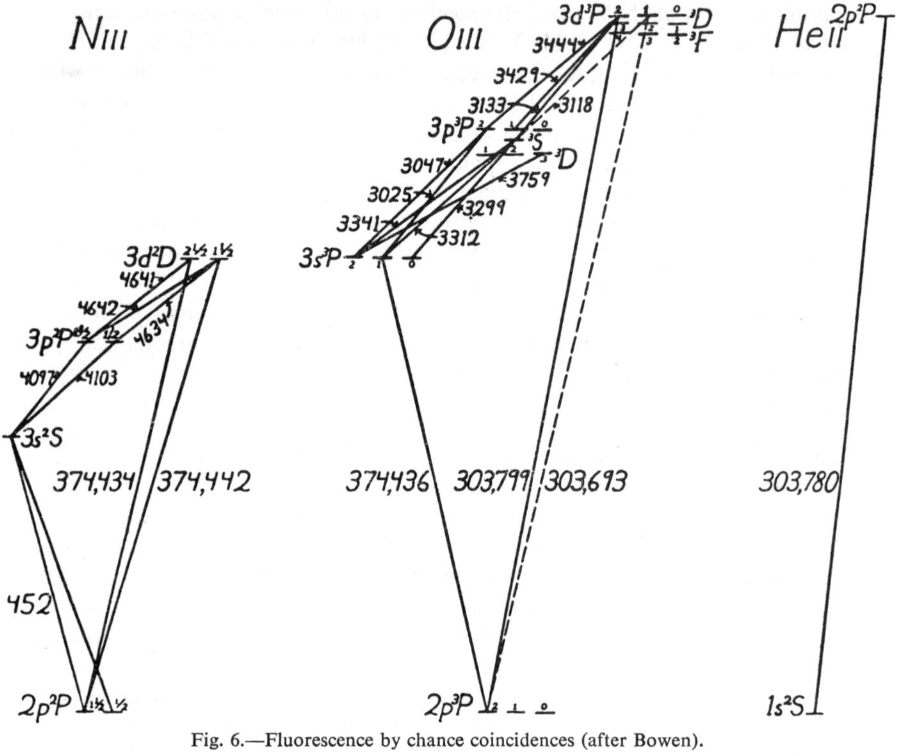

Fig. 6.—Fluorescence by chance coincidences (after Bowen).

We can thus predict the emission of *certain* well-defined lines of O III, and of those only (Fig. 6), and it is just these which appear with great intensity in many of the nebulae.

In its descent by jumps, the O III ion finally comes back to its ground level emitting the resonance line 374·436 A. By a further chance this happens to coincide, very nearly, with resonance lines 374·434 and 374·442 A of N III. The gaps are easily bridged by the small radial velocities of the ions (Doppler-Fizeau principle). N III can, therefore, in its turn, absorb the line 374·436 A of O III and thus finds itself raised to one or the other

γ Cas.

γ Cas.

Plate VI.—Nebulosities near γ Cassiopeiae (NGC 59). 80 cm. telescope (M. de Kérolyr).

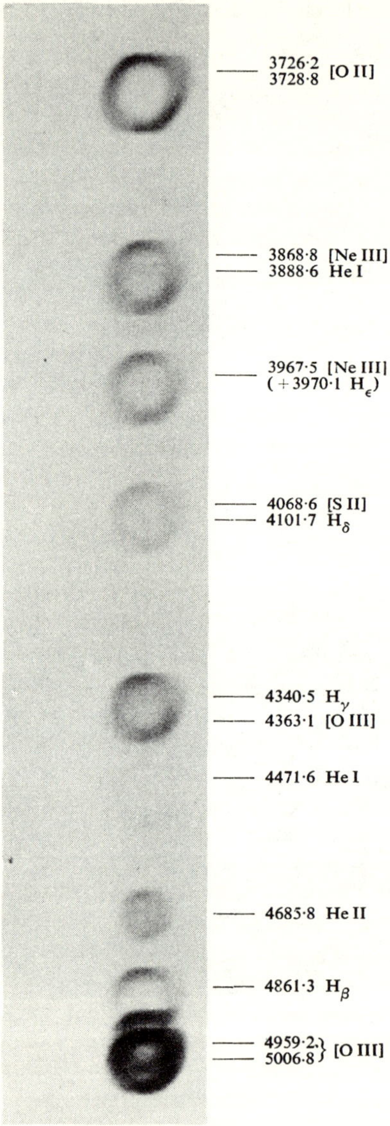

Fig. 7.—Spectrum of the Ring Nebula in Lyra (NGC 6720) obtained by using a wide slit with the single-prism spectrograph mounted on a 120 cm. telescope (Ch. Fehrenbach).

of the two very close levels $^2D_{\frac{5}{2}}$ and $^2D_{\frac{3}{2}}$. The descent is then effected either by the emission of one or the other of the lines 374·434 and 374·442 A or by jumps. In the latter case, we can forecast the emission of certain selected lines of the spectrum of N III, and it is precisely these very lines which we find very reinforced in the nebular spectra (4641, 4634, 4103, 4097 A).

By the action of these two chance coincidences, Bowen gave a perfect explanation of the curious appearance of the O III and N III lines in the spectra of many nebulae.

3. Radiation of Nebulae and Temperatures of Associated Stars

In 1931, Zanstra [387, 388] gave the theory of emission in gaseous nebulae a quantitative form, and the study of their spectra enabled him to determine the temperature of the exciting stars indirectly. We shall try here to give an outline of his theory without entering into the details of his calculations.

Monochromatic images of planetary nebulae by means of a slitless spectrograph

Let us glance at Fig. 7 (opposite) of a spectrum of the Ring Nebula in Lyra obtained by means of a slitless spectrograph.[2] Each line thus gives a monochromatic image reproducing the shape of the nebula. But different images have different dimensions: the line 4686 A of ionized helium gives a smaller image than the other lines, e.g. neutral helium (4472 A) or hydrogen (Hβ).

In order to interpret this curious appearance let us recall that the photoionization of the He^+ ion, which is necessary for the production of the recombination-spectrum, requires radiation of very short wave-length ($\lambda<229$ A, see page 63), whereas radiation of greater wave-length ($\lambda<506$ A) is capable of ionizing a neutral He atom, or a hydrogen atom ($\lambda<912$ A). The shortest wave-lengths ($\lambda<229$ A), can be completely absorbed at a certain distance from the central star. Beyond that, the lines of He^+ are no longer excited. Radiation of wave-lengths 229 and 506 A is effective over longer distances, before it in turn is absorbed. The emission of lines of the neutral helium atom therefore extends over a larger distance from the star. Those of the hydrogen atom extend even further.[3] Thus, we must assume that *ultra-violet radiation of short wave-length is completely absorbed in the interior of the nebula.* This is one of the fundamental hypotheses of Zanstra's theory.

The forbidden lines of [O II], [O III], [Ne III] give images of larger diameter. They are excited by the collision of liberated electrons nearer the nucleus, and by the ionization of atoms of hydrogen and other gases.

Zanstra's Theory: Ionization and recombination

Let us suppose that the nebula, formed almost entirely of hydrogen, absorbs, in the radiation of the star, all those photons whose frequency is greater than the limit ν_0 of the Lyman Series. Let N_0 be the number of these photons emitted by the star in one second. Each one of them ionizes the hydrogen atom in such a manner that the number of hydrogen atoms ionized in one second is also N_0. If we admit that the nebula is in equilibrium there are as many proton-electron recombinations per second as there are ionizations. Thus, the number of recombinations per second is still N_0.

At the moment of recombination, return of the hydrogen atom to the ground level can take place in various ways, shown in Fig. 8, where the wave-lengths of the continuous spectra of Lyman, Balmer, Paschen and Brackett, are represented by L_c, B_c, P_c and Br_c, respectively, and the lines of the corresponding series by the usual symbols (see Fig. 2).

A certain number of atoms return directly to the ground state (case a of Fig. 8) with emission of a wave-length belonging to the continuous Lyman spectrum. The other paths necessarily involve the emission of a line of the same series as a final step (cases b, c, . . . , h). Thus the number N_0 of recombinations per second is the sum of the number n (L) of photons emitted in one second in the Lyman Series and of the number n (L_c) of photons emitted in the adjacent continuous spectrum.

$$N_0 = n(\mathrm{L}) + n(\mathrm{L_c}). \qquad \text{(II.2)}$$

Since all these transitions are unobservable, we endeavour to evaluate their number by means of those observable transitions which precede them.

Let us first consider the processes leading up to the emission of the $\mathrm{L}\alpha$ line. To begin with, all of them require either the emission of a wave-length of the continuous Balmer spectrum (case b of Fig. 8), or of a line of the same series: $\mathrm{H}\alpha$ (case c and d), $\mathrm{H}\beta$ (case e), etc. If we represent by n(B) and $n(\mathrm{B_c})$ respectively, the number of transitions in the Balmer Series and its continuum, the number $n(\mathrm{L}\alpha)$ of photons emitted per second in the line $\mathrm{L}\alpha$ is then

$$n(\mathrm{L}\alpha) = n(\mathrm{B}) + n(\mathrm{B_c}).$$

Similarly, we could evaluate the number of transitions $\mathrm{L}\beta$ by means of the preceding transitions in the Paschen Series and its continuum. But in fact Lyman radiation has very little chance of leaving the nebula, since the absorption co-efficient of hydrogen is much larger for these lines than for

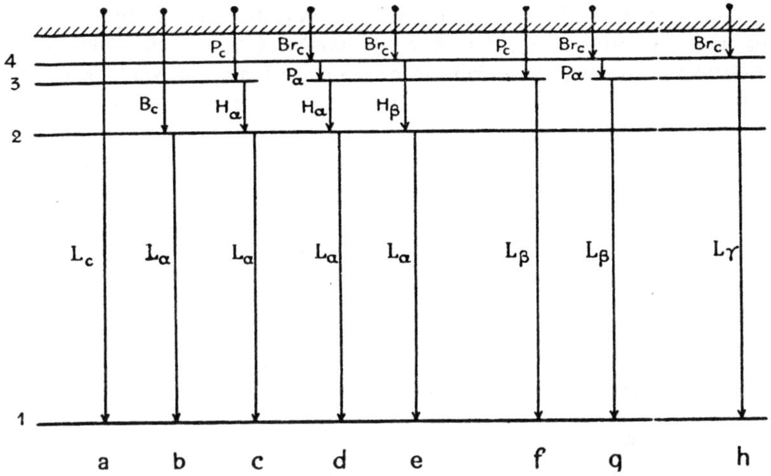

Fig. 8.—Recombination of an electron and a proton; various wave-lengths emitted when the hydrogen atom drops to the ground level.

the adjacent continuous spectrum. The majority of photons Lβ, Lγ, etc., are absorbed by other hydrogen atoms, which are then raised to excited levels. In this case, each emission of any line whatsoever of the Lyman Series is followed by the emission of the line Lα and, in practice, we can write

$$n(\mathrm{L})=n(\mathrm{L}\alpha).$$

By replacing n (L) by that value obtained for n (Lα), in (II.2) we have

$$N_0=n(\mathrm{B})+n(\mathrm{B_c})+n(\mathrm{L_c}). \qquad \text{(II.3)}$$

The number N_0 *of photons emitted by the star for frequencies greater than* ν_0 (equal to the number of ionizations and of recombinations) *is thus the sum of the number of photons emitted by the nebula in the Balmer Series and in the continuous spectra of Balmer and Lyman.*

It is impossible to evaluate the number n (L$_c$) of photons of the Lyman continuum, which actually leave the but nebula, since it must be very small we can, at first approximation, neglect it and the equation (II.3) becomes

$$N_0\simeq n(\mathrm{B})+n(\mathrm{B_c}). \qquad \text{(II.4)}$$

The right-hand side contains measurable quantities only, since we can observe the entire Balmer Series and its continuous spectrum through the terrestrial atmosphere.

Application to diffuse nebulae

We shall endeavour to evaluate the relation between the total photographic brightness of the nebula and the light which it intercepts in the same spectral range.

Let us assume for the moment that we know the curve $N(\nu)$, representing as a function of the frequency ν the variations in the number of photons received by the nebula from the star in each very small interval of frequency $d\nu$. The Number N_0 of photons emitted beyond the Lyman limit ν_0 is represented by the area S_1 which lies between the curve, and the

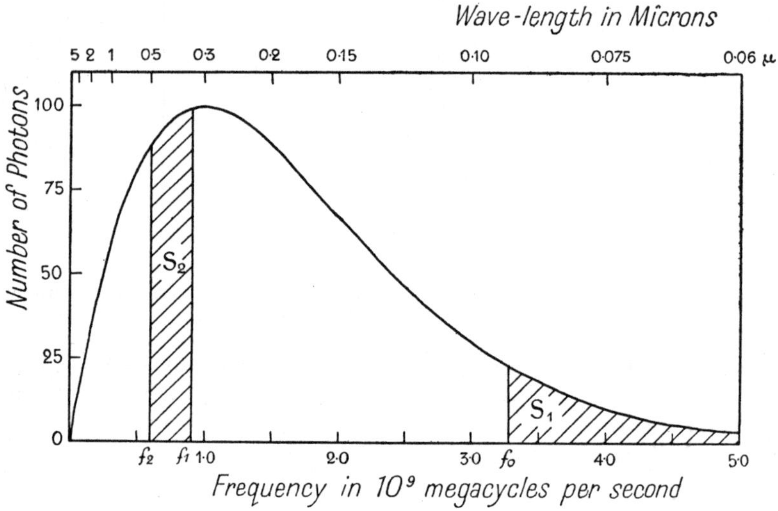

Fig. 9.—Graph as a function of the frequency v of the number of photons N_v emitted per second per square centimetre by a black body at T=30,000°K between frequencies v and $v+dv$ (calculated by means of formula (0.14)). $\frac{S_1}{S_2}=0.985$.

abscissa beyond ν_0 (see Fig. 9). This can be evaluated in the nebula, by means of relation (II.4) if we measure the total intensity of all the Balmer lines and of the adjacent continuous spectrum; that is to say, in practice by measuring the total photographic brightness of the nebula. In the "photographic" region the number N_g of photons emitted by the star and intercepted by the nebula is represented by the area S_2, which is limited laterally by the ordinates corresponding to wave-lengths λ_1 and λ_2 which bound the spectral region which we employ. By using (as has been the custom for some twenty years) a telescope with a silver mirror and so-called "ordinary" photographic plates, we effectively have λ_1=3300 A, λ_2=5050 A.

According to Hubble's statistical researches (page 58), the number N_g' of photons emitted by the nebula in the photographic region is approximately equal to the number N_g of photons which the nebula intercepts in the radiation of the star between the same limiting wavelengths.

$$\frac{N_g'}{N_g} \simeq 1 \qquad \text{(II.5)}$$

The ratio of the areas S_1 and S_2, therefore, ought to be near unity. But it depends uniquely on the temperature of the star. If we suppose that the latter radiates like a black body, we can construct the curves $N(\nu)$ with respect to different temperatures and investigate in which case $S_1=S_2$. This is the case with a temperature in the vicinity of 30,000°K. We find this figure quite acceptable for those stars of the final O type and of the type B0 which are generally associated with diffuse nebulae (see Table I, page 26).

In actual fact, the measurement of total photographic brightness of the nebula does not include the Hα line which has no effect on "ordinary" plates. We must, therefore, write

$$N_g' = N_0 - n(\mathrm{H}\alpha),$$

denoting by $n(\mathrm{H}\alpha)$ the number of photons emitted in the Hα line. On the other hand, total photographic measurement is not affected by hydrogen lines only, since the nebula is not exclusively constituted of that gas. In actual fact, the theory can only give us an idea of the order of magnitude.

Application to planetary nebulae

The application to planetary nebulae leads to more interesting quantitative results.

We still use relation (II.4). To evaluate $n(\mathrm{B})$, Zanstra uses negatives obtained by means of a slitless spectrograph and measures the ratio r_1 between the total illumination H_1 produced by each of the Balmer lines, and the illumination J_1 produced by the continuous spectrum at the same frequency. He obtains the number of quanta emitted by one of the Balmer lines, by dividing H_1 by the energy $h\nu$ of the quantum. Thus this number is

$$\frac{H_1}{h\nu} = \frac{r_1 J_1}{h\nu}$$

and the relation (II.4) becomes

$$\sum \frac{r_1 J_1}{h\nu} = n(\mathrm{B}) + n(\mathrm{B_c}) = N_0. \qquad \text{(II.6)}$$

The summation ought to be extended to include the entire Balmer Series and, in principle, the continuous spectrum beyond it.

Further, by comparing the radiation of the star to that of a black body, we can calculate, for each temperature T, both the intensity J_1, and also N_0, and we observe for which temperature the relationship is satisfied.

Second Mechanism: Excitation by electron collisions

The nebula, completely surrounding the star is, in the main, formed of atoms of hydrogen. These then constitute the principal source of photo-electrons. Since the ionization of an H atom requires energy $h\nu_0$ and the number of atoms ionized per second is N_0, the energy used in ionizing the hydrogen is $h\nu_0 N_0$, per second.

Let W_0 be the energy radiated by the star in the frequency range beyond ν_0. We can then assume that this energy serves to ionize the hydrogen and to communicate to the total of expelled electrons a kinetic energy W_e:

$$W_0 = h\nu_0 N_0 + W_e \qquad \text{(II.7)}$$

We assume that the kinetic energy W_e is entirely used to excite forbidden lines by collisions. It can thus be evaluated by measuring the intensities of all these lines. Zanstra, again, compares the total illumination L_i, for each forbidden line with the intensity J_i of the continuous spectrum of the star for the same frequency, and measures the ratio $r_i' = \frac{L_i}{J_i}$.

Equation (II.7) thus becomes

$$\Sigma L_i = \Sigma r' J_i = W_0 - h\nu_0 N_0 \qquad \text{(II.8)}$$

This time the summation is extended to include all the forbidden lines. J_i, W_0 and N_0 are calculable for a black body at temperature T and we again investigate for which temperature the relation (II.8) is satisfied.

Combination of the two mechanisms

Stoy [312] devised a variant of Zanstra's methods, having the advantage of dispensing with the always difficult and somewhat uncertain photometric comparison of the intensity of the lines of the nebula with those of the continuous spectrum of the star. His procedure, however, necessitates the standardization of negatives by means of a source of known energy distribution, since it is necessary to adjust the intensities of radiation or different wave-lengths.

The method consists of measuring the ratio between the sum of the intensities of all the forbidden lines and the sum of the intensities of all the Balmer lines (including the associated continuous spectrum). By dividing (II.8) by (II.6) we immediately obtain

$$\frac{\sum L_i}{\sum \left(\frac{H_i}{h\nu}\right)} = \left(\frac{W_0 - h\nu_0 N_0}{N_0}\right) \qquad \text{(II.9)}$$

The left-hand side is determined by observation, the right-hand side is calculated for different temperatures of the central star.

Visual magnitudes of planetary nebulae and photographic magnitudes of the nuclei

We have still to give an account of a much more convenient method which is independent of spectrophotometric measurements. In the spectrum of the nebula, almost all the light capable of affecting the eye is derived from forbidden lines and especially from the two lines 5007 and 4959 A of O III, which are prominent in planetary nebulae. We can therefore assume that the total luminosity l'_ν of the nebula roughly measures the total brightness of the forbidden lines, which in turn is equal to the total kinetic energy W_e of the electrons already calculated (relations II.7):

$$l'_\nu \simeq W_0 - h\nu_0 N_0.$$

On the other hand, the photographic luminosity l_g of the nucleus measures, to all intents and purposes, the intensity J_g of the continuous spectrum for a certain frequency, corresponding to about $\lambda = 4210$ A:

$$l_g \simeq J_g.$$

By means of the ratio l'_ν / l_g we obtain

$$\frac{l'_\nu}{l_g} \simeq \frac{W_0 - h\nu_0 N_0}{J_g}$$

If we represent the visual magnitude of the nebula by m'_V, and the photographic magnitude of the central star by m_{pg}, the definition of the differences of magnitude leads to

$$m_{pg} - m'_V = 2{\cdot}5 \log \left(\frac{W_0 - h\nu_0 N_0}{J_g} \right) + \text{constant} \qquad \text{(II.10)}$$

The first term of the right-hand side is a known function of the temperature of the central star and we can determine the constant if the temperature of the nucleus of the planetary nebula is already known.

The calculations involved in Zanstra's theory do not present difficulties, but they utilize certain special integrals which we shall now set out.

The energy radiated by a star of radius R, considered as compared to a black body, between the frequencies ν and $\nu + d\nu$ is

$$W_\nu d\nu = 4\pi R^2 \mathbf{R}_\nu d\nu$$

where $\mathbf{R}_\nu$ is the spectral radiance given by the formula (0.13), whence

$$W_\nu d\nu = 8\pi^2 R^2 h c^{-1} \nu^3 \left(e^{h\nu/kT} - 1 \right)^{-1} \qquad \text{(II.11)}$$

Similarly, the number $N_\nu d\nu$ of photons emitted by the star in the frequency band $d\nu$ is

$$N_\nu d\nu = 4\pi R^2 \mathbf{N}_\nu d\nu$$

$\mathbf{N}_\nu d\nu$ represents the number of photons emitted per square centimetre between ν and $\nu + d\nu$. This is given by (0.14). Thus we have

$$N_\nu d\nu = 8\pi^2 R^2 c^{-2} \nu^2 \left(e^{h\nu/kT} - 1 \right)^{-1} \qquad \text{(II.12)}$$

From this we can easily find the expressions for W_0 by integrating (II.11) between ν_0 and infinity, and for N_0 or N_g by integrating (II.12) between ν and infinity or between ν^1 and ν^2 respectively.

Putting $x=\frac{h\nu'}{k_T}$, $I_1=\int_{x_0}^{\infty}\frac{x^2}{e^x-1}\,dx$, $I_2=\int_{x_0}^{\infty}\frac{x^3}{e^x-1}\,dx$, $I'_1=\int_{x_2}^{x_1}\frac{x^2}{e^x-1}dx$,

the equations (II.5), (II.6), (II.8), (II.9) and (II.10) become respectively

$$\frac{N'_g}{N_g}=\frac{I_1}{I'_1} \qquad \text{(II.5}'\text{)} \qquad\qquad \sum\frac{x^3}{e^x-1}\frac{r_i}{\nu}=I_1, \qquad \text{(II.6}^1\text{)}$$

$$\sum\frac{x^4}{e^x-1}\frac{r'_i}{\nu}=I_2-x_0\,I_1, \qquad \text{(II.8}'\text{)} \qquad\qquad \frac{\sum L_i}{\sum\frac{H_i}{h_\nu}}=\frac{I_2-x_0\,I_1}{I_1} \qquad \text{(II.9}^1\text{)}$$

$$m_{pg}-m'_v=2{\cdot}5\log\left[\frac{e^{x_g}-1}{x_0^4}(I_2-x_0\,I_1)\right]+C \qquad \text{(II.10)}$$

Results

The "hydrogen" and the "forbidden lines" methods were applied to some nebulae by Zanstra [388] and by L. Berman [35]. For three nebulae, which are common to the two lists, their results differ but little. Table III gives those of Berman and those which Stoy [312] extracted from the measurements of J. H. Plaskett and of Berman, combining, as we have seen, the two mechanisms. We see that on the whole, the different methods agree well enough.

TABLE III

Temperatures of the Central Stars

Nebulae	*Berman*		*Stoy*
	Recombination	*Electronic Collision*	
I C 4593	25,000°K	24,000°K	20,000°K
N G C 6826	26,500	28,000	27,000
. . . 6543	36,100	35,000	24,000
. . . 7009	40,000	40,000	51,000
. . . 6572	43,500	41,500	34,000
. . . 7662	43,000	51,000	35,000
. . . 7027	52,000	53,000	

The method based on the comparison of the visual magnitude of the nebula and the photographic magnitude of the nucleus, was applied to a greater number of nebulae by Zanstra, Berman, Vorontzov-Velyaminov [358, 360], etc., and gave comparable results.

Verification by the direct study of the spectrum of the central star

It is particularly difficult to measure the temperature of the colour of the nuclei of planetary nebulae directly by means of spectrophotometry, but Aller [4] meticulously determined the spectral types of some nuclei showing absorption lines. The temperatures resulting therefrom are given in Table IV, where they are compared with those found by Berman (by means of magnitudes) and with those re-determined by Aller himself, using Stoy's method.

TABLE IV

Temperatures of the Central Stars

Nebulae	*Spectrum*	*Temperatures*		
		according to spectral type (Aller)	*according to $m - m^1$ (Berman)*	*Stoy method (Aller)*
I C 4593	0 6·5	34,000	25,000	40,000°K
N G C 6826	0 6	34,600	30,000	48,000
. . . 6891	0 7	32,900	30,000	40,000
. . . 2892	0 6·5	34,000	25,000	40,000
. . . 6210	0 7	32,900	30,000	57,000

The temperatures deduced from the spectral types are of the same order of magnitude as those obtained indirectly from the study of nebulae: in all cases they are intermediate between those obtained by Berman and by the Stoy method.

We notice the greatest divergencies for the first two nebulae, common to Tables III and IV, but it is curious to note that the most important correspond to measurements made by the Stoy method, both by Stoy himself and by Aller. This proves that these determinations are still very uncertain. Further, we must not forget that the temperatures of the very hot stars are *always* poorly determined. By means of spectrophotometry it is easier to measure the colour temperature of a star within some hundreds of degrees when, for instance, it is near to 6,000°, than to measure it within 10,000°, when it is as much as 30,000°.

There is no doubt, however, that the methods of the Zanstra theory lead to a convenient order of magnitude. *The nuclei of planetary nebulae are amongst the hottest of known stars; their temperatures generally appear to range between* 30,000° *and* 50,000°*K, and they are sometimes even hotter.*

NOTES TO CHAPTER II

[1]Some exceptions will be studied at the end of Chapter III.

[2]The simplest of slitless spectrographs consists of a prism followed by an objective in the focal plane of which the plate is placed (prism objective). Actually the spectrum of Fig. 7 was photographed with a spectrograph with its slit as wide open as possible, placed at the focus of the telescope. The image of the nebula slightly overlapped the rectangle of the opening and it is for this reason that most of the monochromatic images are found to be laterally limited by the rectilinear sides of the slit.

[3]It is possible that the varying expansion velocities for the different atoms, measured by O. Wilson, in the case of NGC 3242 (page 60), produce also a real stratification in the nebular envelope.

CHAPTER III

OUTLINES OF THE PHYSICS OF GASEOUS NEBULAE

Quantitative theory of the emission of forbidden lines

We have just seen how photometric measurement of the spectrum of the nebula and of that of the associated star—or, as in the Stoy method, only of the nebula itself—leads to an evaluation of the temperature of the exciting star. Until now, however, we have had no numerical information on the physical conditions of the nebula itself.

This was provided by the continuation of Bowen's study of the excitation mechanisms in precise quantitative form. From 1937 to 1945, Donald Menzel and Lawrence Aller published, together with various collaborators, a series of remarkable papers on this subject (*Physical Processes in Gaseous Nebulae*) [196], which made decisive advances in the physics of gaseous nebulae possible. Because of its great complexity, and also because of the often laborious theoretical calculations which it implies, there can be no question here of going into the details of this work. We shall have to be content with a very brief outline of the general ideas on which the theory is based, and with the essentials of the results obtained.

Let us consider especially the emission of forbidden lines of O III (5007, 4959, 4363 A), the first two of which are, in general, the most intense of nebular lines. They correspond to transitions between the levels ^3P, ^1D and ^1S (see Fig. 4) which, for reasons of convenience, we shall hereafter simply denote by the numbers 1, 2 and 3 in order of increasing energy. Each transition 3–2, for instance (4363 A), gives rise to the emission of a quantum $h\nu_{32}$ (h=Planck's constant, ν_{32}=the frequency of the line 4363 A). The number of transitions 3–2, per second, inside 1 cc. of gas is the product of the spontaneous transition probability A_{32} and the number n_3 of O^{++} ions per c.c., in level 3 (see page 53). The emitted energy per c.c. of gas in the transition 3–2 is then

$$I_{32}=h\nu_{32}.A_{32}n_3.$$

This measures the intensity of energy in the line 4363 A.

We can in a similar way represent the intensity of the lines 2–1, 5007 and 4959 A, but it is sufficient to consider only one of them, or even the sum of the intensities of the two lines. Since they involve the same higher level 2, the ratio of their intensities is always equal to the ratio of the spontaneous transition probabilities, multiplied by the ratio of the frequencies. It is, therefore, strictly constant, and has the value 2·83.

The spontaneous transition probabilities have been calculated theoretically by considering the magnetic dipole and electric quadrupole radiation separately (Pasternack [235]). To calculate the intensity of the lines all that remains to be done is to evaluate the numbers n_1, n_2 and n_3 of ions in each of the levels 1, 2, 3.

The atoms are raised to metastable levels by electron collisions such that the respective population of the levels depends both upon the known spontaneous transition probabilities, and on transition probabilities due to electron collisions. We cannot here give an account of the possible transitions for energy levels higher than 3 (or much greater), which give rise to the absorption or emission of permitted lines. The excitation of levels 2 and 3 by absorption of radiation is almost negligible, because of the extremely small probability of the forbidden *absorption* transitions. Finally, because of the rarefication of the nebulae, we shall neglect the excitation and the de-excitation phenomena caused by collision with particles other than electrons.

The problem is thus simplified. Let us suppose that at a given moment both the transition probabilities due to electron collisions, and also the spontaneous transition probabilities are known. Their sum constitutes the total transition probability. As the intensity of the emitted lines does not vary from one moment to the next, we could suppose the nebula to be in equilibrium, as if the population of the three levels considered remained stationary. The number of ions entering each level is, therefore, equal to that leaving it during the same time $\left(\frac{dn_1}{dt}=\frac{dn_2}{dt}=\frac{dn_3}{dt}=0\right)$. These three equations enable us to calculate the relative populations of the three levels as a function of the total transition probabilities. If their ratios are $\frac{n_2}{n_1}$, $\frac{n_3}{n_1}$ and $\frac{n_3}{n_2}$, then we obtain the relative intensities of the lines 2–1 and 3–2.

Electron temperature and density

The transition probability q due to electron collision depends upon the kinetic energy or velocity distribution of the electrons. For a gas in equilibrium, at a temperature T, we know that the distribution of velocities can be represented by Maxwell's formula, if the collisions are elastic. It depends only on the temperature T and on the mass m of the molecules.[1]

It is natural to assume that the same distribution law holds in the case of a gas formed of free electrons. The mean kinetic energy is then

$$\frac{m_e V^2}{2}=\frac{3kT_e}{2} \tag{III.1}$$

where m_e represents the mass of one electron and k is Boltzmann's constant (see page 27). The parameter T_e, which corresponds to the temperature of a gas in equilibrium, is called the *electron temperature.* The transition probability for electron collision is inversely proportional to $\sqrt{T_e}$.

Also, it is directly proportional to the number of electrons per c.c., that is to say, to the *electron density* n_e. We can write:

$$q=\frac{\beta n_e}{\sqrt{T_e}},$$

where β is a constant depending on the transition considered. Menzel, Aller and Hebb (196, XIII) were able to calculate the constant β for transitions between the levels 1, 2 and 3 of O III.

Thus, generally speaking, the ratio of the intensity of the lines 2–1 to that of lines 3–2 depends on both of the two parameters T_e and n_e. Since, however, the electron density is not too great, it has little influence upon the relative intensity of the O III lines. With the numerical values of the spontaneous transition probabilities, calculated by Pasternack [235], and the transition probabilities due to electron collision, Menzel, Aller and Hebb finally obtained the expression

$$\frac{I_{21}}{I_{32}}=1{\cdot}07\, e^{\frac{33000}{T_e}}, \tag{III.2}$$

where I_{21} represents the sum of the intensities of the lines 5007 and 4959 A, and I_{32} the intensity of the line 4363 A.

This formula can be used to good effect when the electron density is of the order of 10^4 electrons per c.c. and we shall see that this is nearly so in the case of planetary nebulae (Table V). We can thus calculate the electron temperature by measuring the ratio $\frac{I_{21}}{I_{32}}$.

The first spectrophotometric measurements of Aller thus led to temperatures which differed only very slightly from one nebula to the next, ranging between 6,000° and 10,000°K, in most cases they were about 9,000° or 10,000°. However, H. Andrillat, working at the Haute Provence Observatory, found very much higher values, increasing with the degree of excitation of the nebula [9].

These are very difficult measurements, because the ratio $\frac{I_{21}}{I_{32}}$ can exceed 100 when the electron temperature is low. Further, to compare the energies carried by radiation of different wave-lengths, it is necessary to standardize the plates with a source whose energy distribution is well known.

Andrillat deals with these difficulties in the following manner [441]:

On the one hand he measures the intensity ratios of the lines 4959 A and 4861 A (Hβ), and on the other hand the intensity ratios of the lines 4340 A (Hγ) and 4363 A. Since these measurements are made on two pairs of wave-lengths, very near to each other and generally of similar intensities, no serious difficulties are encountered. Since the lines 5007 A and 4959 A start from the same higher level ($^1D_2$), their relative intensity, which is strictly invariable, depends only on the spontaneous transition probabilities, and I_{21} is obtained by multiplying the intensity of the line 4959 A by a constant factor (3·83 or 3·93 depending on whether we use Pasternack's transition probabilities, or those calculated more recently by Garstang [442]). To pass from the ratio I_{21}/I(Hβ) and I(Hγ)$/I_{32}$ to I_{21}/I_{32}, all that need now be done is to determine the intensity ratio of the lines Hβ and Hγ. Now, according to the theory developed by Baker and Menzel, this ratio, practically independent of T_e is equal to 2. Andrillat measures it by comparing each nebula with a neighbouring star of well-known gradient.[2]

Liller and Aller, on their part, continued the measurement of the ratio by means of a photoelectric photometer, and thus improved the determination of I_{21}/I (Hβ) [443].

During this time there were also advances in theory. Seaton [444], [445] determined the collision cross-sections by a new method, and calculated the transition probabilities due to electron collisions in a much more correct manner. With Garstang's transition probabilities he arrived at

$$I_{21}/I_{32}=8{\cdot}74e^{\,33{,}000/Te}$$

which must henceforth replace (III.2), and leads to higher temperatures.

Table V compares the temperatures evaluated by this formula according to the measurements of Andrillat (column 2), and according to those of Liller and Aller (column 3). It will be seen that the two series of values are in good agreement. Nevertheless, some divergencies will still be found, notably in the case of the strongly excited nebula NGC 7662. Thus there is an effective difference in the ratios I_{21}/I (Hβ) between Andrillat and Liller and Aller.

The latter, working with a prism objective, made measurements on the entire nebula, whereas Andrillat compared the lines emitted in the central region. Now, as was seen on page 58, different wave-lengths are not weakened equally with increase of distance from the exciting star. It is precisely when the dimensions of the images corresponding to the lines 5007, 4959 and 4861 A differ most, that the authors no longer agree. Since the local electron temperature certainly varies from one region to the next of the nebula, Andrillat's results seem preferable. The increase of electron temperature with degree of excitation should therefore be considered as real.

TABLE V

Electron temperatures and densities

Nebulae *NGC* (1)	T_e *Andrillat* (2)	*Liller-Aller* (3)	n_e *Seaton* (4)	(5)
7662	28,900°K	15,000°K	50,000	17,000
7027	22,000	16,800	40,000	14,000
2392	19,000	19,600		
6803	16,800	12,800		
2440	15,500		40,000	
7009	15,400	13,900		
6572	14,600	13,500	50,000	14,000
3242	14,300	17,700		
6818	13,800	15,700		
6210	12,800	10,800		
1535	12,300	14,300		
I 418	12,200	18,300	8,000	7,000
6826	11,400	11,200		
II 4593	11,000	7,700		
II 2149	10,600	14,200		
6543	8,000	9,500		

For ions other than O III (for instance O II, N II, or S II) the ratio I_{21}/I_{32} depends both on electron temperature and density. The measurement of the ratio leads to a relation between T_e and n_e. If we plot T_e against n_e for O III, this is obviously a horizontal line of ordinate T_e.

Different measurements made for many ions lead to a set of curves, intersecting approximately at a point whose co-ordinates determine n_e and T_e. It is thus that Seaton [445], using the observations of many investigators, evaluated the electron densities, appearing in column 4 of Table V.

If we already know the value of the electron temperature approximately, n_e may be obtained quite differently by measuring the *absolute* intensity of a hydrogen line, i.e. the number of photons emitted per c.c. per sec. in the form of Hα or Hβ lines (Menzel and Aller; Ambarzumian). These absolute measurements are delicate, and the method requires that the distance of the nebula be known, so that its true dimensions can be calculated. The numbers calculated by Seaton, and given in the fifth column of Table V, show that the densities thus obtained are generally smaller than those resulting from the study of forbidden lines. The difference might still result from the fact that absolute intensity measurements must be made over the entire nebula, whereas the method of forbidden lines is applied to a region where the lines are most intense (Seaton). In support of this point of view, it can be noted that the agreement between the two methods is satisfactory in IC 418, an object of very uniform brightness, free of any local condensations.

When applied to diffuse nebulae, the measurement of the absolute intensity of the hydrogen lines has led to very much smaller electron densities than in planetary nebulae. Thus in measuring the intensity of the Hα line in the Rosette nebula (Plate XII), Minkovski [207] found $n_e=32$ electrons per c.c., and Shain and Gaze $n_e=28$ electrons per c.c. [429], [430]. The agreement of these two numbers is remarkable. In NGC 6523 (M8, Plate XXI) and NGC 6618, Shain and Gaze obtain $n_e=53$ and 110 electrons per c.c. respectively.

We shall soon see that the study of the continuous spectrum of the nebulae at the limit of the Balmer series can also tell us something about electron temperature and density.

Chemical composition

The discovery of forbidden lines belonging to multiple-ionized atoms, like those of [Fe II], [Fe III], [Fe VI] and [Fe VII] (Bowen and Wyse, 1939; Wyse, 1942) has proved that gaseous nebulae contain many more elements than one would have thought possible less than fifteen years ago. Those elements which are characterized by their permitted lines are relatively few in number: they are hydrogen, helium, carbon, nitrogen and oxygen (perhaps silicon and magnesium). But we can identify the heavier atoms—fluorine, neon, sulphur, chlorine, argon, potassium, calcium, iron and perhaps nickel—by their forbidden lines.

Although the spectra of gaseous nebulae always have the same general appearance, there exist between them very marked differences which are illustrated on Plate VII. These must not be attributed to a difference in chemical composition, but rather to differences in the degree of excitation. Amongst the planetary nebulae, NGC 7672 and NGC 7027, showing

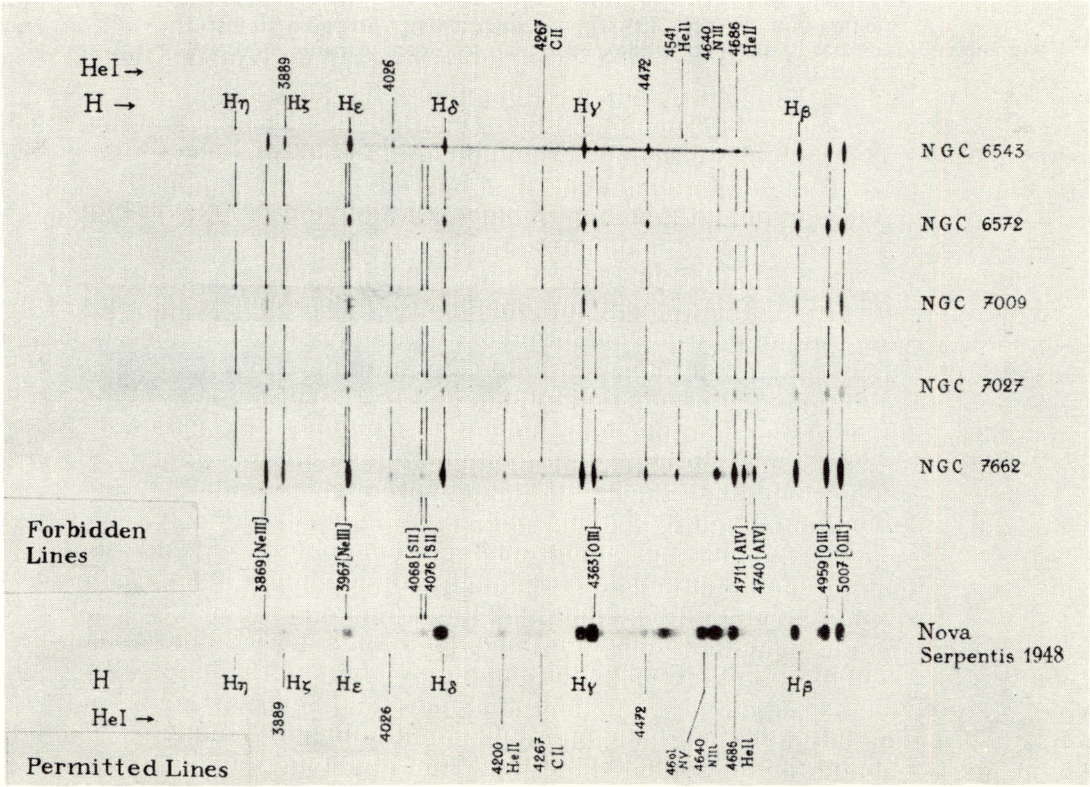

Plate VII.—Spectra of some planetary nebulae and of Nova Serpentis 1948. Spectrograph having one prism mounted on the 120 cm. telescope of the Haute-Provence Observatory. NGC 6543, 6572, 7006 and 7662 (J. Dufay); NGC 7027 (H. Andrillat).

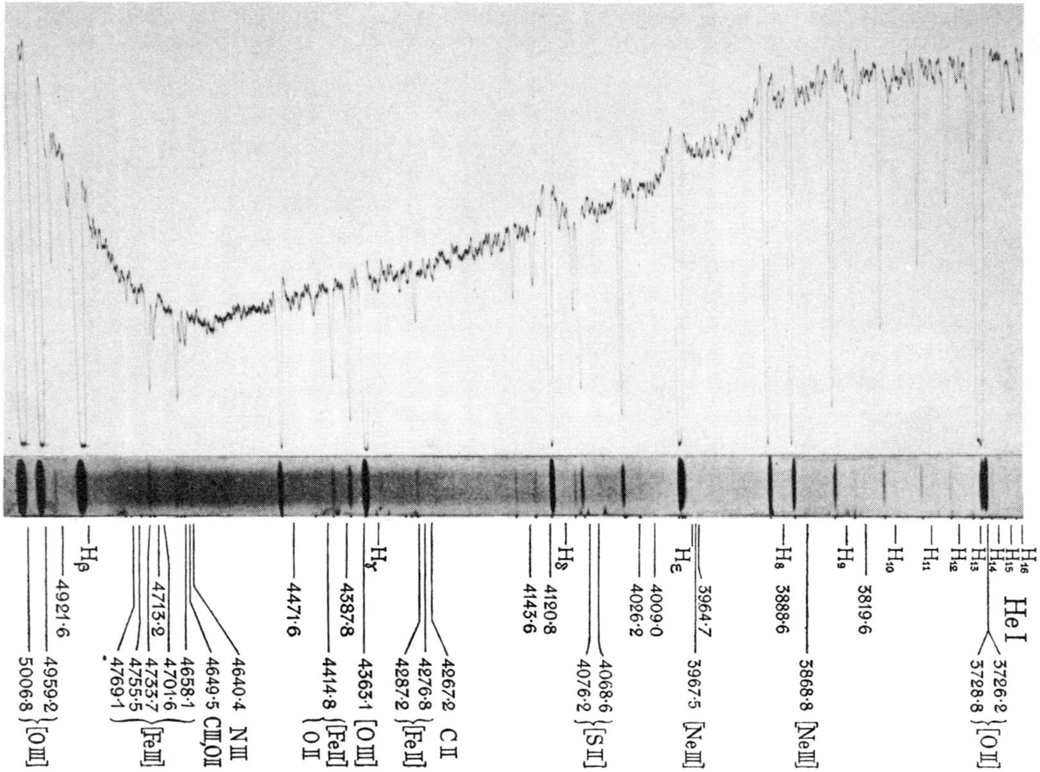

PLATE VIII.—Microphotometric record of the Great Nebula in Orion. Spectrograph having one prism mounted on 120 cm. telescope (Tcheng Mao-Lin and Dufay).

lines of He II, C IV [Fe VI], [Fe VII], and [Ca V], bear witness to a very much stronger excitation than, for instance, NGC 6572, where we can only detect weak traces of H II lines with difficulty, and where the other ions enumerated above are practically absent. The excitation is still weaker in the case of irregular nebulae with emission lines. Thus, in the nebula in Orion (Plate VIII) helium only appears in the neutral state (lines of He I and not of He II), iron is only once or twice ionized (lines of [Fe II] and [Fe III], but not of [Fe VI] and [Fe VII]). This increase in the degree of excitation is clearly associated with the temperature of the exciting stars.

Measurement of the line intensity gives us an idea of the relative abundance of the various elements. We must bear in mind the fact that the lines are emitted by various mechanisms. We must also be able to pass from the number of atoms in one particular state of ionization to the total number of neutral and ionized atoms of the same element. The results of these calculations are necessarily very uncertain, and can lead to no more than an order of magnitude.

TABLE VI

Relative abundance of light atoms

Element	*Planetary Nebulae* (*Menzel*)	*NGC* 7027 (*Aller*)	*Normal stars* (*Aller*)	*Interstellar Gas* (*Strömgren*)
Hydrogen	1,000,000	1,000,000	1,000,000	1,000,000
Helium	200,000	90,000	50,000 to 100,000	
Carbon	165		90	
Nitrogen	1,650	330	170	
Oxygen	2,000	900	540	
Fluorine	6	2·3?	2·7?	
Neon	320	275	590	
Sodium		2·6		4·1
Sulphur	40	150?	18	
Chlorine	20	22?	11	
Argon	160	15?	55	
Potassium		0·14?	0·13	0·4
Calcium		0·36?	2·8	10
Titanium			0·1	0·2

The first column of Table VI gives (according to Menzel [195]) the relative abundance of light atoms in the planetary nebulae. The second column gives the more recent results of L. H. Aller [446] for the nebula NGC 7027. Aller has assumed that the electron temperature in it was T_e= 15,000°K (a value which is perhaps a little low), and that there were 7.10^4 ions per c.c. in the bright filaments. For purposes of comparison, the third column is made to show the relative abundance of the same elements in the atmosphere of normal stars, also according to Aller [447]. Finally, the last column gives some indications of the composition of interstellar gas, according to Strömgren (see Chapter V).

As far as we can judge at the moment, *the chemical composition of gaseous nebulae differs neither quantitatively nor qualitatively from that of the Sun and of the stars.* This conclusion was extended by Bowen and Wyse to include heavy atoms like iron (NGC 7027).

Density of gaseous nebulae

We can now attempt to evaluate the total density of gas in the nebulae.

The number of atoms must be of the same order of magnitude as the number of electrons. Therefore, the planetary nebulae ought to contain 2,000 to 15,000 atoms per c.c. (half this value perhaps according to Greenstein [229]). If we recall that under conditions of normal temperature and pressure each c.c. contains $2{\cdot}7.10^{19}$ molecules, we shall have some idea of the extreme rarefication of nebular matter.

Hydrogen is the principal constituent of the nebulae, as also of the stars. Taking account of the relative abundance of the elements by means of Menzel's evaluations (Table VI), we find that in the nebulae the mean atomic mass is approximately 1·6. Since the mass of the hydrogen atoms is $1{\cdot}67.10^{-24}$ gm., the mean mass of an atom can equally be taken as $1{\cdot}6\times 1{\cdot}67\times 10^{-24}=2{\cdot}67.10^{-24}$ gm. With 10^3 to 10^4 atoms per c.c. we thus arrive at specific masses of the order of $2{\cdot}7.10^{-21}$ to $2{\cdot}7.10^{-20}$ gm. per c.c. The mass contained in 1 cubic-kilometre should range between 1/3,000 and 1/300 of a milligram, and a volume equal to that of the Earth ($1{\cdot}08.10^{12}$ km^3) would contain from 3 to 30 tons of gas.

These figures, let us recall, concern planetary nebulae. The density of diffuse nebulae is much smaller. According to the electron densities given previously (page 77), we obtain specific masses of the order of 10^{-22} gm./c.c.

In spite of their low density, gaseous nebulae have a significant mass, by reason of their great dimensions. A planetary nebula of radius 0·1 parsec ($3{\cdot}08.10^{12}$ km.), with density of 2.10^{-20} gm./c.c., ought to have a mass $2{\cdot}45.10^{33}$ gm. This is almost the order of mass of the Sun ($\simeq 2.10^{33}$

gm.).* Diffuse nebulae, which are much more rarefied, have much greater masses by reason of their enormous dimensions. For NGC 6618, 6523 and 2237, Shain and Miss Gazé [229, 230] have given values of about 260, 3200 and 5800 solar masses respectively.

The continuous spectra of gaseous nebulae

Bright line nebulae often emit, with fairly large intensity, the continuous spectrum which extends the Balmer Series towards the ultra-violet. In that region there is, however, much more blue of longer wave-length. Long exposures or even those made with a large slit spectrograph also show a more or less intense continuous spectrum. We call this continuous spectrum "visible" in order to distinguish it from the Balmer continuum. It is very apparent in the spectrum of the Nebula in Orion (see Plate VIII). In this case, according to Greenstein and Henyey [115], it would carry a little more energy than would be the case if the aggregate of the lines corresponded to a colour temperature of about 11,500°K; this is a little higher than the electron temperature and a little lower than the colour temperature of the exciting star θ_1 Orionis (12,500°K, according to Greenstein).

The continuous "visible" spectrum of the planetary nebulae has mainly been studied by Page [226], who has shown its presence in twenty-nine out of fifty-eight nebulae, that is, in half the cases. The distribution of energy in it appears to be very uniform between 3900 and 5000 A, and its intensity is such that an interval of 100 A (towards 4000 A) carries an energy comparable to that emitted in the Hδ line.

There is a strong possibility that certain bright-line nebulae scatter the light of the stars also, in the same way as true diffuse nebulae in which more or less only the continuous spectrum appears. Thus the colour temperature of the star θ_1 Orionis, which is abnormally small for its spectral type, is certainly the result of absorption in the interior of the Nebula in Orion, which is almost opaque to the stars situated behind it. There are cases, however, where absorption in the nebula appears negligible. In those planetary nebulae like NGC 7662, whose expansion is apparent from the splitting of the bright lines (see page 83), the violet component is emitted by those atoms situated nearest to the observer and which are approaching him, the red component by the furthest atoms which are receding. If then the nebula brings about a notable absorption, the red component would, as a rule, be weaker, which is not the case (Zanstra [390]).

The study of the polarization of the continuous spectrum of gaseous nebulae seems to bear witness to the existence of an appreciable scattering.

*The total mass could be much greater than that of the visible region.

Hall [123] has recently shown that the light of the Nebula in Orion is slightly polarized: the privileged vibration seems to be perpendicular to the plane of scattering and the proportion of light polarized is less than 4 per cent (the reader is referred to page 198 for an account of polarization). Since both the bright lines and the continuous spectrum were measured, it is possible that for the latter the proportion of polarized light may be larger. Analogous observations made on a number of planetary nebulae by Page [228] have only shown that the proportion of polarized light is certainly less than 5 per cent in the case of NGC 6210 and less than 10 per cent in the case of IC 3568. Nor has any greater degree of polarization been found in the case of NGC 6543. It is desirable that these most interesting observations be completed in the near future.

In actual fact, it is generally held that scattering only plays an accessory role in the production of the continuous spectrum of many gaseous nebulae. An attempt was made to attribute this to recombination spectra of atoms ("free-bound transitions") and the transitions between non-quantized states ("free-free transitions").

We have already considered for the case of hydrogen, the continuous spectra associated with the Lyman and Balmer Series, which are emitted by electron and proton recombinations on levels 1 and 2 (Fig. 2). Similarly, the Paschen Series (limit 8204 A) is extended on the short-wave length side by a continuous spectrum whose emission accompanies the capture of electrons in level 3.[3] The continuous spectra of the Nebula in Orion suddenly increases in intensity, in the ratio 1–4, when passing the Balmer limit towards the ultra-violet (Barbier [19], Greenstein [110]). It is, therefore, certain that the continuous Paschen spectrum (always weaker than that of Balmer) is emitted with notable intensity in the visible region, and it is possible that the recombination spectra of hydrogen at levels above 3 would not be entirely negligible.

A photon can also be emitted when a hydrogen atom passes from one non-quantized state to another (Cillié, 1932 [58]). Since the recombination spectra in the Bohr scheme correspond to the change of an electron from an hyperbolic orbit to an elliptic orbit of order 1, 2, 3, etc., transitions between non-quantized states correspond to changes from a hyperbolic orbit to another of lesser energy.[4] The continuous spectrum resulting therefrom necessarily extends over the entire range of frequencies. Its contribution to the continuous spectra of the nebulae seems to be weaker than that of the recombination spectra, at least in the case of not very small frequencies.

In any case, it would seem that the intensity of the spectrum emitted by recombination, and transitions between non-quantized states, decreases more rapidly towards the short wave-lengths than that of the continuous spectrum of planetary nebulae. According to Greenstein and Page [116] we can only just account for 5 per cent of the intensity observed at 3900 A in this way. These authors have shown that towards the limits of the region in which hydrogen is ionized by the radiation of the central star, very favourable conditions exist there for the formation of *negative* hydrogen ions H^-. The electrons attach themselves to the neutral atoms as follows:

$$H+e \rightarrow H^- + h\nu,$$

with emission of radiation which, for an electron temperature of the order of 10,000°K, would nearly give a very uniform distribution of energy from 3000 to 7000 A. The ratio of H^- ions to H atoms could be 10 to 100 times that in the solar atmosphere, where, as we know, absorption by negative hydrogen atoms is very important. Unfortunately the total number of H^- ions must be very small indeed, and calculations show that the radiation intensity accompanying their formation is possibly 10,000 times smaller than the observed intensity.

Simultaneous emission of two photons by metastable hydrogen atoms

Spitzer and Greenstein [293] have put forward an entirely new interpretation, based on the *fine structure* of the levels of the hydrogen atom[5] (Kipper [169], however, had already considered the problem). The level $n=1$ on which the Lyman lines converge, is a singlet,* but the level $n=2$

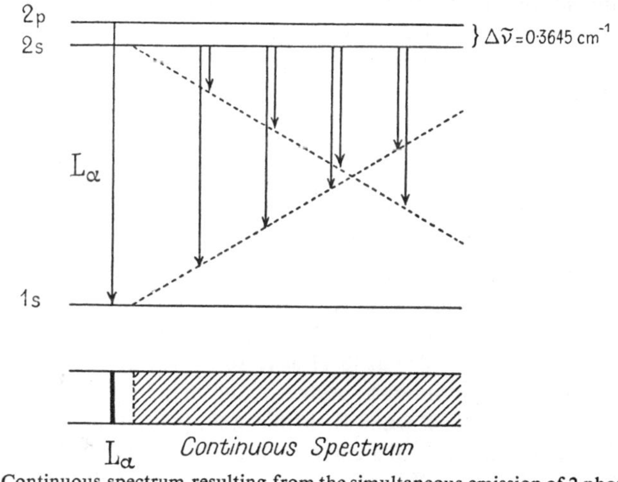

Fig. 10. Continuous spectrum resulting from the simultaneous emission of 2 photons from the 2*s* level of the hydrogen atom.
The lengths of the arrows are proportional to the wave numbers of the photons. The sum of the wave numbers of a pair is constant. (The separation of the levels 2*p* and 2*s* is shown very much exaggerated.)

*In terms of *fine structure* only. We shall see further on that it is a doublet in terms of *hyperfine structure* due to nuclear spin.

is a very close doublet, whose two components, denoted by $2p$ and $2s$ respectively, are at a distance of 0·365 cm.$^{-1}$ (see Fig. 10). The transition from level $2p$ to the levels $1s$ corresponds to a Lα line emission, but that from $2s$ to $1s$, which is strictly forbidden, cannot take place spontaneously with the emission of a photon.

We can, therefore, imagine that the atom on level $2s$ loses its energy by *simultaneously emitting two photons*. The sum $\tilde{\nu}_1+\tilde{\nu}_2$ of their wave numbers is constant, since it is necessarily 0·365 cm.$^{-1}$ less than the wave number of the Lα line

$$\tilde{\nu}_1+\tilde{\nu}_2=\tilde{\nu}(\mathrm{L}\alpha)-0{\cdot}365 \text{ cm.}^{-1}.$$

With the sole condition that this relation be satisfied $\tilde{\nu}_1$ and $\tilde{\nu}_2$ can take on any values whatsoever. We can thus forecast the emission of a continuous spectrum (Fig. 10).

Now, the metastable level $2s$ can be heavily populated. Certain atoms are directly brought to it by electron and proton recombinations; others indirectly, by the capture of an electron in a level higher than 2, followed by the emission of lines in a series of jumps, of which the last drops to the level $2s$. According to Spitzer and Greenstein's calculations, 32 per cent of the recombinations would finally bring the atoms to the level $2s$. Further, an atom raised from the level $1s$ to the level $2p$ can, by absorption of the line Lα, give up a part of its energy to a free electron which collides with it, and so find itself left in the level $2s$. This new mechanism leads to an increase in the population of the level.[6]

Because of the rarity of collisions, all the atoms in the state $2s$ end by losing their energy with the simultaneous emission of 2 photons. The transition probability is, in fact, relatively large ($A_{2s,\ 1s}=8{\cdot}23$ sec.$^{-1}$) and corresponds to a mean life time of 0·12 sec. for the metastable level.

Spitzer and Greenstein finally studied the distribution of energy thus produced in the continuous spectrum at various temperatures. If we suppose that 32 per cent of the atoms pass through the level $2s$, we find that, when the temperature is near to 10,000°K, towards 4800 A the intensity is already comparable to that of the continuous spectrum of Paschen. This intensity increases constantly towards the ultra-violet.

D. Barbier and H. Andrillat [8.1] have since studied, in the violet and the ultra-violet, the continuous spectra of the nebulae NGC 6543, 6572, 6720, 6826, 7009 and 7662, while R. Minkowski [449] did similar work on IC 418. The quantitative results can be represented graphically by plotting the logarithm of the intensity against the wave number $\tilde{\nu}$. If the continuous spectrum is due uniquely to free-free transitions, and to the recombination spectrum of Hydrogen, we ought to find, on either side of the limit of the Balmer series, two *parallel* straight lines, vertically displaced towards each other. The straight line on the ultra-violet side is naturally higher up, since the continuous Balmer spectrum is then superposed on that of the free transitions. The theory of Cillié then enables us to calculate the electron temperature, either from the *magnitude* of the Balmer discontinuity, or from the common slope of the straight lines.

If, on the other hand, the continuous spectrum is due uniquely to the emission of photon pairs by the hydrogen atoms $2s$, the slope observed on the red side of the discontinuity should be nearly horizontal. In this case the electron temperature can be evaluated from the theory of Spitzer and Greenstein, either from the magnitude of the discontinuity, or from the slope of the ultra-violet straight line.

Now, in general, observation has led to straight lines of different slope on the respective side of the discontinuity—smaller on the red side, but never zero. The explanation for these results was provided by Seaton [450].

Spitzer and Greenstein thought that the disexcitation by collisions from the metastable level $2s$ could be neglected. Seaton shows that, on the contrary, the density in the planetary nebulae is great enough for the collisions of protons and electrons with the hydrogen atoms on the level $2s$, to depopulate this level considerably. With the temperatures and densities assumed today, he evaluates the probability of disexcitation by collisions, and calculates the magnitude of the Balmer discontinuity in IC 418 and in five of the six nebulae observed by Barbier and Andrillat. The agreement with the observations is excellent, and the slopes of the two straight lines obtained (much more difficult to determine) nearly agrcc with the theoretical predictions.

If we account for the recombination spectra of hydrogen and ionized helium, the emission of photon pairs by the hydrogen atoms on the metastable level $2s$ thus accounts for the continuous spectrum of the planetary nebulae.

A particular case: The Crab Nebula

The Crab Nebula (see Plate IV) has been studied from different points of view by Duncan, Mayall [189], Baade [10], Minkowski [204, 422] and Barbier [20]. It shows a continuous spectrum of much greater intensity than that of the lines, and without doubt it carries 90 per cent of the total energy. The emission lines, few in number, belong to O I, O II, O III, N II, S II, Ne III, He I and H. Highly split, and with a width corresponding to an expansion of velocity above 1,100 km./sec., they are mainly emitted in the external filaments of the nebula. The strongest are the red lines of N II, but those of S II are also quite intense. The amorphous central mass essentially gives a continuous spectrum.

Scattering may be entirely left out of our considerations, since it would not appear that the red component of the rays is perceptibly weakened. On the other hand, no bright star exists in the interior of the nebula, whose photographic brightness surpasses by seven magnitudes that of the only star which could have been responsible for the excitation. We, therefore, have a true emission spectrum, which could be the result of recombinations and transitions between non-quantized states. But it would seem that hydrogen plays hardly any part, since the Balmer lines are very weak ($H\alpha$ is weaker than the neighbouring line [N II] 6583 A) and, at the limit of the series, the discontinuity is either very small (Minkowski) or even non-existent (Barbier). This very peculiar nebula could, therefore, be very poor in hydrogen indeed.

The colour temperature of the continuous spectrum is only between 6,000 and 9,000°K (Minkowski, Barbier), but the electron temperature is certainly much higher. According to Minkowski it would be of the order of 30,000°K with an electron density near 10^3 per c.c. The only star capable of producing this excitation shows a complete continuous spectrum, without either emission or absorption lines. Minkowski's calculations show that this could be a white dwarf with a radius 0·02 that of the Sun, but 30,000 times more brilliant than the Sun. Its temperature could reach 500,000°K, which would make it, in effect, a most extraordinary object.

Oort [224] is of the opinion that the electron temperature is even higher still (500,000° or 1,000,000°). In these conditions the continuous spectrum ought to be extremely intense because of hydrogen and helium recombination lines and it would no longer be necessary to imagine that the central mass of the nebula is nearly free of hydrogen.

Astronomers see, in the Crab Nebula, the residue of a *supernova* whose explosion, recorded in Chinese annals, was produced in the year A.D. 1054 [85]. The star became visible to the naked eye by day, for more than three weeks and by night for twenty-two months. A discussion of the ancient observations led Mayall and Oort [199] to the belief that this supernova must have been one of the most brilliant ever observed. Its absolute magnitude ought to have attained the value of $-16{\cdot}5$, such that its luminous intensity ought to have been greater than the total intensity of all the stars in the Milky Way.

The identification of the nebula and of the supernova whose position is obviously only known in a rough manner, is based principally on the actual expansion velocity of the nebula, whose apparent diameter increases by 0·238″ per annum. If we calculate the mean expansion velocity since the explosion, and if we assume that this took place in 1054, we get 0·201″ per annum. The difference with the observed value is not very large; it may be due to experimental errors or to a variation in the expansion velocity during the last nine centuries, since the gases could have been expelled with some appreciable acceleration. The comparison of the annual increase of the apparent diameter with radial expansion velocity (which is well within the order of magnitude expected for a nova or a supernova) leads to a value of 1,500 pc. as the distance of the nebula, and consequently the actual linear diameter is of the order of 2 pc.

According to Oort [224], the high electron temperature could be that of the interior of the star at the moment of explosion, since a gaseous mass of such weak density would cool very slowly, and since its temperature could not have varied perceptively in the course of nine centuries. Further, if the density of the surrounding medium were about 0·2 hydrogen atoms per c.c., the work of expansion of the nebula would suffice to maintain this radiation temperature. It is interesting to note further, that the position of the Crab Nebula coincides with that of one of the strongest sources of radio waves of the Milky Way (Bolton and Stanley [46]). Finally the study of the diurnal variation of cosmic rays of great energy led Sekido, Yoshida and Wada [271] to the belief that the Crab Nebula was itself an intense source of cosmic rays.

A new and fascinating theory on the excitation of the Crab Nebula has recently been put forward by Ramsey [425]. The energy necessary to cause the complete breaking up of a star must surely be of nuclear origin. It is, without doubt, liberated by the synthesis of heavier elements from helium, for it is generally admitted following Chandrasekhar, that the star would have already exhausted the major part of its hydrogen before its explosion. Now, the explosion of this gigantic "helium bomb" would have produced both stable isotopes and unstable radio-active isotopes. Most of the latter would disappear rapidly, but the former would have a much longer lifetime, disappearing slowly.

Of those radio-active isotopes whose life is longer than thirty years, and whose atomic mass is less than seventy, the element C^{14}, an isotope of carbon, is of particular interest. Its mean life is 7,300 years and the mass of C^{14} need only be 5/10,000 that of the nebula for its destruction to maintain the radiation of the latter. This proportion appears reasonable, if we remember that the nebula must be very poor in hydrogen and, without doubt, in helium. The decrease in brightness of one magnitude in seventy centuries would be quite imperceptible, in a human lifetime.

C^{14} becomes transformed into N^{14} with the emission of β rays whose energy attains 160,000 eV. The electron temperature corresponding to this would be of the order of milliards of degrees—in so far as we may speak of temperature at all when the velocity distribution does not follow Maxwell's law. This would explain the extremely high electron temperature of the central body.

Finally, the abundant production of stable nitrogen would seem to account for the great intensity of the [N II] lines in the outermost filaments of the nebula.

Very close to the position of Kepler's supernova, which appeared in 1604, Baade [11] discovered a feeble nebulosity whose spectrum, formed exclusively of some emission lines, is very similar to that of the Crab filaments (Minkowski [205]). Ramsey's theory would still apply in this case.

Excitation of certain filamentous nebulae

The neighbourhood of a group of the bright-line nebulae NGC 6960, 6992 and 6995, known by the names of Veil and Lacework Nebulae in Cygnus (Plate IX), also seems devoid of all stars capable of exciting the luminescence of the gas. The general appearance of the filaments, arranged in arcs of circles as they are, also suggests that the phenomenon can be attributed to the explosion of a star. Oort [223] thought that the excitation could be the result of the expansion of the envelope of a nova, or rather a supernova, through the interstellar gas. The absence of appreciable radial expansion velocity, however, noted by Humason [425], renders this hypothesis somewhat improbable. The phenomenon could be due to the propagation of a shock wave due to the explosion (J. M. Burgers), which would be capable of raising the temperature of the gas sufficiently for the atoms to be excited by electron collisions.

A recent study of Chamberlain [399] is not opposed to this point of view. Chamberlain measured the brightness of the filaments of the Lacework Nebula for the lines Hα, [O III] 5007–4959 A and [O II] 3737 A, which were isolated with appropriate filters. Knowing the approximate dimensions of the filaments, he deduced from his measurements the emission intensity of the three wave-lengths thus studied, in ergs per c.c. per second. The results are capable of interpretation by the theory of shock waves, with an electron temperature of order of 20,000°K, and by a density of 500 hydrogen atoms per c.c. However, at an electron temperature of 10,000°K, excitation by radiation could equally well account for these observations so that the actual data do not definitely permit a choice of one or the other explanation. It would be very useful to measure the relative intensities of the Balmer lines. These ought to decrease much more rapidly starting from Hα in the case of excitation by collisions; an attempt to determine the electron temperature should also be made.

In the meantime, the absence of an identifiable exciting star makes excitation by fluorescence somewhat improbable. Here again Ramsey's theory may be considered as favourable.

Furthermore, this group of nebulae does not constitute an isolated case today. In fact Shain and Miss Gazé have recently discovered two very similar systems. Near o^1 and o^2 Cygni, there are no O and B filaments either, at least not up to the ninth magnitude. Near the third system situated in Auriga towards $\alpha=5^h35^m$, $\delta=+27°$ we find three weak B0 stars, but these are not sufficient to account for the luminescence of all the very fine filaments, still arranged in arcs of a circle. The excitation mechanism could, therefore, be the same as in the case of the Lacework Nebula. Shain and Miss Gazé have spoken, although somewhat vaguely, of a phenomenon of electromagnetic origin. Turbulence could produce an appreciable increase in the magnetic field and the nebular matter would then tend to flow out along the lines of force.

Are planetary nebulae vestiges of former novae?

Since the Crab Nebula is the residue of an old supernova, certain authors have been of the opinion that planetary nebulae represent the vestiges of former novae. The explosion of a nova is a much rarer phenomenon than that of a supernova, and neither the number nor the galactic distribution of planetary nebulae would contradict such an hypothesis. The authors base their arguments on analogies which, at first sight, would seem to be very striking.

1. The spectra of novae at a certain phase of their evolution are, to all intents and purposes, similar to those of planetary nebulae. Thus the spectrum of the Nova Serpentis 1948 (Plate VII) very much resembles those of planetary nebulae shown on the same plate, even though their emission lines are much broader than those of the nova.

2. The spectra of former novae are similar to those of the nuclei of the planetary nebulae. In both cases we are dealing with hot stars which are dense and of small radius.

3. Many of the planetary nebulae, expanding just as the gaseous envelopes ejected by the novae, are responsible for the bright-line spectrum of the nebular appearance. These envelopes can be seen developing in the form of weak nebulosities around the star in the course of a given time.

Deeper study will show that this analysis remains somewhat superficial. A study of Aller's on the nuclei of five planetary nebulae with absorption lines, has confirmed the fact that these were stars of radius smaller than that of the Sun, but considerably larger than that of white dwarfs. Their density appears to be only ten times that of the Sun, and their absolute bolometric magnitude would be of the order of -2 and -3. It would therefore seem that the nuclei of planetary nebulae are stars which are less hot; greater in volume, and less dense than former novae.

Two even more decisive arguments have been invoked by Minkowski [206]:

1. The expansion velocities of the planetary nebulae are of the order of 20 km./sec., with moderate dispersion. The greatest known expansion velocity, that of NGC 2392, is only 55 km./sec. The expansion velocities of the envelopes of novae are, on the average, fifty times as great and range roughly between 300 and 1,700 km./sec. It had been suggested that in the case of planetary nebulae, the expansion velocity could have been slowed down by the interstellar medium. The different velocities measured by means of lines however (see page 60), and also the fact that the greatest velocity appears in the case of lines of weak ionization, emitted in the external regions of the nebula, would contradict the proposed explanation.

2. The most typical bright planetary nebulae have a diameter in the neighbourhood of 0·1 parsec. With an electron and proton density of the order of 5.10^3 (which is very small), we arrive at a total mass of 0·1 of the Solar mass, excluding the nucleus. (We have found on page 82 a mass of the order of magnitude of that of the Sun by taking a radius twice as large and a greater density, and we shall soon see (page 103) that the total mass is probably even greater still). The nebular envelopes of the novae are much less dense, and their mass is evaluated as being 10^{-4} the Solar mass only. This difference in mass is shown by the fact that most of the envelope of the novae seem to disappear the moment their radius attains some seconds of arc.

By reason of their great expansion velocity and of their weak mass, the envelopes of the novae have only an ephemeral life. With very few exceptions they disappear at the end of some twenty years. The more massive planetary nebulae whose expansion is slower, are, at least in appearance, permanent objects. Their life can be counted in at least some tens of thousands of years.

If it is permissible to conclude that planetary nebulae are not vestiges of former novae, it is yet probable that the gaseous mass has been ejected by the central star in the manner of an envelope of a nova. In both cases, the ejection doubtless results from the internal instability of the star. In other words, the formation of a planetary nebula and the explosion of a nova would appear to be somewhat analogous phenomena, without, however, any direct connexion.

NOTES TO CHAPTER III

[1]If n is the total number of particles, and dn the number of those whose velocity is between v and $v+dv$, then the Maxwell Distribution Law is

$$\frac{dn_v}{n}=2\sqrt{\frac{m}{2\pi k^3T^3}}\,e^{-\frac{1}{2}mv^2/kT}\,v^2\,dv,$$

where k is Boltzmann's constant.

[2]In general $I(H\beta)/I(H\gamma)$ is found to be fairly close to 2. Interstellar absorption can intervene in certain cases, weakening $H\gamma$ more than $H\beta$. Observations would then give a higher ratio than 2. In taking the ratio 2, this absorption effect is automatically corrected.

Nevertheless, for a small number of nebulae Andrillat found ratios clearly below 2. This anomaly has not yet been explained satisfactorily.

[3]Naturally the continuous Lyman, Balmer and Paschen spectra are observed as absorption spectra when the atoms are ionized from levels 1, 2 and 3 upwards. The continuous Balmer spectrum appears as a very intense absorption spectrum in the A stars.

[4]Conversely, we have the absorption of a photon when the atom passes from one non-quantized state to another of higher energy, or, in other words, when the electron passes from one hyperbolic orbit to another of higher energy.

[5]The fine structure corresponds to the fact that, for a principal quantum number n, the azimuthal quantum number l may take all integral values between 0 and $n-1$ (page 46). For $n=1$, we necessarily have $l=0$ and the level is a singlet ($1s$). For $n=2$, we can either have $l=1$ (level $2p$), or $l=0$ (level $2s$). The selection rule $\Delta l=\pm 1$ (page 48) shows that the transition $2p \longrightarrow 1s$ is permitted ($\Delta l=-1$) and the transition $2s \longrightarrow 1s$ is forbidden ($\Delta l=0$).

We neglect here the splitting due to electron spin. At the level $n=2$, $l=1$, $s=\frac{1}{2}$ a quantum number J may take two values $J=l+s=\frac{3}{2}$ and $J=l-s=\frac{1}{2}$. There exist, therefore, two sub-levels $^2P_{\frac{3}{2}}$ and $^2P_{\frac{1}{2}}$. Elementary considerations will show that the latter coincides with the level $2s$, for which $J=l+s=0+\frac{1}{2}$. In fact, it differs from it by 0·035 cm^{-1}.

[6]When the density is of the order of 10^3 to 10^4 electrons per c.c., one in every 10^{10} Lα absorptions carries the atom to the level $2s$. In the region H I which surrounds the nebula, free electrons are much rarer and this fraction becomes $\frac{1}{10^{13}}$. Nevertheless, it is in this H I region that the process is most effective in populating the level $2s$.

CHAPTER IV

EXCITATION OF ATOMS IN SPACE: EMISSION LINES IN THE MILKY WAY

1. Emission in the Visible and Ultra-violet Spectrum

Discovery of Galactic emission areas

We can proceed in two ways in order to photograph the spectra of galactic nebulae. We use a slitless spectrograph if we want to obtain a series of monochromatic images of the nebula (see page 65 and also Plate VII of the spectrum of the Nebula in Lyra). In all other cases an image of the nebula is projected on the slit of an ordinary spectrograph by means of the mirror of a reflecting telescope, or the objective of a refracting telescope. The spectra on Plate VII were obtained in this manner.

A variant of this method, due to Otto Struve [321], allows us to dispense with the mirror or the objective (Fig. 11).

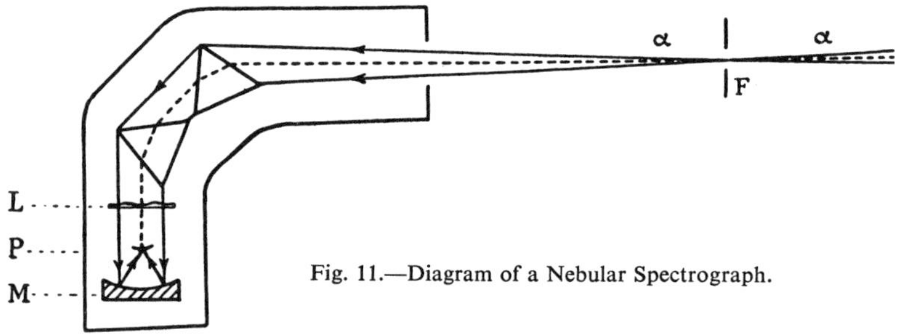

Fig. 11.—Diagram of a Nebular Spectrograph.

The spectrograph is pointed straight at the sky and its objective produces a spectral image of the distant slit on the film P. In these conditions a star will only send, through the prism, a narrow beam of parallel light with section equal to that of the slit. A nebula, however, emits a cone of rays of angular aperture equal to the apparent diameter of the nebula. If the diameter subtends an angle greater than α at the slit F, the first face of the first prism is completely covered by the beam, and the apparatus is working at full aperture. The spectrograph of Struve and his collaborators consists of two quartz prisms of 60° and of a Schmidt chamber of approximately 10 cm. focal length, with aperture $f/1$.[1]

Fig. 12 is a diagram of the mounting at the McDonald Observatory in Texas [328]. The light coming from the nebula is reflected from the first plane mirror M_1, whose shutters V permit one to limit its useful width, and then from the second distant plane mirror M_2, before entering the spectrograph S. The chamber is focused on the point of the mirror M_1 which acts as a slit. The distance M_1M_2S is 46 metres and the angle α is 6′. The mirror M_1 and the spectrograph are placed symmetrically with respect to the hour axis H H′ of an equatorial mounting. The continuation of this axis, parallel to the axis of rotation of the Earth, is normal to the mirror M_2 and allows the diurnal movement to be followed during the exposure.

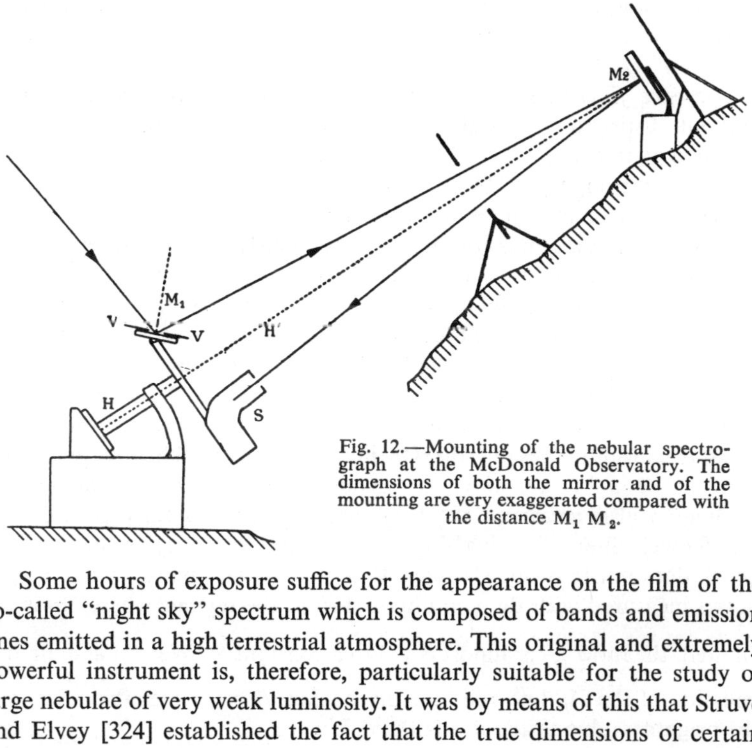

Fig. 12.—Mounting of the nebular spectrograph at the McDonald Observatory. The dimensions of both the mirror and of the mounting are very exaggerated compared with the distance M_1 M_2.

Some hours of exposure suffice for the appearance on the film of the so-called "night sky" spectrum which is composed of bands and emission lines emitted in a high terrestrial atmosphere. This original and extremely powerful instrument is, therefore, particularly suitable for the study of large nebulae of very weak luminosity. It was by means of this that Struve and Elvey [324] established the fact that the true dimensions of certain known nebulae (near γ Cygni for instance) were much greater than was indicated by direct photographs taken in total light. The dimensions of the nebulae are revealed by the presence of the Hα line (6563 A) and of the forbidden doublet O^+ (3726, 3729 A). This, however, is not all: the same lines are frequently observed along the whole length of the slit of the spectrograph (more than 1° of the Sky), in some regions where direct photographs show no traces of nebulosity (for instance around 25 Cygni) (Struve and Elvey, 1938). These emission lines can be observed with very varying intensities, in different regions of Cygnus, of Cepheus and of Cassiopeia. Moreover, we can very often distinguish Hβ (4861 A), and the forbidden doublet of O^{++} (5007, 4959 A). With a spectrograph of higher dispersion the forbidden lines of ionized nitrogen (6584 A) and of ionized sulphur (6717, 6731 A) have been identified (Thorton Page, 1948). The doublet 3727–3729 of O II is relatively intense in Monoceros and in Canis Major, but is weak in Sagittarius and in Scorpion.

Strömgren's photoelectric measurements

Struve and Elvey's method of detecting galactic emissions is extremely sensitive but also very laborious, since each examination of a celestial area of 1° by 6′ requires an exposure of some hours. The invention of *interference filters* permitting the isolation of narrow spectral regions, without dispersion apparatus, has led to more expedient yet equally sensitive methods.

Bengt Strömgren [315] has placed behind the focal plane of a telescope an electron multiplier tube preceded by an interference filter approximately isolating an emission line. It is obviously essential to ensure that each of the small chosen areas of the order of 1′ in diameter does not contain any stars brighter than the sixteenth or seventeenth magnitude. So far measurements have been made on the Hβ line for which sensitive photomultipliers and easy methods of manipulation exist (caesium-antimony cathode which does not require refrigeration). The interference filter has a transparent band of approximately 200 A, centred on 4861 A. Thus it transmits, besides the Hβ line, a not negligible fraction of the light of the night sky, which is of complex origin. To account for this, Strömgren also measured the brightness of the area through a second interference filter of which the maximum transparency is 4610 A and which does not transmit Hβ lines. If B_1 and B_2 are the respective brightnesses measured with the two filters, then in the regions where Hβ is not emitted, the ratio $\frac{B_1}{B_2}$ is almost equal to a constant R. We can assume that emission of the Hβ line will take place when B_1 is greater than RB_2 and the brightness due to the line is the difference

$$B_\beta = B_1 - RB_2.$$

The photomultiplier having an almost instantaneous response, measurement is very rapid and, within only two months (January and February 1948) in the McDonald Observatory, Strömgren was able to

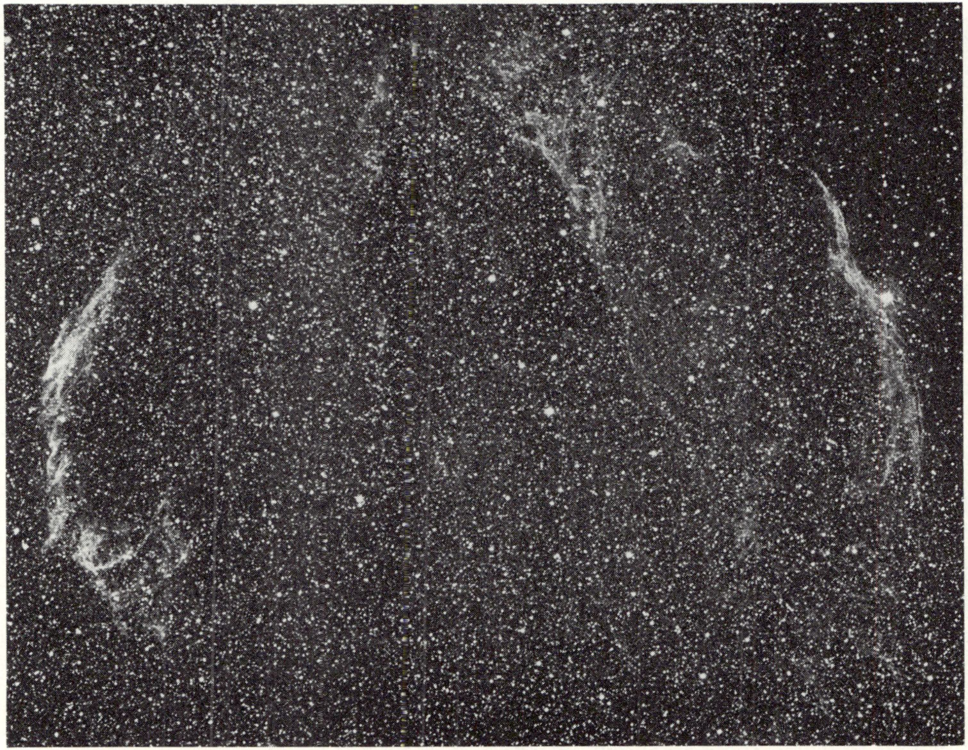

PLATE IX.—The Lacework Nebula in Cygnus. Schmidt telescope of the Haute-Provence Observatory (J. Texereau).

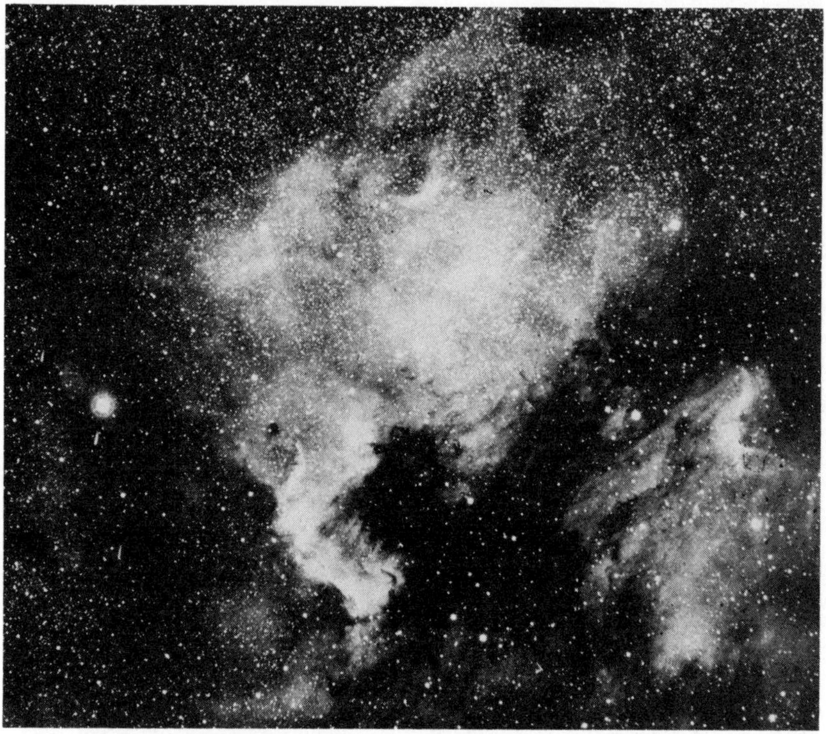

PLATE X.—The nebula 'North America' and the Pelican Nebula, in Cygnus, photographed in the light of the H α line. Schmidt telescope of the Haute-Provence Observatory (P. Berthier).

explore the galactic plane over a longitude of approximately 150°, by measuring the galactic latitudes $-4°$, $0°$ and $+4°$ in intervals of 4° of longitude. He also went on to make a more detailed study of the regions of Orion (1 point per degree) and of Monoceros (4 points per degree) and he compiled a first chart of galactic emission.

Photographic observations

Emission areas can also be revealed by photographing the same region of the sky first with a filter which isolates, for instance, the $H\alpha$ line, and then with one which only passes a spectral band not containing this bright line. Plate X shows the region which is situated to the west of the star ξ Cygni, photographed by means of the $H\alpha$ line with Schmidt telescope of the Haute Provence Observatory, constructed by A. Couder. The nebulae, particularly "North America" and "The Pelican", appear in far greater contrast than on ordinary photographs taken without filter and which are sensitive both to emission lines and to a large region of the continuous spectrum. On the negative reproduced on Plate XI, the same region was photographed in the near infra-red spectral band 7500–8800 A, devoid of all prominent emission lines. All the nebulosities have disappeared there; there is no longer a trace of "The Pelican" and of "North America"—these are only represented by their stellar skeletons.

The photographic method naturally has much more to teach us about the structure of the nebulae than photoelectric measurements. It has the added advantage of allowing excellent modifications: very powerful objectives covering a large field, such as certain cinematographic objectives, or better still, the Schmidt or Matsukov chambers.

The very large field of Henyey and Greenstein's optical apparatus [434] also allows the rapid recognition of extended regions of weak emission, and was used by Sharpless and Osterbrok [432] for this purpose. A spherical mirror of convenient radius gives a real image of the sky which is photographed with a wide open objective of short focal length. It is thus possible to obtain clear images over a circle of 140° in diameter, or even, as Courtès [401] has shown, of 180°.

Thus, at the Haute Provence Observatory, G. Courtès and A. Duflot [60] photographed the entire accessible part of the Milky Way with two identical small objectives of 75 mm. focal length and of aperture $f/1{\cdot}4$ used together. Each one of them was preceded by an interference filter which allowed a band of 150 A to pass, the first one centred on $H\alpha$, and the second, not passing this line, on a wave-length of 6150 A. In this manner a list of 149 emission areas was compiled, two-thirds of which did not appear in Cederblad's catalogue of diffuse nebulae [54]. Particularly in the region of Orion, results agree well with those of Strohmeier [311], who used a similar method, but with ordinary colour filters instead of interference filters.

At the Siméiz Observatory, the important researches of Shain and Miss Gazé [274, 429, 430], mentioned on more than one occasion, was undertaken by a similar process, but with objectives of a relatively large aperture ($f/1{\cdot}4$) and of much greater focal length (450 and 650 mm.). These were used together with interference filters of which one was still centred on the Hα line but the other a little further towards the red ($\lambda \simeq 6600$ A). The accounts of these two Russian astronomers, illustrated by a number of photographic reproductions and recently completed by a very beautiful atlas [431], give us details of more than 300 gaseous nebulae, two-thirds of which were previously unknown (many of these new objects, however, also figure in the list of the Haute Provence Observatory). The study of these negatives suggests new ideas on the structure of the nebulae, their mode of excitation and their relationship with certain groups of stars. We have already mentioned (page 81) the discovery of nebulae whose excitation did not appear to be the result of ordinary fluorescence. Another important factor is that the nebulae often manifest very clear tendencies to associate themselves with groups of hot stars, rather than with isolated hot stars. Herein the authors see an indication of a genetic, rather than an accidental link between the nebulosities and those stars responsible for their luminescence.

A very beautiful atlas of gaseous nebulae photographed in the light of the Hα line, with the Matsukow telescope of the Alma-Ata Observatory (f=120 cm., aperture=$f/2{\cdot}4$) has recently been published by Fessenkov and Rojkovsky [451].

Interpretation

Observation of emission lines shows that, apart from those bright gaseous nebulae previously catalogued, certain regions of the Milky Way contain hydrogen atoms and ions of O^{+}, O^{++}, S^{+}, etc. Since the observed spectrum is very similar to that of other gaseous nebulae, it is natural to suppose that the excitation mechanisms are the same. Thus, the emission of bright lines requires both the presence of atoms and of ultra-violet radiation capable of exciting them. Must we then conclude that when no emission takes place there are no atoms capable of being excited, or rather that the flux of ultra-violet radiation is too feeble to excite these atoms present? The study of interstellar absorption lines (Chapter V) will show us that atoms are relatively abundant in the neighbourhood of the galactic plane, but that their distribution is far from uniform. There is, however, no doubt that the localization of emission lines results essentially from the variation in the density of ultra-violet radiation from one point to another in the Milky Way. Observation does in fact show that the bright lines appear in the neighbourhood of the class O stars, and especially near those stellar clusters containing an appreciable proportion of O stars. This is markedly the case in the clusters of Cygnus and Cepheus, near which the emission lines are very strong, whereas near the bright clouds of Sagittarius, containing very few hot stars, the lines are generally weaker.

Strömgren has made a theoretical study of the ionization of interstellar hydrogen by the ultra-violet radiation of stars [313, 314]. On calculating, by means of Saha's formula, the degree of ionization as a function of the distance of a star, he found that ionization is practically total up to a certain critical distance r_0, and then diminishes rapidly to become quite inappreciable. *There are, therefore, well-defined areas in space in which hydrogen is almost completely ionized* ("*H II regions*"), *and others in which it is neutral* ("*H I regions*").

If we suppose hydrogen atoms to be uniformly distributed in an extended portion of space, the critical radius r_0 of the ionized sphere around a star depends both on the number n_H of (neutral or ionized) hydrogen atoms contained in 1 c.c. and on the spectral type of the star. Strömgren arrives at

$$r_0 = R_0 n_H^{-\frac{2}{3}}, \tag{IV.1}$$

with

$$R_0 = 0{\cdot}67.10^{-6}.\bar{a}_u.\ T_e^{\frac{1}{6}}.\ T_s^{\frac{1}{3}}.\ 10^{\frac{-2280}{T_s}}.\ R_s^{\frac{2}{3}} \tag{IV.2}$$

$\bar{a}_u$—effective absorption co-efficient below 912 A of the hydrogen atoms;

T_e=electron temperature;

T_s=effective temperature of the exciting star;

R_s=radius of the exciting star.

Fig. 13 represents the variations of the critical radius R_0 as a function of the spectral type, from O5–A0, when the density n_H is 1 atom per c.c. The temperature T_s and the absolute visual magnitude adopted by Strömgren corresponding to each spectral type are shown on the diagram as well.

The O5 and O6 stars ionize gas at enormous distances, but R_0 diminishes rapidly from type O5 to type B0, such that the O stars are almost the only ones effective. To obtain the radius r_0 of the ionized sphere, when the value of hydrogen atoms per c.c. has a fixed value, one simply divides the ordinates of the curve by $n_H^{\frac{2}{3}}$. Thus, if we take n_H=10 cm.$^{-3}$, we find r_0=30·2 pc. around an O5 star, r_0=14·2 pc. around an O8 type star, and r_0=5·6 pc. around a B0 type star. We would obtain no more than r_0=0·8 pc. around a B5 star.

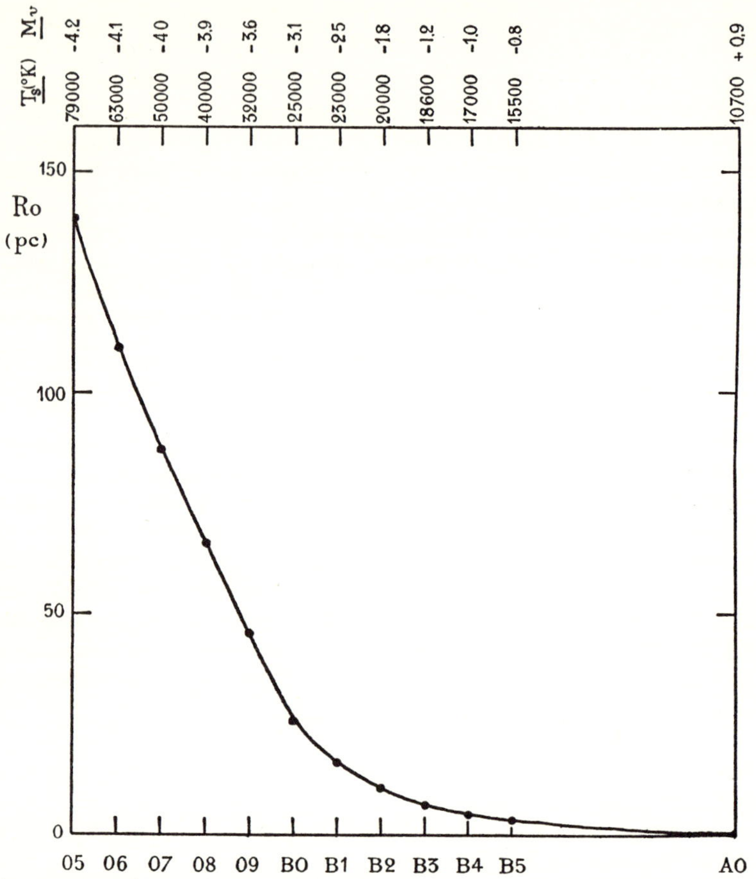

Fig. 13.—The radius R_0 of the Strömgren sphere where hydrogen is ionized for a density of one atom of hydrogen per c.c. On the abscissa are shown the spectral types and also (on the top scale) their temperatures and absolute visual magnitudes.

One can see how a measurement of the extension of a galactic area in the vicinity of a star of known distance, permits the evaluation by means of formula (IV.1) of the density n_H of the hydrogen atoms. We can also compare the monochromatic brightness of a Balmer line in these areas, to the brightness observed in a typical nebula, such as that in Orion, where the electron density is already known. By this means, Struve and Strömgren originally found that the density of hydrogen atoms, supposed to be uniform near the Galactic plane, ought to be of the order of 2 to 3 atoms per c.c. The study of interstellar absorption lines (Chapter V), however, will show us that gases frequently form discrete clouds, where the mean density can be of the order of 10 atoms per c.c. within a continuum of much smaller density.

Sharpless and Osterbrok [432] recently obtained comparable values in the emission areas by studying Hα lines around the stars λ Orionis (type O8) and ζ Ophiuchi (O9·5). The first has an apparent diameter of 7° and without doubt, a linear diameter of 71 pc., the second extends over 10°, and should have a diameter of 24 pc. Both resemble Strömgren's ideal H II regions, for they appear to be homogeneous and almost spherical around the exciting star. The proton densities calculated by the relation (IV.1) are 2·5 and 5·2 per c.c. respectively, if we suppose the gas uniformly distributed in the H II regions. They could reach 11 and 24 protons per c.c. if the protons were concentrated in small clouds occupying only 5 per cent of the total volume of the H II region. The Nebula "North America" (NGC 7000) (see Plate X) would be excited by the star HD 199579, of sixth magnitude and of spectral type O6, and not by α Cygni (spectrum cA2), as was previously supposed. Its diameter would be of the order of 33 pc. and on the hypothesis of uniform distribution, the mean proton density would be 15 per c.c.

The weak emission areas discovered by Struve and Elvey do thus differ but little in density from the brightest diffuse nebulae. In other words, bright nebulae would merely represent local condensation of interstellar gas in the neighbourhood of a star which is sufficiently hot to ionize it. They are H II regions of great density, to which, in certain cases, Strömgren's calculations can be applied as well.

Here is an example. The open cluster NGC 2244 in Monoceros is surrounded by a vast nebulosity, whose bright regions are known as NGC 2237, 2238 and 2246 (see Plate XII). The symmetrical form suggests that both the nebula and the dark clouds surrounding it are part of a large interstellar cloud whose density is uniform enough for the relation (IV.1) to be applied. The excitation is produced by a group of four stars: One of type O6, and three of type O8, whose distance was evaluated as 760 pc. by Minkowski [207] (allowing for interstellar absorption). The apparent diameter (80′) of the nebula gives a linear diameter of approximately 17 pc. Now, we can calculate the volume ionized by the four stars together: it is the sum of the volumes ionized by each one of them. The critical radius is then $r_0=130\ n_H^{-2/3}$ pc. By taking it to be 17 pc. we find n_H to be 60 atoms per c.c.

But as we saw on page 80 the measurement of the absolute intensity of the Hα line permits the evaluation by an entirely different procedure of the electron density—practically equal to that of the hydrogen ions—and we found $n\text{H} \simeq 30$ cm.$^{-1}$. The agreement between the two determinations may be taken as sufficiently satisfactory and to confirm the idea that the nebulae represents a Strömgren sphere within a relatively dense interstellar cloud.

In the case of the nebula NGC 6523 in M8 (see Plate XXI) Strömgren's theory leads to a density of 60 to 140 hydrogen atoms per c.c. depending on whether a temperature of 50,000 or 10,000°K has been attributed to the exciting star (type O5) [452], whereas in measuring the absolute intensity of the Hα line Shain and Gazé found nH=53 cm.$^{-1}$ [429].

Extension of the theory to planetary nebulae

In the past it seemed natural to attribute the abrupt outlines of planetary nebulae to a discontinuity in gas distribution, but today this has become an untenable idea. With an expansion velocity of 20 km./sec., the radius of the mass of gas ejected by the star ought to increase by $6{\cdot}3.10^{13}$ cm. (more than 4 A.U.) per century, which figure is more than $\frac{3}{1000}$ of the actual radius of a typical planetary nebula (estimated to be 2.10^{17} cm. (0·065 pc.). The age of the nebula could therefore not exceed 3,000 years, and this is ridiculous.

The edge of a planetary nebula is therefore not due to a discontinuity in the density, but to one in the ionization of the gas. The visible portion of the nebula would then represent H II region of ionized hydrogen around the exciting star. The radius r_0 of the Strömgren sphere is very large around a hot star when the number n of atoms per c.c. is small (page 99). It can, however, reach an order of magnitude of the radius of a planetary nebula, when the density is much greater, since r_0 is proportional to $n^{-\frac{2}{3}}$.

Page and Greenstein (1951) [229] have compared the linear radius R, evaluated from the distance and the angular radius, with the radius r_0 of the ionized sphere for nine planetary nebulae on which we have detailed information. The calculation of r_0 involves, as we have seen (page 99) the mean absorption co-efficient in the region of the continuous Lyman spectrum ($\lambda<912$ A), the temperature, and the electron density of the nebula, the temperature, and the radius of the star (the latter is calculated by equation (0.24) from the radiation temperature of the star and the absolute bolometric magnitude). Although these quantities can only be obtained in a very uncertain manner from observation, the comparison of the values of R and r_0 is nevertheless very suggestive: in all cases they agree with each other within a factor 2. It is further obvious that certain of the data are inexact: by correcting them in a reasonable manner, the equality between R and r_0 becomes almost rigorous.

For seventeen other planetary nebulae of unknown distances, Page and Greenstein compared the angular values corresponding to R and r_0 and the agreement is still very satisfactory. This time the apparent diameters range from 2″ to 71″. The authors conclude, therefore, that the visible parts of the planetary nebulae are identical with the H II regions. Thus we find one of the fundamental hypotheses of Zanstra's theory confirmed: the nebula absorbs all those photons emitted by the exciting star of frequencies higher than the Lyman limit.

This reasoning, however, is not proof against criticism. Gourzadian [407] notes that the calculation of the radius of the H II region utilizes the effective temperature T_S of the star (formula IV.2). Now this was only obtained by the application of Zanstra's method; the identity between the dimensions of the ionized region and of the nebula follows automatically and cannot be considered as a verification of the theory. The Russian astronomer thinks that Zanstra's method furnishes temperatures which are too low, so that the radius of the ionized regions (which increases with the temperature T_S) can readily exceed that of the nebula.

For Page and Greenstein's calculations to be convincing, it would be essential to obtain the temperature of the exciting star independently, by the study of its spectrum. The radius r_0 however, only increases slightly with the temperature T_S (proportional to $T_S^{\frac{1}{3}}$). In order to change the dimensions of the ionized region markedly, we must assume that the temperature deduced from Zanstra's theory is grossly in error. This is not the case. Aller arrived at values drawn from the spectral type of the stars (Table IV) which differ very little.

The identification of the visible part of the planetary nebulae with the H II region surrounding the central star seems quite probable. In consequence, the gas which is ejected by the star ought to extend far beyond the bright nebulosity. The total mass of the nebula must be very much larger than was previously imagined. With a density of the order of 10^{-20} gm./cm.^{3}, a radius of 2.10^{17} cm. and an expansion velocity of 20 km./sec., the mass which annually passes out of the bright internal H II region into the invisible external H I region is approximately 10^{-4} times the solar mass. If the nebula had only existed for 10^5 or 10^6 years, it would be surrounded by a mass of gas of the order of 10 to 100 solar masses i.e. 100 or perhaps 1,000 times greater than the mass of the bright region. These results would suggest that the planetary nebulae have a relatively short life, and possibly that they contribute appreciably to the increase in atoms scattered through interstellar space.

2. Emission of Radio Waves

Radiogalactic emission and interstellar gas.

In studying the direction of the arrival of radio noise on a wave-length of approximately 14·6 m. (frequency $2{\cdot}06.10^6$ sec.$^{-1}$, i.e. 20·6 megacycles per second, or—for short—Mc./sec.), Jansky [154] showed in 1932, the arrival of waves of extra-terrestrial origin. Soon he was able to show that they came from the Milky Way and that the most intense source was found approximately in the direction of the centre—in the constellation of Sagittarius [155]. This important discovery was very soon confirmed by other observers. Reber [241], in particular, by operating on a wave-length eight times shorter (1·85 m., or ν=162 Mc./sec.), was able to make use of a directional aerial and traced lines of equal intensity of galactic emission. Today we have a great number of similar observations made on wave-lengths ranging between 14·6 m. and 0·625 m. (i.e. between 20.6 and 480 Mc./sec.) [339]. Near the centre of the Milky Way the emission has been measured between λ=31·6 m. (ν=9·5 Mc./sec.) and λ=10 cm. (ν=3,000 Mc./sec.).

Do radio-frequency waves come from interstellar space or from the stars? The question is difficult to decide because of the extremely small resolving power of "radio-telescopes". Initially the first hypothesis seemed to be the more reasonable, at a time when radio waves from the Sun had not yet been detected. Whipple and Greenstein [374] have shown that the thermal emission by the solid particles of space is much too weak to account for the observed intensity. Emission by interstellar gas demands a closer examination (Reber, Henyey and Keenan, Van de Hulst, etc.). This would still be due to transitions between none-quantized levels ("free-free") produced between protons and electrons. With an electron temperature of 10,000°K, and a density of 1 proton and 1 electron per c.c., Henyey and Keenan [134] were soon able to account for the intensity of the wave-length 1·85 m. measured by Reber. They supposed, however, that all the interstellar gas could participate in the emission. Today we know that it ought to be limited to the H II regions where hydrogen is ionized (Woolley).

Today it would appear that galactic radiation for the lower frequencies is much too intense to be due to interstellar gas. The energy falling on unit surface (energy of illumination) per unit solid angle $d\omega$, between frequencies ν and $\nu+d\nu$, equals

$$dW_\nu=\mathbf{B}_\nu d\nu d\omega,$$

so that it gives us a measure of the spectral brightness $\mathbf{B}_\nu$. It is convenient to characterize the latter by means of the temperature of a black body which would have the same spectral brightness for the frequency considered [175]. According to Planck's equation (0.13), the spectral brightness is

$$\mathbf{B}_\nu=\frac{\mathbf{R}_\nu}{\pi}=2hc^{-2}\nu^3\left(e^{h\nu/kT}-1\right)^{-1}.$$

But, for the relatively very small frequencies of radio waves the exponent $\frac{h\nu}{kT}$ is very small and, without appreciable error, we take

$$e^{h\nu/kT}-1 \simeq \frac{h\nu}{kT},$$

so that the spectral brightness becomes

$$B_\nu = 2c^{-2}\nu^2 kT = \frac{2kT}{\lambda^2}$$

(the Rayleigh formula). For a fixed temperature it is proportional to ν^2 and, for a given frequency, it is proportional to T.

Townes [338] has shown that in order to account for the emission of low frequency, interstellar gas must have a temperature of the order of 150,000°K. More recently, Piddington [237] calculated the temperatures of relative brightness for various frequencies, in numerous galactic regions, using all the measurements made and published before 1951. Generally T diminishes rapidly with increase of frequency ν. Thus, for the region of co-ordinates $b=-2°$, $l=330°$, in the neighbourhood of the Galactic centre, T is, at first, of the order of 2.10^5°K while the frequency is between 9·5 and 18·3 Mc./sec., then diminishes progressively as ν becomes greater. For $\nu=200$ Mc./sec., we have $T=1{,}190°$ and for $\nu=3{,}000$ Mc./sec., the temperature drops to 2·8°K. In the region of the co-ordinates $b=0°$, $l=180°$, studied in a much smaller range of frequencies, the temperature of brightness diminishes regularly from 75,000°K, for $\nu=18{\cdot}3$ Mc./sec., down to 16·6°K, for $\nu=480$ Mc./sec. The curve representing the variation of T as a function of ν has the same appearance as a point situated outside the Milky Way, of co-ordinates $b=-30°$, $l=200°$, with temperatures which are systematically much lower ($T=50{,}000$°K with $\nu=18{\cdot}3$ Mc./sec.).

Interstellar gas, whose electron temperature can never exceed 10,000°K in the H II regions (see page 241) can hardly account for the brightness measured for the lower frequencies ($\nu<60$ Mc./sec., $\lambda<5$ m.). But this is not all: Piddington shows that thermal radiation from ionized gas generally gives a temperature of brightness which increases with frequency, and this is contrary to all observations. For the temperature to diminish with increase of frequency, it is necessary to have either an intense magnetic field in the gas or else to have its dielectric constant differing appreciably from unity. Neither of these conditions can be realized: the first would require a magnetic field of the order of tens of gauss, the second an extremely high electron density (perhaps of the order of 10^8 electrons per c.c.). An optically thin layer of gas (that is to say, one which only exercises negligible absorption) would give a temperature of brightness independent of frequency.

We must therefore return to the stellar or nebular origin of radio-galactic emission. The discovery of radio waves from the Sun (Southworth 1945, Appleton 1945, Appleton and Hey 1946) has shown that this is probable. Galactic emission, however, is so intense in certain stars, that we must imagine emission which is incomparably stronger than in the Sun in the "calm" state (perhaps 10^{n} times stronger). Thanks to the great improvements in detectors it has been possible to make measurements with increasingly smaller solid angles. Thus, Hey, Parsons and Phillips (1946) managed to show the existence of discrete radio sources of small apparent diameter, soon to be confirmed by the observations of Bolton and Stanley (1948) and those of Ryle and Smith (1948).

In the same way that the Michelson interferometer permits an increase in resolving power of optical telescopes, so does interference apparatus lead to more precise localization of radio-galactic sources (Bolton, Ryle). Ryle, Smith and Elsmore [246] have published a list of fifty discrete radio sources, which were detected in the northern celestial hemisphere by means of a very ingenious interferometer. Their positions have been determined with a precision proportional to the intensity of the source. Thus, the right ascensions of two of the strongest sources, situated in Cassiopeia and in Cygnus are known within five or ten seconds of time (1·25′ to 2·5′ of arc) and their declinations within 4′ or 7′. For weaker sources, however, the uncertainty may be as much as a few degrees. The brightness of the sources ranges over more than seven magnitudes. Their directions, distributed almost at random, do not show any galactic concentration. A priori, this fact may be given two quite different interpretations: either the sources are situated outside the Milky Way, or else they are very near to the Sun. We know, in effect, that near stars do not show the very galactic concentration of the far stars, which is easily explained. Of these two hypotheses the second seemed to be the more acceptable at first, but it appears today that neither corresponds with reality.

Identification of some radio-sources

The radio-sources of Ryle can neither be identified with bright stars, with novae, nor with planetary nebulae. They might be stars of a new type characterized by strong radio emission together with weak luminous emission.

The most recent interferometric research [413] seems to indicate that the most intense sources have an appreciable apparent diameter. Thus, that of Cygnus extends over at least 1′ (Mills) and that of Cassiopeia over a number of minutes of arc (F. G. Smith). Bolton also shows sources whose diameter attains some degrees, like that of Centaurus.

Today it is, therefore, certain that the majority at least of the strong radio-sources are not stars. They correspond to numerous classes of galactic and extra-galactic objects.

The sources of a first group coincide with nebulae more or less similar to the Crab Nebula. We have already seen (page 89) that the Crab Nebula itself corresponds to a radio-source of Bolton and Stanley. At less than 5′ from the position of the source in Cassiopeia, Baade [395], on negatives from Mount Palomar, showed the existence of a filamentous nebula of a sufficiently similar structure, whose expansion velocity is of the order of 1,500 km./sec. Hanbury Brown identified another source in Tycho-Brahe's supernova and it is possible that a third corresponds to Kepler's supernova, surrounded by a nebulosity whose spectrum recalls that of the Crab Nebula (page 90). At Mount Palomar, an object of a similar kind was photographed in Puppis, near an extended source of Bolton.

A second group corresponds to very special extra-Galactic Nebula. Very near to the intense source of Cygnus, Baade [395] showed the presence of two colliding galaxies with a relative velocity of 1,600 km./sec. Here, the Hα line is excited with such intensity that in the visible it gives more light than both galaxies together. M 87, already identified as a radio-source of Virgo, would represent a partial collision. NGC 1275 would also have radio emission.

Finally, in the third group we find normal galaxies, relatively near, such as M 51, M 101, and especially the Great Nebula in Andromeda (M 31). The emission of the latter has been studied in detail by Handbury Brown and Hazard [125] on a wave-length of 1·89 m. The energy received is of the same order of magnitude as that which would come from the Milky Way if the latter were situated at the same distance. By basing themselves on the absolute mean magnitude of extra-Galactic nebulae, and on their spatial distribution, the authors arrived at the conclusion that the radio emission of all the external galaxies only represents $\frac{1}{100}$ of the radiation received by the earth.

A new account of radio emissions of Galactic or extra-Galactic origins was published in 1954 by the *Journal de Physique* [453], and a monograph on discrete radio sources has been published by the Union Radio-scientifique Internationale [454].

The role of interstellar gas

Thus we have only been able to identify some of the most intense radio sources. It is probable, however, that a great number of weak sources, of small extension, largely account for the total emission of the Milky Way. Ryle, Smith and Elsmore think that interstellar gas can never contribute more than 1 per cent to the production of total radiation.

The problem of the origin of the latter was re-discussed by Piddington [237] and by Westerhout and Oort [370]. Piddington's analysis, of which we have already given some account, leads to the conclusion that the radiation received is of complex origin and is partly due to interstellar gas, and partly to what he believed was the atmosphere of the stars. For frequencies greater than 500 Mc./sec. ($\lambda < 0{\cdot}60$ m.), the energy distribution curve of Galactic radiation is, in effect, similar to that of an optically rare ionized gas. On the other hand, for frequencies lower than 100 Mc./sec., the curve could correspond to the emission from stellar atmospheres, which would not be of thermal origin. Interstellar gas, however, would exercise appreciable absorption on the radiation of radio sources, and ought to modify the total energy curve especially for frequencies lower than 60 Mc./sec. ($\lambda > 5$ m.).

Westerhout and Oort have shown that the emission of ionized gases is much too weak to account for Bolton and Westerfold's observations on a frequency of 100 Mc./sec. These, however, could be interpreted satisfactorily by supposing that the radiation is due to radio-sources whose spatial distribution would be similar to that of ordinary dwarf stars of the classes G and K. It is rather paradoxical that the distribution of radiation resembles that of the common stars, while the individual identifications of the most characteristic radio-sources makes use of large proportions of singular stars.

The line F (1)–*F* (0) *of the hydrogen atom, of wave-length* 21 *cm.*

If some doubt still exists about the contribution which interstellar gas makes to radio-galactic emission in general, it is at least certain that a particular monochromatic radiation can be attributed to the hydrogen atoms of space.

We know that in any atom each electron rotating around the nucleus spins on its own axis and possesses an angular momentum $\frac{1}{2}\left(\frac{h}{2\pi}\right)$, the electron spin (page 46). It is thus that we were able to give a simple interpretation of the multiplicity of levels, e.g. the double levels of the alkali metals. But to account for the *hyperfine structure* of the spectrum lines, we have to conclude that the nucleus is also in rotation and itself possesses an angular momentum of $i\left(\frac{h}{2\pi}\right)$, where i is a new quantum number, which can take only integral or half integral values. The total momentum $\vec{J}$ of the extranuclear electrons is obtained by compounding, in accordance with the known quantum rule, the vector $\vec{L}$, which is a resultant of the azimuthal moments of all the electrons and the vector $\vec{S}$, resultant of their spins (Russel-Saunders coupling, page 48). We must now similarly compound the vectors $\vec{J}$ and $\vec{i}$ in order to obtain *the total moment of the atom, including the nuclear spin.*

The case of the hydrogen atom is the simplest, since it has only one planetary electron and since the nucleus reduces to a proton. Let us consider the atom in its ground level. The electron is then in the first quantum orbit of Bohr; the principal quantum number is $n=1$, and the azimuthal quantum number, at most, equal to $n-1$, is $l=0$. The total moment of the electron is reduced to the spin $\frac{1}{2}\left(\frac{h}{2\pi}\right)$. On the other hand, the study of the H_2 molecule has shown that the spin of the proton is also $\frac{1}{2}\left(\frac{h}{2\pi}\right)$. We must therefore compound two vectors $\vec{s}$ and $\vec{i}$, each of value $\frac{1}{2}$, such that their resultant F would have an integral or half-integral value. There are only two solutions possible: either the vectors $\vec{s}$ and $\vec{i}$ are parallel and of the same sense, or they are parallel and of opposite sense. In the first case $F=1$, in the second $F=0$.

Thus the ground state of the hydrogen atom is a hyperfine doublet, whose two components correspond respectively to $F=1$ (electron and proton with the same spin) and $F=0$ (opposite spin).

The transition from the level $F=1$ to the level $F=0$, which corresponds to a reversal of electron as compared with proton spin, ought to be accompanied by a magnetic dipole emission of a line, but the lifetime of the level $F(1)$ is estimated to be of the order of 11,000,000 years. As the two sub-levels are very near each other, the frequency of the line is small and falls into the range of decimetre waves. In the laboratory we have been able to resolve the lines in a magnetic field (Zeeman effect) and to obtain a very precise value of the frequency. By the *atomic beam* method due to Stern and Gerlach's classical experiments, Prodell and Kusch [240] found that $\nu=1{\cdot}420405.10^9$ sec.$^{-1}$$=1420{\cdot}405$ megacycles per second ($\tilde{\nu}=0{\cdot}047$ cm.$^{-1}$) The corresponding wave-length is 21·1 cm.

In 1945, Van de Hulst [341] suggested the possibility of observing this line in the radio spectrum of interstellar emissions. In fact, in the H I regions almost all hydrogen atoms ought to be on the ground level. The possibility of detecting the line depends on the difference between the temperature characterizing the distribution of atoms on levels $F(0)$ and $F(1)$ ("spin temperature") and the temperature characterizing the continuous radiation in that region of the spectrum. It may be either an emission or an absorption line according to whether the spin temperature is higher or lower than that of the continuous background. It is certainly invisible if the temperatures are equal. Provided that the source is sufficiently thick to be practically opaque, the intensity of the line ought to depend on the difference of the two temperatures only.

The first results

In 1951, various observations led to a positive result. It was at Harvard University that the line was first discovered by Ewen and Purcell [95] at the end of March, then in Holland by Muller and Oort [211, 347] in May, and finally in Australia by Christiansen and Hindman [236] in July. Detectors were used so as to allow the direct comparison of the intensities of two very narrow spectral bands, one of which is centred on the line $F(0)-F(1)$, 75 to 110 kilocycles per second apart. At Harvard, use was made of a directional aerial, at Kootwijk of a swivelling paraboloid of 7·5 m. aperture and of 1·7 m. focal length. The half width of the admitted beam is, in the first case, 12° and in the second case 2·8°.

In Ewen and Purcell's first experiments, the aerial was installed in a fixed direction with regard to the Earth at a declination of $-5°$, and owing to the rotation of the Earth, various regions of the sky passed successively into the field of the detector. Under these conditions, an emission line was seen during six hours and its greatest intensity corresponded to a right ascension of approximately eighteen hours, or nearly in the Galactic plane. Its frequency was approximately 80 kc./sec. and the displacement of its frequency (which can be up to 150 kc./sec.) could be explained by the Doppler effect due to the orbital movement of the Earth and the movement of the solar system towards Hercules.

In the centre of the line the temperature reached a maximum which was greater by 25° than the temperature of the continuous background. The extrapolation of the results to lower frequencies leads one to think that towards a wave-length of 21 cm., this will not exceed 10°K. The spin temperature of hydrogen would, therefore, not exceed 35°K, if the source were really dense, a hypothesis which seems to be well confirmed. The probable excitation mechanism is, according to Ewen and Purcell, mainly the collision of two hydrogen atoms. In support of this point of view, we might remark that, in the regions H I, the kinetic temperature of the gas seems hardly greater than the spin temperature (see Chapter XIII).[2] It is possible, however, that energy exchanges with the continuous radiation take place.

Muller and Oort also let the sky pass over the field of their paraboloid kept stationary. They always observed, in following the declination, a maximum intensity near the Galactic plane but sometimes some degrees north or south of it. The distribution of intensity in the Galactic plane is very irregular; in Cygnus it is greater than in the direction of the centre of the Milky Way. The great extension in latitude of that region in which the line can be seen, shows that emission comes from relatively near regions, perhaps at a distance of 300–400 pc. generally, but of 600–800 pc. in Cygnus. Towards the Galactic centre where rotation does not affect the radial velocity, the line has a symmetrical profile and its magnitude corresponds to a mean gas velocity of the order of 5 km./sec. along each co-ordinate axis. If we get further away from the direction of the centre, the rotation of the Milky Way comes into play. The frequency of the line becomes modified by the Doppler effect (see page 60), and its displacement, in a given direction, is proportional to the distance of the source. Since the emission comes from clouds of a certain thickness, the line is also enlarged. Thus, at 30° from the centre ($l=355°$), the magnitude of the line attains 350 kc. In operating on a frequency lower by 250 kc. than the normal value, corresponding to a velocity of separation of +55 km./sec., Muller and Oort managed to observe the line emitted in regions so distant (some thousands of parsecs), that it can only be detected in a very narrow interval of latitude.

These first experiments thus revealed the rotation of the internal parts of the Milky Way, ordinarily hidden from view by absorbing clouds. A new method, full of promise, may henceforth be applied to the study of the structure of the Milky Way. In the same way that the emission of the Hα line gives us an idea of the nature of the H II regions, the emission of the line $F(1)$–$F(0)$ permits the study of vast regions of space, in which hydrogen is not ionized.

Since the first publications, work has been actively pursued [413]. Christiansen and Hindman observed the splitting of the line over a sufficiently extended section of the Galactic equator, and have thus detected the emission from two gaseous masses of distinct radial velocities, and therefore situated at entirely different distances. They certainly are two arms of the Galactic spiral which are situated one behind the other. (We shall, in fact, see in Chapter XVII that interstellar matter is generally concentrated along the arms of the spirals.)

In the same way Oort, Muller and Van de Hulst [423, 413] explored that entire portion of the Galactic equator which is observable in Holland, and they have traced a chart which represents, in this plane, the distribution of the emitting gaseous masses. At least three spiral arms appear on their diagram. One of them is very near the Sun, in the opposite direction to the centre of the Galaxy; another, at 2,000 pc. approximately at its nearest, partially overlaps the first arm in Monoceros. Finally a third one, which is the longest, extends over more than 70° of longitude. It is situated behind the second in the region of Cygnus and extends very far on the other side of the Galactic centre, around which it describes an almost circular arc of radius more than 10,000 pc. It has been traced up to approximately 20,000 pc. from the Sun.

The detailed results of the Dutch astronomers were published in 1954 in two papers, one concerning the external regions of the Milky Way [455], and the other the internal regions [456].

It is interesting to compare these results with those of Sharpless and Osterbrok's large-scale study of the H II region. This latter gives much more fragmentary results since the H II regions are few and far between. In its outlines, however, a model of the Galaxy, constructed in accordance with the data of W. W. Morgan, Sharpless and Osterbrok [433] is in sufficient agreement with the Dutch astronomers' map.

NOTES TO CHAPTER IV

[1]The Schmidt chamber consists of a spherical mirror M, of which the aberrations are corrected by means of a transparent plate L, placed near the centre of curvature and having a non-spherical cross-section of special design. It gives a much clearer field than a parabolic mirror and permits the use of relatively much larger apertures up to approximately $f/0{\cdot}7$.

[2]Spitzer and Savedoff [294] arrived at a kinetic temperature between 30° and 100°K, most probably 50°–60° (see Chapter XIII).

Plate XI.—Regions of the 'North America' and Pelican Nebulae, photographed in the near infra-red. Schmidt telescope of the Haute-Provence Observatory (J. Dufay and P. Berthier).

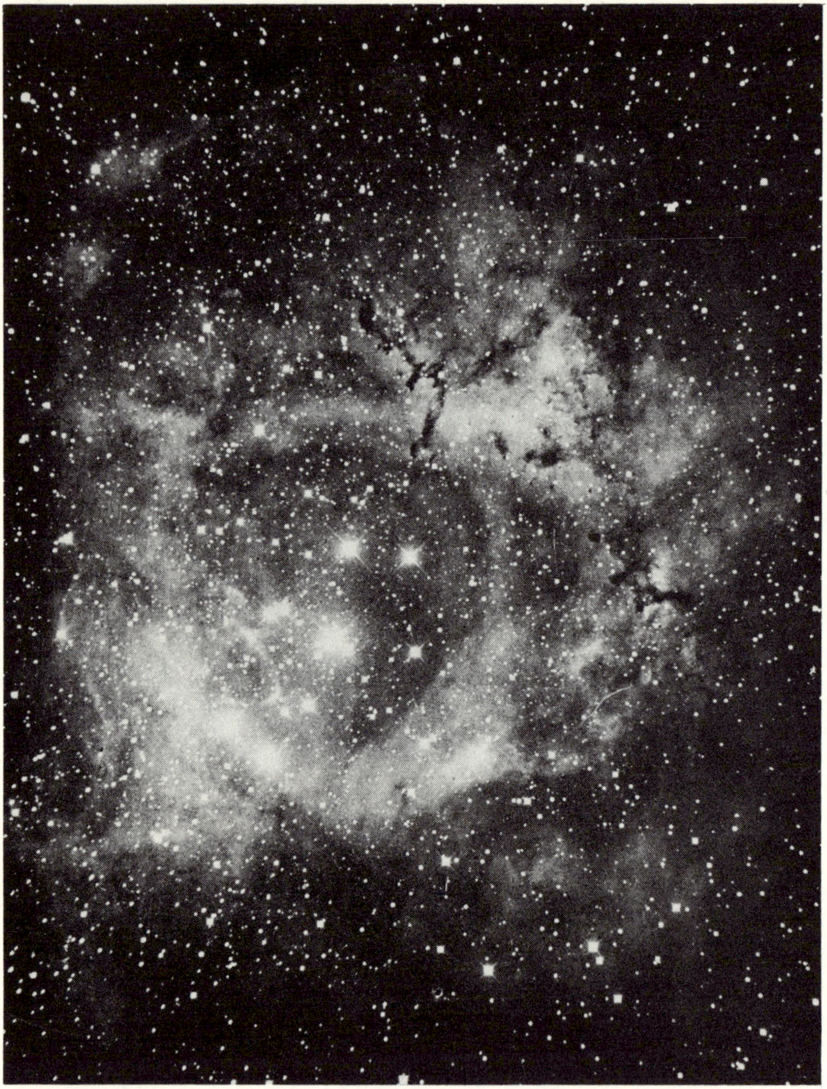

PLATE XII.—The Nebulae NGC 2237, 2238, 2239, around the open cluster NGC 2244, in Monoceros. Note the absorption marks in the bright nebulosities and around them. 80 cm. telescope (M. de Kérolyr).

CHAPTER V

SELECTIVE ABSORPTION BY THE ATOMS AND MOLECULES OF SPACE: INTERSTELLAR LINES AND CLOUDS OF GAS

"Stationary" or "detached" lines of ionized calcium

Long before the discovery of galactic emission, observation of interstellar absorption lines had established the presence of atoms in space. In fact, in 1904, J. Hartmann [126] had discovered the very singular behaviour of the K line of ionized calcium (3933·7 A) in the absorption spectrum of the star δ Orionis (type B1). This is an *eclipsing binary*, i.e. a pair, too compact to be resolved by the telescope, but whose binary nature is clearly established by spectroscopic and photometric observations. The former show that the spectral lines oscillate around their mean position, with a period of 5·7325 days, the latter that the light is regularly weakened twice per period. This is very well explained by the presence of two neighbouring stars, both of which turn around the centre of gravity of the system in 5·7325 days. The two light minima correspond to the mutual eclipses of the two components.[1] The oscillation of the positions of the spectral lines of the brighter star which alone is observable in this case, results from the fact that in describing its orbit the star alternately approaches and recedes from the observer. The displacement of the spectral line permits the measurement of the radial approach or separation velocity in km./sec., in each case.[2]

Now, since the radial velocity measured over all the other lines (H, He, Mg^{+}) of the spectrum of δ Orionis varies from $+135$ to -79 km./sec., Hartmann thought that the K line—and only this line—has a fixed radial velocity of $+16$ km./sec. (separation velocity). "*A careful study*," he says, "*has led me to the altogether surprising conclusion that the calcium line* 3934 *A does not participate in the periodic displacement of the other lines, caused by the orbital movement of the star.*"

After having shown that the "stationary" line can neither be produced in the terrestrial atmosphere, nor be attributed to the spectrum of the weaker component of the double system, Hartmann concludes: "*We are thus led to suppose that somewhere in space along the line of vision between the Sun and δ Orionis, a vapour cloud of calcium exists which produces this absorption and which recedes with a velocity of* 16 *km./sec.*" In order to determine the lateral extension of the absorbing cloud, Hartmann proposed to observe the spectra of neighbouring stars which had different radial velocities, such as ϵ and ζ Orionis. But the radial velocity of these stars varies little and does not differ from that of the calcium cloud. It is therefore, at the moment, impossible to decide whether the K line belongs to the spectrum of the star, or to that of the cloud.

The "stationary" K line has been found in other double spectroscopic stars of the first B types. But, if it is produced by a cloud of calcium interposed between the star and the observer, there is no reason why it should not appear in the spectrum of simple stars also. Hartmann himself, in discussing his observations on δ Orionis, recalled those previously made on Nova Persei in 1901. After its explosion, lines of hydrogen and of other elements became greatly increased, strongly displaced, and underwent frequent changes of structure. Conversely, the two lines K and H (3968·4 A) of ionized calcium and the doublet D of neutral sodium (5890, 5896 A) remained sharp and fixed, with an unvarying separation speed of +7 km./sec. Hartmann had already expressed the opinion that it would be

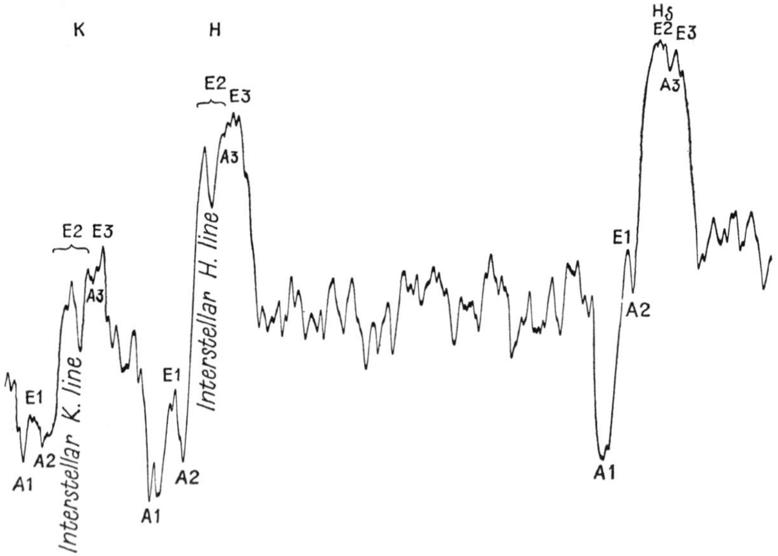

Fig. 14.—Interstellar H and K lines in the spectrum of Nova Cygni 1948. The very complex structure of the lines shows three absorption components (A_1, $v_r = -1240$ km./sec.; A_2, $v_r = -620$ km./sec.: A_3, $v_r = +230$ km./sec.) and three emission components (E_1, $v_r = -800$ km./sec., E_2, $v_r = -60$ km./sec. and E_3, $v_r = +420$ km./sec.). The sharp interstellar H and K lines cut into the strong emission components E_2 of the corresponding lines. Spectrograph with one prism mounted on the 120-cm. telescope of the Haute Provence Observatory (M. Bloch).

necessary to look for the origin of this, not in the atmosphere of the nova itself, but in a nebular mass across the line of vision.

It is easy to understand why we cannot observe the "detached" calcium lines more frequently. It has been established that the K line does not usually belong to the spectrum of stars of the first B types. Normally, it begins to appear in the B5 stars and becomes progressively stronger in the coldest stars. It is, therefore, almost impossible to observe the "detached" K line in stars of spectral types higher than B3 or B4, for it is masked in them by a stronger and particularly broader K line, produced in the atmosphere of the star itself. As for the H line of the same multiplet, even with the stars of the first B types, it is always represented by the strong $H\epsilon$ line of hydrogen, whose wings extend sufficiently far on both sides of 3970 A. It is only in the spectra of novae that we can generally distinguish between very fine sharp lines of Ca II and Na I amidst the complex structure of the corresponding stellar lines, which are strongly enlarged and displaced. Fig. 14 shows an example thereof.

We, therefore, have no chance of detecting the K line in the calcium cloud, except amongst stars of types O5 to B3. Now a certain number of them effectively show a feeble K line which is much sharper than the other spectral lines. On every occasion that exact measurements have been possible, it has been found that this K line does not lead to the same radial velocity as all the other lines. It is, therefore, an *interstellar* line.

The interstellar line of Na I (5890, 5896 A), whose intensity is comparable, can be observed in a greater number of stars, for normally neutral sodium only appears in the atmosphere of relatively cold stars.

Distribution of calcium ions in space: First approximation

Some authors have suggested that the O and B stars could be surrounded by a calcium cloud and might be displaced towards the centre. A very attractive theory of Milne shows that the radiation pressure in the solar chromosphere ought to expel Ca^+ ions into space. But the hypothesis does not explain why only *certain* stars of the O5 and B5 type should be surrounded by such a cloud. It would have been much more rational to suppose, following Eddington, that the atoms of calcium are almost uniformly distributed in space.

This idea seemed to have been confirmed when a sufficiently large number of good spectrograms of O and B stars were available. Otto Struve [318, 319] has shown that the intensity of the detached K line increases roughly with the distance of the stars (1929). The distances of the B stars were still, it is true, very uncertain. Few of them are sufficiently close for us to measure their parallax directly; in their spectra the detached line is weak or absent. On the other hand, the evaluation of the distances of the B stars from their absolute spectroscopic magnitude only gave provisional estimates. But the proper movements could, to some extent, serve as criteria of distance. They are roughly the smaller the more distant the stars are. Now, there exists a clear negative correlation between proper movement and the intensity of the K line. The intensity of the line increases with apparent magnitude, which is, on the average, the higher the further away the star is. Finally, the distance of certain groups of B stars was determined fairly accurately by indirect procedures (for instance in the case of stars "in group movement"). With the aid of all of these data, Struve was able to construct a graph showing that the intensity of the detached K line increases constantly with distance of the star.

By immediately reversing the terms of the problem, he could put forward an audacious idea which, by means of the intensity of the interstellar K line, determined the distance of the B stars. Although today we know that the distribution of calcium atoms in space is only uniform as a first approximation, this new method applied preferably to groups of these stars rather than to individual stars, has decidedly improved the determination of their distances.

The research of J. S. Plaskett and Pearse (1935) [239] has also confirmed the fact that the distribution of Ca^+ ions could be considered as being roughly uniform. Now statistical study of the proper movement of radial velocities had led J. H. Oort (1927) to establish beyond doubt the rotation of the Milky Way, predicted by Lindblad. The Milky Way does not revolve as a solid body, and there is every reason to think that the angular velocity of rotation is the greater the nearer the stars are to the centre. From this results the possibility of the movement being one of *differential rotation.* A simple calculation shows that this imparts to a star, situated at a distance r from the Sun, at a Galactic longitude l and at latitude b, a radial velocity

$$v_r = Ar \sin 2\,(l - l_0) \cos^2 b. \qquad \text{(V.1)}$$

In this formula, l_0 represents the Galactic longitude of the centre of rotation (near 325°) and the constant A (Oort's constant) measures in km./sec., the radial velocity imparted to a star situated at 1 parsec from the Sun (for $l - l_0 = 90°$ and $b = 0°$). A is approximately 0·017 km./sec.

Naturally, to the velocity v_r is added that velocity resulting from the known translational movement of the solar system, and also an unknown "particular velocity", belonging to each star. If, however, we consider a group of stars of determined longitude, latitude and distance, we can assume that the particular movements are random and cancel each other on the average. The radial velocity, corrected for the translational movement of the Sun ought, therefore, to satisfy equation (V.1). Oort's statistical studies made the determination of the two constants l_0 and A possible.

Plaskett and Pearse evaluated the mean product $\overline{Ar}$ (amplitude of the Oort effect) of a number of B star groups situated at various distances from the Sun, and they concluded that it is roughly proportional to the mean distance $\bar{r}$ of the group, as is required by equation (V.1). Their measurements were naturally made on absorption lines produced in the atmospheres of the stars. But in measuring the average radial velocities from the interstellar K line, Plaskett and Pearse found that the product $\overline{Ar}$ was, for each group of stars, approximately half that found for true stellar lines. All this is as if the mean distance of the calcium cloud were, in each case, half the distance of the stars. Calcium therefore appears to be distributed in a roughly uniform manner.

Further research: The equivalent magnitudes of the interstellar lines of Ca II and Na I

In the original investigations, the intensity of the K line was "estimated" in a rather arbitrary manner. Later on, correct photometric measurements were made on the interstellar lines H, K, D_1 and D_2, notably by Beals [25, 26], E. G. Williams [378], Wilson and Merrill [382], Merrill, Sanford, Wilson and Burwell [200], Sanford and Wilson [247, 249] and more recently by Sanford [248], Spitzer, Epstein and Li Hen (1950) [292].

By measuring the intensity of the spectrum at a certain number of points in the line and in its neighbourhood, the photometric outline

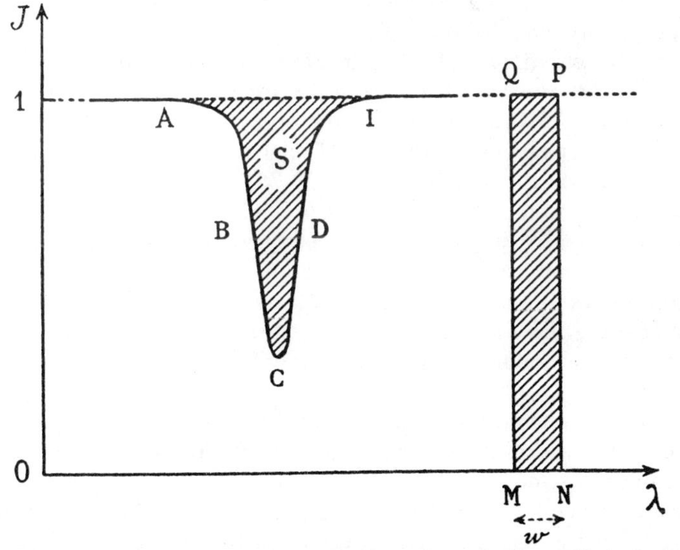

Fig. 15.—Definition of the equivalent breadth of an absorption line. This is the breadth ω=MN of the rectangle MNPQ, whose area is equal to the area S of the line.

A, B, C, D, I is traced (Fig. 15). The area S lying between this profile and the line A I of the continuous spectrum, measures the total absorption. We generally express this by the breadth M N of the rectangle M, N, P, Q, whose height is equal to the intensity of the continuous spectrum (taken as unity) and whose area is equal to S. This is what we call *the equivalent breadth of the absorption line*. It is measured in ångströms.

The discussion of these measurements has, on the whole, confirmed the results of Struve and those of Plaskett and Pearse. Evans [94] particularly studied the relation between the distance of the stars (evaluated by means of the radial velocity due to galactic rotation) and the equivalent breadth of the K and D lines. His main conclusion is that the intensity of the interstellar line is probably the best criterion of the distances of the B stars: even if applied to individual stars there is never a greater mean error than 20 per cent. Van Rhijn thinks, however, that this error could attain 40 per cent. Schilt (1947) also concludes that in the first approximation there is a uniform distribution of Ca^+ ions. Finally, Sanford (1949) [248], measuring the radial velocities of the stars in fifteen open Galactic clusters, also finds a sufficiently close correlation between the equivalent breadth of the K line and the amplitude $\overline{Ar}$ of the Oort effect (but the increase in the equivalent breadth as a function of $\overline{Ar}$ is not exactly linear). The amplitude of the Oort effect is twice as small in the case of interstellar lines as that of absorption lines produced in the atmosphere of the stars.

Second approximation: Clouds of gas

Nevertheless, in 1936, Beals [26, 27] announced that the interstellar lines H, K, D_1 and D_2 were split in the spectrum of numerous stars as if there existed, between them and the observer, two clouds of calcium (and of sodium) having different radial velocities. This fact was confirmed by Sanford, Merrill and Wilson (1938–39) [249] and, with a highly dispersive spectrograph (2·9 A/mm.), Adams [2] proved that the interstellar H and K lines were, in fact, very frequently complex (1943, 1949). Of the 300 stars which he studied, 87 showed double lines, 17 triple lines, and 4 quadruple lines. If one adds to these 108 stars, 40 others where the breadth of the H and K lines suggests a complex structure, we find that the interstellar lines are double or multiple in approximately half of the studied spectra.

Today it is, therefore, certain that the Ca^+ ions are often grouped in distinct clouds possessing different velocities. But the sharpness of each of the components indicates that the velocities of the ions vary but little inside any one cloud. We conclude that there are marked differences in the behaviour of the lines in various regions of the sky. Complex lines are particularly frequent in Cygnus, Sagittarius and Orion but in the latter, their intensity is moderate. The structure of the lines in the spectra of neighbouring stars is very often the same. On the average, the radial velocities of the clouds are small (arithmetic mean 4·5 km./sec.), but there exist, nevertheless, some which are greater than 30 or even 50 km./sec. With few exceptions, the clouds with great velocity produce weak lines, and the strongest lines come from slow clouds.

The results of the first measurements made by Adams on only fifty stars have been analysed by Whipple [373] and by Melnikov [421]. Whipple found that the mean square velocity of opaque clouds producing strong lines was approximately 7 km./sec. and that of more transparent clouds, giving weak lines, 23 km./sec. A straight line of the Galactic plane, 1,000 pc. in length, should roughly traverse six to ten distinct clouds. Melnikov also found a mean square velocity of 7 km./sec., by supposing a Maxwellian velocity distribution.

In a more recent work, based on all the measurements of Adams, Blaauw [397] has shown that the frequency of the observed velocities is better represented by the expression $e^{-v/\eta}$ than by a Gaussian distribution. For stars situated at least 500 pc. from the Sun (120 in number), the mean velocity η of the clouds is 5 km./sec. and the light of each star traverses three to six clouds on the average. For the more distant stars (forty-three in number) the mean velocity is approximately 8·2 km./sec. and the number of clouds traversed can be between four and seven. A straight line of 1,000 pc., in the Galactic plane, traverses eight to twelve clouds on the average. This study leads one to believe that a greater number of multiple lines would be observed if we could further improve the resolving power of the spectrograph.

The dimensions of the clouds probably vary between rather large limits. As a working hypothesis, we could still take the provisional scheme of Strömgren [314], in which a straight line of 1,000 pc. traverses on the average, seven clouds, whose mean diameter is of the order of 5 parsecs. (The study of the spatial distribution of interstellar gas will be completed by that of solid particles in Chapter XIII.)

With the 200-inch telescope of Mount Palomar, G. Münch [457] has observed K, H and D lines in the very disperse spectra of distant B stars of low Galactic latitudes and of longitudes between 65° and 130°. He finds that they show two strong components of comparable intensities. The violet component of the K line is often broader and shallower than the red, and it is sometimes resolved into a certain number of fine lines (up to seven or eight). The red component obviously occupies the same position as the unique interstellar line of the stars situated in the same region of the sky, but at only about 1,000 parsec from the Sun.

We can thus interpret the large splitting of the interstellar lines in this region as a result of the superposition of two groups of clouds, situated at different mean distances.

The radial velocities of the red and violet components respectively, correspond to Galactic rotation velocities at distances of the order of 400 and 3,000 parsec, and the two groups of clouds seem to align themselves along the two arms of the spiral, demonstrated by Morgan, Osterbrock and Sharpless [433] in that region, by observing emission areas.

In the spectra of near stars also, such as 56 Eridani and γ Eridani, a large splitting of the H and K lines can be observed, which cannot be explained in the same way. For instance, in the case of 56 Eridani, studied by E. M. Burbidge and G. R. Burbidge [458], the radial velocity of the red component corresponds fairly well with the Galactic rotation, but the violet component must be attributed to a cloud of the same arm of the spiral, but having a velocity of 40 km./sec. with respect to the principal body of the latter.

Other interstellar atomic and molecular absorption lines

By means of spectrographs of great resolving power, new and weaker interstellar absorption lines have been discovered since 1936. Table VII gives a list of the lines and the corresponding transitions which have been identified, mainly according to the indications of Beals [28, 29].

We immediately see that all the known atomic lines are *resonance lines*. The absorbing atoms and ions of space are, therefore, all in the ground state, as is the case at very low temperatures. In the case of Ti^+, the lowest level (^4F) is quadruple. Now, the two observed lines originate from the lowest sub-level. It is, therefore, probable that, in space, the sub-levels immediately below are depopulated by emission of forbidden lines (Dunham [80]).

The identification of other lines presented a little more difficulty. Thus, the line 4300·3 A, discovered by Adams and Dunham, does not seem to belong to any known atom. Swings and Rosenfeld [332] proposed to attribute it to the CH molecule, already known in the absorption spectra of the Sun and of stars, and in the emission spectra of comets. In the laboratory, as in the stars, a molecular band consists of a great number of lines, for, at high temperatures, the molecules are distributed on numerous rotation levels. The number of lines diminishes when the temperature becomes lower and, at very low temperatures, each band can be reduced to some straight lines almost similar to atomic lines, since the major part of the molecules only have a very weak or completely absent rotational energy. Thus McKellar [193] predicted the observation in interstellar space, of the first rotation lines of the band 3883 A of the CN molecule, and

TABLE VII

List of Interstellar Absorption Lines

	Atom or Molecule	Transition	Discovered
		A. *Atomic Lines*	
3302·4 3303·0	Na I	$3^2S - 4^2P^0$	Miss Heger (1921) [129]
5890·0 5895·9	Na I	$3^2S - 3^2P^0$	Miss Heger (1919) [129]
7664·9 7699·0	K I	$4^2S - 3^2P^0$	Dunham (1937) [79]
4226·7	Ca I	$4^1S - 4^1P^0$	. . . . [79]
3933·7 3968·5	Ca II	$4^2S - 4^2P^0$	Hartmann (1904) [126]
3073·0	Ti II	$a^4F - z^4D^0$	Dunham and Adams (1936) [83]
3229·2 3242·0	Ti II	$a^4F - z^4F^0$	. . . . [83]
3283·8	Ti II	$a^4F - z^4G^0$	. . . . [83]
3719·9	Fe I	$a^5D - z^5F^0$	Dunham and Adams (1941) [84]
3859·9	Fe I	$a^5D - z^5D^0$	. . . . [84]
		B. *Molecular Lines*	
4300·3	CH	$A^2\Delta \leftarrow X^2\Pi$, (0.0), R (1)	Dunham (1937) [79]
3890·2	CH	$B^2\Sigma \leftarrow X^2\Pi$, (0.0), $P_{Q1\cdot 2}$ (1)	Adams (1941) [1]
3886·4	CH	. . . ., Q (1) + $Q_{R\,2\cdot 1}$ (1)	. . . . [1]
3878·8	CH	. . . ., $R_2(1)$	. . . . [1]
3875·8	CN	$B^2\Sigma+ \leftarrow X^2\Sigma+$, (0.0), P (1)	Adams (1941) [1]
3874·6	CN	. . . . , R (0)	. . . . [1]
3874·0	CN	. . . . , R (1)	. . . . [1]
4232·6	CH+	$A^1\Pi \leftarrow X^1\Sigma+$, (0.0), R (0)	Dunham (1937) [79]
3957·7	CH+	. . . . ,(1.0), R (0)	. . . . [79]
3745·3	CH+	. . . . , (2.0), R (0)	Adams (1941) [1]
3579·0	?	?	. . . . [1]
		C. *Diffuse Lines*	
4430·6 (10)	?	?	Merrill (1936) [197] Beals and Blanchet (1938) [30
4760	?	?	Merrill (1940) (Miss Heger, 1921) [129]
5780·5 (3)	?	?	. . . . [129]
5797·1 (1)	?	?	Merrill (1934) [197]
6203·0 (1)	?	?	Merrill and Wilson (1938) [201]
6270·0 (1)	?	?	. . . . [201]
6283·9 (6)	?	?	Merrill (1934) [197]
6613·9 (2)	?	?	. . . . [197]

these were, in fact, discovered by Adams. Other lines found by Adams have been attributed to the ionized CH molecule (Douglas and Herzberg [69]). All the molecular lines whose production in space has been discovered up to now, only involve the two lowest rotation levels of the molecules CH^+, CH, CN (and also the two lowest vibration levels).

Do molecular lines come from the same clouds as those of ionized calcium? According to the most recent research of Adams (1949), it seems that, in general, we could reply to this question in the affirmative. The molecular lines—and especially CH^+ 4232 A—are found in a quarter of the studied stars and the radial velocity which they provide almost invariably coincides with that of the strongest K component. In one star, the line 4232 A shows the same splitting as the H and K lines.

On the other hand, the degree of ionization and the apparent chemical composition of the clouds appear to vary. In general, the line CH^+ 4232 A is the most intense after K and H—of all the interstellar lines in the violet, but we know some stars where the line 4300 A of the neutral CH molecule is even stronger (ζ and χ Persei, ρ Ophiuchi for instance). In this case, the Ca^+ lines are only of a moderate intensity.

The bright stars of the Pleiades (with the exception of 20 Tauri) behave in a most singular manner; here the line 4232 A is stronger than the H and K lines, and the corresponding intensities show a considerable variation from one star to the next, in spite of their proximity. It is possible that the interstellar lines come from nebulosities which envelop the Pleiades and that radiation of the stars is involved in some way or other (Adams).

Diffuse lines

Some of the interstellar lines discovered by Merrill, which differ profoundly from all the preceding ones in their diffuse character, have not as yet been identified (Table VII, C). They have, for the most part, a breadth of 4 to 10 A and resemble stellar lines; the strongest (4430 A) is a veritable band from 40 to 60 A [30]. These lines are, nevertheless, well produced in space; their intensity increases with the distance of the star, and that of the "band" 4430 A is connected with the intensity of the interstellar K line (Beals and Blanchet [30]; Miss Sherman [282]). The "stationary" character of that band has further been demonstrated in the spectrum of a spectroscopic double star whose radial velocity varies with an amplitude of 700 km./sec. (Merrill and Humason [199]).

The origin of these lines remains a riddle. An attempt has been made to attribute them to absorption by solid particles (Merrill), but we have to admit that we do not actually know similar lines in the spectra of solidified gases [330]. Swings also thought of homopolar molecules, such as C_2 or O_2, which could be distributed on their rotation levels so as to give non-resolvable bands.

Ionization, composition and density of interstellar gas

Our knowledge of interstellar absorption lines, fortunately completed by the study of emission lines which, amongst other things, gives us information about hydrogen and other atoms, is today capable of providing a sufficiently clear indication of average chemical composition, degree of ionization, and density of gases in space.

Numerous attempts to solve these problems had already been made with somewhat vague data, but it was re-examined and treated in a very complete manner by B. Strömgren in 1948 [314]. Without entering into the details of his calculations here, we shall attempt to give a general idea of the consequent progress. We must proceed step by step and at first establish the relation between the intensity of an interstellar line (expressed by its *equivalent breadth*) and the total number N of atoms or absorbing ions contained in a column of unit section joining the stars with the observer. We can then evaluate the mean spatial density of the absorbing atoms (or ions). To obtain the total density of the element considered, we must further calculate the degree of ionization. To obtain the total density of calcium or of sodium for instance, from the intensity of the Ca^+ or of the neutral Na atom lines, it is necessary to know the proportion of ionized to neutral atoms.

The relation between the number N of the absorbing atoms and the total intensity of an absorption line is given by the *curve of growth* traced by plotting the equivalent breadth $\omega(\lambda)$ of the line against N, or more often, $\log \omega(\lambda)$ against $\log N$. Since the number of absorbing atoms is sufficiently small, the intensity of the line increases proportionately to N and at first we obtain a straight line. But, for the greater values of N, $\omega(\lambda)$ increases more and more slowly. We say that the line becomes "saturated". Such curves of growth have been constructed for the study of stellar atmospheres, by supposing that the velocity distribution among the absorbing atoms is represented by Maxwell's law (page 67). We use the same hypothesis in the case of interstellar atoms, but the question is complicated if the atoms are distributed in distinct clouds having differing total velocities. These clouds produce lines displaced with respect to one another and capable of partial detection. Strömgren, agreeing with Lyman Spitzer, shows that the relation which holds for a particular cloud can be applied to an aggregate of clouds if we modify a certain Doppler effect constant.

It we are dealing with a doublet whose two lines belong to the same lower level, the relative intensity of the two lines in the case of absorption only depends on the ratio of the transition probabilities, so long as the number N of the absorbing atoms is sufficiently small. Thus the D_2 line (5890 A) is twice as strong as the D_1 line (5896 A), the K line two times stronger than the H line. But the most intense line becomes more quickly saturated as the number of absorbing atoms increases, such that the ratio of the intensities of the two lines diminishes progressively and finally tends towards unity. This fact is very clear in the case of the interstellar D lines, from the measurements made on Mount Wilson, and we can see that the measurement of the ratio of the intensities of the two lines would be very useful in the determination of N (Wilson and Merrill).

The proportion of ionized atoms in the stellar atmosphere is governed by the Saha equation, which involves the ionization potential of the atom, the temperature and the number of free electrons per c.c. Eddington (1926) has shown that the same equation can be applied to atoms in space by taking as the temperature that of those stars whose radiation brings about the ionization ($T \simeq 10{,}000°K$), and by introducing a *dilution factor* ($\simeq 10^{-14}$) to account for the separation of the stars. Rosseland applied another correction factor because the kinetic temperature of the particles (notably of the free electrons) differs from the temperature T of the stars.

Since the atoms and the ions are all practically at the ground level, it is always on this level that ionization takes place. But Strömgren observed that this is not so in the case of the recombination of ions and of electrons which could give rise to excited atoms. From this results a new correction sometimes of the greatest importance.

The relative intensity of the Na I and the Ca II lines shows that the number of neutral Na atoms is of the same order of magnitude as that of the Ca^+ ions (1 Na atom for 1 to 3 Ca^+ ions). Now, the ionization potential of sodium (5·11 volts) being less than that of calcium (6·08 volts) most of the sodium atoms ought to be ionized. This would result in a preponderance of neutral and ionized sodium atoms over neutral calcium atoms in space. Since, in the solar atmosphere or in that of the stars, almost as many calcium atoms as sodium atoms are found (Table VI, page 81), the first calculations based on the equations of Eddington, would indicate 100 sodium atoms to 1 calcium atom, in space. Applying the new correction factors, Strömgren has found only 25 atoms of sodium to 1 of calcium. The difference from the atmospheres of the stars is therefore reduced, but remains very important. It is most surprising that as far as the other elements are concerned, the chemical composition of interstellar gas does not seem to differ appreciably from that of the Sun, the stars or the planetary nebulae.

A new discussion by Seaton [269] confirmed this general conclusion. The ratio of relative abundances of sodium and of calcium alone remains anomalous. But Seaton suggests a very probable explanation of this anomaly. To evaluate the ionization degrees of the different atoms we assume a certain energy distribution in the radiation of all stars. It is very possible that this energy distribution based on Dunham's statistics page 238) is false. The radiation which ionizes the Na atoms and the Ca^+ ions belongs to quite different spectral regions: the first, between 2400 and 1800 A approximately are largely (80 per cent) due to stars whose temperature is lower than 16,000°K; the second, between 910 and 1040 A, are due to very much hotter stars (near the B0 type). Now there are a great number of indications to show that we have attributed too large a value to the continuous energy in the first spectral region, and too small a value to the continuous energy in the second region. The possible magnitude of these errors can almost account for the anomalous abundance ratio of Na to Ca.

The composition of interstellar gas given in the last column of Table VI (page 81) was obtained by Strömgren from the dense clouds situated in front of χ^2 Orionis, containing approximately 50 atoms of hydrogen per c.c. The spectrum of χ^2 Orionis, rich in interstellar lines, in this case permits the more certain determination of the composition of the cloud. The ratio of sodium to calcium is, this time, almost normal.

As in the stars and the nebulae, hydrogen is always preponderant and Strömgren concludes that inside clouds, where hydrogen is not ionized (H I regions), there are approximately 10 atoms of hydrogen per c.c. If one further assumes that the mean atomic mass is of the order of 1·6 (page 104), we arrive at a specific gravity near $1{\cdot}6\times1{\cdot}67\times10^{-23}=2{\cdot}7.10^{-23}$ gm./c.c., in the case of a typical cloud.

In the H II regions, the density of the electrons is practically equal to that of the atoms; it ought to be some hundreds of times, or even a thousand times, smaller in the H I regions, since free electrons are only due to atoms more easily ionized than hydrogen (and helium), and which are very much less abundant (Seaton obtained 0·02 electrons per c.c.). There is, on the contrary, no reason for any differences in atomic densities between the two types of regions.

Strömgren [316] estimated that in the neighbourhood of the Sun, the volume occupied by the H II clouds is hardly 1/10 that of the H I clouds. It is possible that the H II regions occupy a larger fraction of space between the clouds. It is thought that the density of matter can here be 50 to 100 times smaller than inside typical clouds, with 1 atom of hydrogen per 5 or even 10 c.c.

NOTES TO CHAPTER V

[1]The spectroscopic binary stars only show eclipses if the plane of their orbit is inclined slightly to the line of vision. The amplitude of the 2 minima is weak in the case of δ Orionis (0·15) and 0·12 magnitudes respectively), and such variations in luminosity were unknown when Hartman made his discovery.

[2]When the two components have comparable luminous intensities and identical spectra, the lines are split, and undergo oscillations of different amplitude, always in opposite phase.

Part Two

SOLID PARTICLES IN SPACE

PLATE XIII.—The dark nebulae of Barnard 92 (right) and Barnard 93 (left), in Sagittarius. 80 cm. telescope (M. de Kérolyr).

PLATE XIV (A).—The dark nebulae around ξ Ophiuchi. 17 cm. Zeiss objective (F=120 cm.) (M. de Kérolyr).

PLATE XIV (B).—Map of the same region showing the main nebulae catalogued by Barnard. In the north the globular cluster M 9 is shown.

CHAPTER VI

THE DARK NEBULAE

General

In regions of great stellar density, photographs of the Milky Way often show dark patches almost devoid of stars.

The American astronomer Barnard (1857–1923) devoted almost his whole life to the systematic photographing of various Galactic regions with objectives of rather short focal length (most frequently between 78 and 128 cm.). He discovered a great number of empty regions. His observations, first collected in 1919, in a catalogue of 182 objects [22], were completed in 1927 by the publication of *The Photographic Atlas of Selected Regions of the Milky Way*, which even today, together with the photographs of Max Wolf and the more recent atlas of Ross-Calvert, remains one of the most valuable sources of information.

The dark areas have extremely varied shapes. Many are almost irregular, oddly serrated or filamentous. Others, mainly the smallest, are oval or near-circular. Their apparent dimensions also are very varied: some extend over many degrees of the sky, others have an apparent diameter smaller than 1′ of arc. (The latter can naturally only be discovered with long-focus instruments.)

Plate XIII shows two small typical Barnard nebulae, photographed with a rather long-focus telescope; Plates XIV and XV represent, on a very much smaller scale, dark nebular fields photographed with a refracting telescope of 120-cm. focal length. The most characteristic, around ξ Ophiuchi (Plate XIV), consists of objects of very varied form and extension.

Two hypotheses could account for these appearances. Either the black patches correspond to true lacunae in the distribution of the stars, or else they represent dark masses, interposed before the stellar clouds, as if they were opaque screens.

A study of their distribution on a celestial planisphere leads to the discarding of the first hypothesis. The dark areas meet mainly in the neighbourhood of the Galactic plane (although certain of them still appear at a latitude of more than 20°) and they are particularly abundant near the centre of the Milky Way, in the constellations of Ophiuchus and of

Sagittarius, i.e. in the directions where the Galaxy shows its greatest thickness. It would therefore be necessary to imagine the existence not of simple *cavities*, localized in the stellar clouds, but of real *tubes*, devoid of stars for many millions of parsecs. If on the other hand they were dark masses, distributed almost at random in the Milky Way, it is natural that we should see a greater number of them in the direction of the greatest thickness.

The idea that they are *true dark nebulae* is further confirmed by other observations. On large-scale photographs of diffuse nebulae taken with a telescope of long focal length, one can often see, projected on to the luminous background, thin filaments or small black globules visibly representing opaque bodies. This shows that dark nebulae are very often associated with bright Galactic nebulae, continuous-spectrum nebulae, or irregular line-spectrum nebulae. Thus, the Great Nebula in Orion (Plate I), the Trifid Nebula (Plate XXII) and M 8 in Sagittarius (Plate XXI) bear clear marks of absorption. The nebula in Monoceros (Plate XII) is surrounded by a large absorbing cloud, as is the nebula NGC 5146 in Cygnus (Plate XVI). Finally, the nebula "North America", in Cygnus (Plate X) and the region of ζ Orionis (Plate XX) present a complex mixture of dark and bright nebulosities.

The objects catalogued by Barnard generally have sufficiently well-defined contours measuring some degrees at most. But there also exist more diffuse clouds with more or less blurred contours sometimes veiling vast regions of the sky. Such are the great clouds which extend over a part of the constellation of Auriga and of Taurus, and those which spread from Cygnus to Cepheus.

Thus, the dark nebulae, in the most general sense of the word, have the most varied appearances and degrees of opacity, from the minute "globules" which speck the brilliant mass of the nebula M 8, up to the immense gathering of absorbing clouds which, in the constellation of Aquila, divide the Milky Way into two branches.

Distances, depths and opacities of absorbing clouds. Max Wolf's method

Actually, the only method capable of telling us something about the distance, the depth, and the opacity of the absorbing clouds consists of counting stars of each magnitude in the dark area and in equal areas in its neighbourhood, but outside of it. Absorption is held responsible for the observed differences, since we suppose that without it, the distribution of stars would be practically uniform in the entire region considered.

We thus determine, for chosen areas, the photographic magnitude of all the observed stars. Let us designate by $N(m)$ and $N'(m)$ respectively the average number of stars brighter than magnitude m contained per unit solid angle, outside and inside the dark region respectively. Obviously $N'(m)$ is smaller than $N(m)$. *Let us suppose that all the stars are of the same absolute magnitude*, i.e. of the same luminous intensity. When no absorption takes place, each apparent magnitude m will have a corresponding well-determined distance r. If, at a distance r, the nebula exercises an absorption δm, the stars whose apparent magnitude would be m, in the absence of absorption, become $m+\delta m$ and, since their number is not changed, we necessarily have

$$N'(m+\delta m)=N(m). \tag{VI.1}$$

Let us then construct a graph by plotting $N(m)$ and $N'(m)$ against m or, more conveniently, the logarithms of these numbers, $\log N(m)$ and $\log N'(m)$. The preceding equation indicates that the horizontal separation of the two curves measures the absorption δm exercised by the nebula, for each distance. The graph is represented by Fig. 16. The two curves which at first coincide, begin to separate from a certain magnitude m_1 onwards (point A). Their horizontal separation (such as BC) then increases up to another magnitude m_2 (point D on the curve $\log N(m)$). Finally, from there on, the two curves become parallel and can be derived from each other by a translation parallel to the abscissa (DE=FG=KL= Δm).

The interpretation is immediate. At the distance r_1, which corresponds to the magnitude m_1 (point A), the line of vision meets the front face of the dark nebula. From that point onwards, the horizontal distance of the

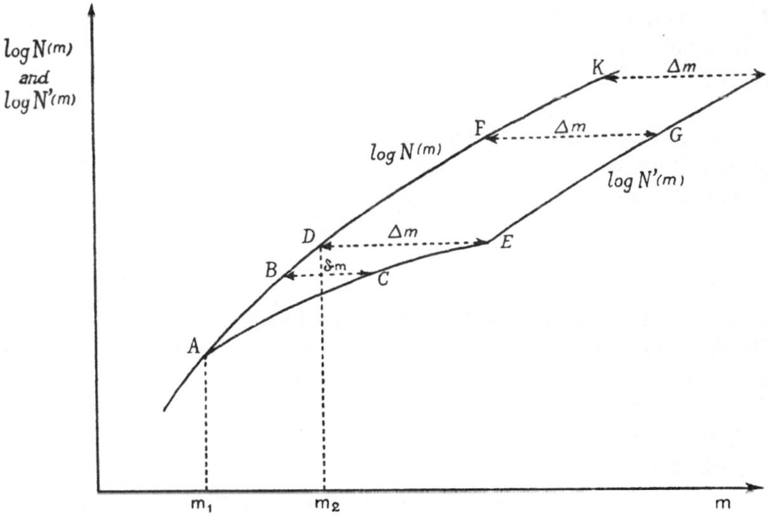

Fig. 16.—Illustration of Max Wolf's Method.

two curves (such as BC) represents the absorption in magnitude δm, and this increases as one penetrates more deeply into the nebula. At the distance r_2, corresponding to magnitude m_2 (point D), the line of vision crosses the back surface of the nebula. All the observed stars in the dark area are then behind the nebula and the light of each of them undergoes *constant* absorption Δm.

Thus, we immediately know the total opacity of the nebula, since it corresponds to the absorption Δm in magnitude. As for the distances r_1 and r_2, they are deduced easily from the magnitude m_1 and m_2, by using the absolute magnitude M, supposed to be common to all the stars. We therefore make use of the known relation

$$5 \log r = m - M + 5, \tag{0.9}$$

which is valid in the absence of absorption, that is to say, along the curve $\log N(m)$. The depth of the nebula is obviously $r_2 - r_1$.

Such is the method of Max Wolf and applied by its author to the first studies of dark nebulae (1923) [384]. Let us remark at once that since it involves the enumeration of stars in the interior of the dark area, absorption must not be too strong, and the apparent surface not too small (so that statistical counts should be significant). Thus, we are much better informed on large clouds producing moderate absorption (1 to 2 magnitudes for example) than on small dark and much more opaque nebulae.

Max Wolf's method [230, 40] is often applied somewhat differently, by considering the number of stars $n(m)$ of magnitude m (in the interval between $m-\frac{1}{2}$ and $m+\frac{1}{2}$), instead of the number of stars $N(m)$ brighter than magnitude m. While the same reasoning is still applicable to these differential curves, the procedure, however, is theoretically less satisfactory (Malmquist).

In Von der Pahlen's method [230], $N(m)$ is taken as the average number of stars brighter than the magnitude m contained in unit solid angle in a zone of the same Galactic latitude, rather than making a count in small areas bordering on the dark nebula.

Improvements of the method (*Malmquist*) [183]

Up till now we have supposed that the stars considered have the same luminous intensity. Applied to all the stars this hypothesis is quite false. The dispersion of their absolute magnitude is enormous and can seriously affect the results deduced from the Max Wolf method in its original form. Even if the total absorption Δm, independent of the dispersion, is correctly evaluated, the distance and the depth of the nebula can be affected by very important errors [183].

Let us now suppose instead that we had a means of determining the *individual* absolute magnitude of all the stars contained in the studied areas.

Equation (0.9) gives us the true distance of each star situated outside the dark cloud, but it is not suitable for the evaluation of the *true* distances of the stars situated in the interior of this area. It is as if their absolute magnitude M were affected by a certain absorption $A(r)$ (in magnitude), corresponding to the distance r, which we are trying to determine precisely. It is, therefore, necessary to replace (0.9) by the new relation

$$5 \log r = m - M - A(r) + 5$$

which can also be written

$$5 \log r + A(r) = m - M + 5. \quad \text{(VI.2)}$$

If we continued to apply the original relation (0.9) without any modification, we could say that we were determining an *apparent distance* r', which is larger than the true distance r and is related to the latter by

$$5 \log r' = 5 \log r + A(r).$$

It should be understood that, in those regions where there is no absorption, the apparent distance coincides with the true distance ($r' = r$).

Let us now count the stars, no longer by their magnitude, but by their apparent distances, and let us construct the two graphs representing, as a function of $y = 5 \log r'$, the logarithm of the number of stars of apparent distance smaller than r' contained per unit solid angle: first, outside the darkened zone, let thus be $\log N(y)$, and secondly, in its interior, let this be $\log N'(y)$. The fundamental relation (VI.1) then becomes

$$N'(y) = N[(y) - A(r)] \quad \text{(VI.4)}$$

and the graph in Fig. 16 is not changed. It is only necessary to replace m by y on the abscissa and $\log N(m)$ and $\log N'(m)$ by $\log N(y)$ and long $N'(y)$ respectively on the ordinates.

The interpretation remains the same, but, this time, we obtain besides the total absorption Δm, the distances r_1 and r_2 of the front and back faces of the nebula. These are in effect, simply related to the abscissae y_1 and y_2 of the points A and D by

$$y_1 = 5 \log r_1, \; y_2 = 5 \log r_2,$$

since, along the curve $\log N(y)$, there is no absorption ($A(r) = 0$, $r^1 = r$).

Actually the precise determination of the absolute magnitude of all the stars in the field, considered individually, is almost impossible. The use of spectroscopic methods, which alone would serve the purpose, would in fact necessitate obtaining spectra of very weak stars with rather large dispersion. But it is possible to approach the ideal case sufficiently by determining the spectral types of all the stars (which can be done by means of spectrograms of very small dispersion), and by clearly distinguishing the giants from the dwarfs. We can even improve the discrimination further by determining the *class of luminosity* of each star, by means of certain criteria of absolute magnitude applicable to spectra of low dispersion. We treat separately each one of the groups of stars in which the dispersion of absolute magnitude can be reduced to 1/2. The results are then very much more certain than those of the original method. The stars of spectral classes B and A are always those whose study presents the greatest interest, since, apart from some rare giants, the major part of the stars of other classes are dwarfs and in front of the absorbing clouds.

The best determined quantity is always the total absorption Δm. The distances r_1 and r_2 are more uncertain, for it is difficult to fix the exact position of the points A and D on the graph. Further, the residual dispersion of absolute magnitudes generally leads to an under-estimation of r_1 and an over-estimate of r_2 (and even more of the depth r_2-r_1). If this dispersion, supposed to conform to a Gaussian distribution, does not exceed ± 0.5 magnitudes, Malmquist shows that sufficiently correct results are always obtained by taking the distance as r_1 when the absorption is 1/10 of the total absorption of Δm and, as r_2, when it becomes 9/10 of Δm.

More exact values of the distance and of the depth of the nebula may be obtained by a method involving more laborious calculations—"the method of moments", introduced in a particular case by Gyllenberg [409] and developed by Lyttkens [414] in an entirely general manner.

Methods of Pannekoek and Bok

In the study of dark nebulae, an analytical method devised by Pannekoek (1921) [231] and "numerical methods" of which the most practical is that of Bok [39, 40], are also employed.

They only require the enumeration of the apparent magnitudes of the stars inside and near the dark areas, but we account for the dispersion of absolute magnitudes by introducing the function $\phi(M)$ representing their relative distribution. This can be defined so that

$$\int_{-\infty}^{+\infty} \phi(M)dM=1.$$

The function $\phi(M)$ is very well known in the neighbourhood of the Sun, and we take it as independent of the distance. On the other hand, the number of stars per unit volume is a function $D(r)$ of the distance, which we do not know. We can, however, take it that this function is the same in the darkened area as it is in the neighbouring area of comparison. Thus, in the absence of absorption the number $n(m)$ of stars of apparent magnitude m contained in solid angle ω leads directly to the expression:

$$n(m)=\omega\int_0^{\infty} r^2D(r).\phi(M).dr$$

or, by replacing M by its value as a function of m and of r (0.9),

$$n(m)=\omega\int_0^{\infty} r^2D(r)\phi(m+5-5\log r)dr. \qquad \text{(VI.5)}$$

The determination of the function $D(r)$ requires the solution of this integral equation, known as the *fundamental equation* of stellar statistics.

If we now suppose that there exists, at a distance r_1, a *thin* layer of nebula producing absorption of magnitude A, we can represent the distribution of the stars in the darkened area by

$$n'(m)=\omega\int_0^{r_1} r^2.D(r).\phi(m+5-5\log r)dr$$

$$+\omega\int_{r_1}^{\infty} r^2.D(r).\phi(m+5-5\log r-A)dr,$$

which Pannekoek writes

$$n'(m)=\gamma_1,n(m)+\gamma_2.n(m-A),$$

by putting

$$\gamma_1=\frac{\omega\int_0^{r_1} r^2.D(r).\phi(m+5-5\log - r)dr}{n(m)}$$

$$\gamma_2=\frac{\omega\int_{r_1}^{\infty} r^2.D(r).\phi(m+5-5\log r-A)dr}{n(m-A)}$$

By varying r_1 and A (and therefore γ_1 and γ_2), we try, by successive approximations, to represent in the best possible way, the distributions $n'(m)$ and $n(m)$ observed in the darkened area and in the area of reference respectively.

The principle of the "numerical method" consists of dividing space into a certain number of concentric spherical shells, with the Sun as centre. Between two successive spheres, we can assume that the stellar density is constant and that all the stars are at the same distance from the Sun. Thus we replace the integral in the equation (VI.5) by the sum of a finite number of terms.

It is convenient to consider, as Kapteyn has proposed, spheres of a radius such that

$$\log r=0{\cdot}1,\ 0{\cdot}3,\ 0{\cdot}5,\ 0{\cdot}7, \text{ etc.}$$

We then take, for all stars contained in the jth shell (between the jth and the $(j+1)^{\text{th}}$ sphere) the distance r as given by

$$\log r_j=0{\cdot}2j=\frac{j}{5}.$$

The relation between the absolute magnitude and the apparent magnitude can then be written

$$M=m+5-5\log r=m+5-j.$$

By supposing that the function $\phi(M)$ is known, Bok constructs a table giving for each integral value of j and m, the product

$$a_{j,m}=V\phi(M)=V\phi(m+5-j),$$

where V is the volume cut off by a cone of solid square angle of 1 square degree between spheres of orders j and $j+1$.

If the stellar density were constant, this product would represent the contribution of the layer j to the total number of stars per square degree of magnitude comprised between $m-\frac{1}{2}$ and $m+\frac{1}{2}$. We should then obtain the total number of stars $n(m)$ having these magnitudes by adding all the numbers in the same column (m=const.) of the table. But, in reality, $D(r)$ varies with the distance. Let $D(j)$ be the value in the shell j. The equation (VI.5) can be written as

$$n(m) = \sum D(j)a_{j,m}$$

and the problem consists in finding those values of $D(j)$ which gives a satisfactory representation for each magnitude of the numbers $n(m)$ observed. These values are obtained by trial and error.

In the darkened area, the distribution $n'(m)$ is different, since at a distance r_1 absorption of magnitude A is produced. With some pairs of values (r_1, A) conveniently chosen new tables are constructed and several attempts (generally three according to Bok) would suffice to determine the distance r_1 and the absorption A which conveniently represent the counts of stars by magnitude.

A comparative study of the methods of Wolf, Pannekoek and Bok has recently been published by Velghe [356], who finally adopts a procedure which is similar to that of Malmquist.

Some results of observation

The study of absorbing clouds by variants of the Wolf method, or by the methods of Seeliger, Pannekoek and Bok, has been the object of many works, notably in Sweden (Malmquist, Schalén [251, 253], Vanäs [340], Wernberg, etc.), in Holland (Pannekoek [232]), in Germany (Wolf, Koppf [171], H. Müller [212], H. Müller and Hufnagel [213], Von der Pahlen, von Kluber [170]), in Russia (Messrs. Shain [275], Berg [34], Lehman-Balanowskaja [174]) and in the United States (Struve [317], Bok and his collaborators, etc.).

The work of the Swedish school is particularly meticulous, and in order to give an account of the technique of the measurements, we shall cite as an example, Wernberg's work at Uppsala [369] on absorbing clouds of the Cepheus region, situated in the immediate neighbourhood of the galactic plane (1941).

A prism objective of small angle (9·7°), of 15 cm. aperture, and 148 cm. focal length gives spectra which only measure 1·4 mm. between Hγ (4340 A) and Hϵ (3970 A), so that 18,000 of them can be observed. The spectra are nevertheless sufficiently large for the determination of spectral type by a microphotometer. The photometric scaling of the negatives permits, amongst other things, the measurement of the *monochromatic magnitude* for a chosen wave-length (which varies with spectral type) and permits the division of stars of the same type into various classes of luminosity. We use Lindblad's criteria which depend, in the case of the spectral classes B and A, on the intensity of the Hγ and the Hδ lines and, for the classes G to M, on the intensity of the band (4216 A) of the CN molecule. We finally construct separate Wolf diagrams for each one of the groups of stars thus constituted.

This work has shown that absorbing matter is distributed in an irregular manner in the region studied. Total absorption varies in different areas, from 0·8 to 1·3 magnitudes. The cloud starts at 200 parsecs from the Sun and extends toward 600 pc. Perhaps a second cloud exists at a greater distance.

In Auriga, next to Taurus, Wernberg (1938) has measured an absorption which is as much as 2 magnitudes in places, in a cloud which starts at a 100 pc. and which is approximately 360 pc. thick. These results confirm those obtained previously by Schalén by a similar method.

In the "Coal-Sack" of Cygnus, Vanäs (1939) has shown two superimposed clouds at 250 and 630 pc. distance, each absorbing 1 magnitude.

The dark nebulae near ρ and θ Ophiuchi have been studied by R. Müller [214], using the original Wolf method. The first extends from 150 to 850 pc., with a photographic absorption of 2·8 magnitudes; the second includes two clouds, impossible to separate completely, at distances of 100 and 250 pc., with a total absorption of 4 magnitudes. However, in treating the different spectral classes separately, Wallenquist [362] finds, in the latter case, a single cloud starting at 250 pc. and some hundreds of parsecs deep, with Δm=2·8 magnitudes.

The considerable thickness attributed to certain absorbing clouds is a little surprising and does not perhaps always correspond to reality. Malmquist (1945) has shown that by taking a count of the residual dispersion of absolute magnitudes, one could often interpret the diagrams by the superposition of numerous clouds at different distances. According to Hartwig [127], in the region of Auriga there ought to be, instead of the single cloud of Schalén and Wernberg, two distinct clouds at about 100, or 350 to 400 pc., absorbing 1·4 and 0·7 magnitudes respectively. Malmquist [183], however, places the first cloud (layer) at about 200 pc., with absorption Δm=1·25 magnitudes, the second at 400 to 500 pc., with Δm=0·75 magnitudes. In the regions of Cepheus, where Wernberg indicated a thick cloud between 160 and 600 pc., absorbing 1·3 magn., he speaks of two clouds, towards 220 and 420 pc. respectively, each absorbing 0·6 magn.

Thus, the actual tendency appears to be the fragmentation of deep clouds. The complex structure of the Milky Way, in the neighbourhood of its equatorial plane, ought to account for the frequency of such superpositions. These are more probable than the occurrence of single clouds, whose elongation just in our line of vision seems unlikely.

Dark nebulae considered as a scattering medium

The absorption produced by the dark nebulae differs profoundly from that of the atoms and molecules of space. As we have seen in the preceding chapter, absorption by the latter is strictly *selective*, in the sense that it affects only certain well-determined monochromatic radiations, characteristic of the absorbing atoms or molecules. On the other hand, the absorption of nebulae makes itself felt in the entire spectral region accessible to observation, from the infra-red to the ultra-violet (but this does not mean that it affects all radiation *equally*).

This type of absorption, independent of the wave-length, or rather varying in a continuous manner with it, always accompanies the scattering of light by various kinds of particles: solid grains, liquid globules, molecules or atoms of gas, and even free electrons. We say that we are dealing with *apparent absorption*, for, in reality, radiation does not disappear. It is not transformed into potential energy or into kinetic energy but is only distributed differently in space. Light transmitted through the cloud of particles is weakened, since part of the energy which it carries is scattered in all directions and, against a perfectly dark background, the cloud appears more or less luminous. It is thus, for instance, that solar radiation, scattered by the terrestrial atmosphere, is seen as the brightness of the blue sky.

The phenomena of scattering, distribution of the scattered flux into various directions around the incident beam, polarization of the scattered light, variation of the apparent absorption as a function of wave-length depend on the dimensions and on the nature of the particles. With certain types of particles, *true* absorption can also be superimposed on apparent absorption.

It is almost natural to attribute the absorption of the nebulae at least partly to a scattering phenomenon, and the problem of determining the nature and the dimensions of the scattering particles immediately poses itself. Observation of the light of the stars transmitted through the Galactic clouds only permits the study of the polarization of the emergent light and of the variations in absorption as a function of wave-length. This study is of paramount interest.

*Differential absorption of the dark nebulae**

Most of the measurements made up till now have only been made on two monochromatic radiations λ_1 and λ_2, or rather on two more or less narrow spectral regions centred on these wave-lengths.

One can, let it be stressed, construct Wolf diagrams for these two radiations separately, and determine the total corresponding absorptions Δm_1 and Δm_2. Their difference

$$\Delta C = \Delta m_1 - \Delta m_2 \qquad (\lambda_1 < \lambda_2)$$

is, by definition, the *differential absorption* of the nebula for the pair of radiations considered.

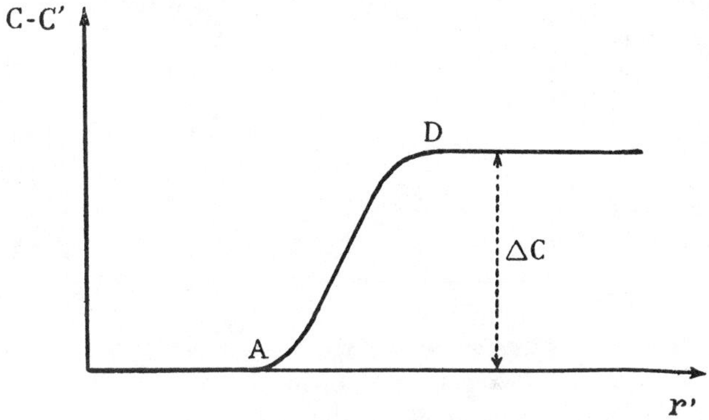

Fig. 17.—Colour excess of a dark cloud as a function of the apparent distance.

It is often advantageous to work in a more direct manner. A group of stars, of determined spectral type and class of luminosity is characterized, in the absence of all absorption, by the *colour index*

$$C = m_1 - m_2,$$

i.e. by the difference of apparent magnitudes appropriate to the two wave-lengths. Although m_1 and m_2 vary from star to star, their difference is constant if the group is sufficiently homogenous.

Seen through the nebula, stars of the same group have their magnitudes m_1 and m_2 increased by Δm_1 and Δm_2 respectively. Their colour index then becomes

$$C' = (m_1 + \Delta m_1) - (m_2 + \Delta m_2) = (m_1 - m_2) + (\Delta m_1 - \Delta m_2) = C + \Delta C.$$

We can directly measure the respective colour indices C' and C of the stars situated behind the nebula and outside of it. Their difference gives the differential absorption or the *colour excess* ΔC.

*This is frequently called "selective absorption". This term ought strictly to be used in the most general sense of "absorption varying with wave-length", and not in the more restricted sense adopted in this work and conforming with its usage by physicists.

TABLE VIII

Total Absorption and Colour Excess of some Absorbing Clouds

Object	*Author*	Δm (4400)	ΔC $\frac{(3950)}{(4400)}$	$\chi = \frac{\Delta m}{\Delta C}$
Auriga Cloud . .	Schalén [253]	1·9 magn.	0·21 magn.	9·05
Cygnus Cloud . .	. . . .	1·5	0·2	7·5
1st Cepheus Cloud	Wernberg [369]	1·3	0·16	8·12
2nd Cepheus Cloud	. . . .	0·8	0·8	10·0
3rd Cepheus Cloud	. . . .	1·1	0·09	12·2
Cepheus, 2 superimposed clouds	. . . .	2·5	0·26	9·61
	Simple average			9·40
	Weighted average			9·27
	(Weights proportional to Δm) By eliminating the last measurement which is less certain			9·16

In practice we plot against the apparent distance r' the difference $C'-C$, where C' is the colour index of the stars situated in the darkened area at an apparent distance r', and C is that of the stars of the same group, at the same apparent distance (which is now equal to the true distance r), in the area which ostensibly has no absorption (Fig. 17). The difference is zero until we enter the nebula. It increases while it crosses it (from A to D) and finally takes the constant value ΔC once the nebula has been crossed. The transformation of apparent distances into true distances naturally involves the study of absorption for one of the two chosen wave-lengths, by the Wolf method.

In all cases positive colour excesses have been found. In other words, the stars appear redder when they are seen through a dark cloud: absorption by the nebulae increases from the red to the violet. The astronomers at Uppsala take as colour index, the difference between monochromatic magnitudes measured at two points chosen from their short spectrograms (most frequently 3950 and 4400 A). In Table VIII measurements made on absorbing clouds of at least 0·8 magn. at 4400 A are collected. The ratio χ between total absorption Δm and colour excess ΔC (between 3950 and 4400 A) varies very little from cloud to cloud. Since the errors are compatible with the degrees of accuracy of the measurements, we can take it that the ratio χ is practically constant ($\chi \simeq 9{\cdot}2$), *as would be the case if the different clouds were composed of similar particles.*

Other authors make use of the colour indices of the international scale, i.e. the difference between the photographic magnitude m_{pg} of a star and its photovisual magnitude m_{pv}

$$CI' = m_{pg} - m_{pv}, \qquad (0.22).$$

If we discard some obviously faulty results due perhaps to systematic errors, we find that the ratio between photographic absorption and colour excess is then between three and four.

These results can already give us an indication of the probable dimensions of the scattering particles. But it would be better to leave the final discussion for a little later; the study of general absorption in the neighbourhood of the Galactic plane and that of light scattered by the nebulae will, in fact, give us other data of very great importance.

CHAPTER VII

ABSORPTION OF LIGHT IN THE NEIGHBOURHOOD OF THE GALACTIC PLANE. TOTAL ABSORPTION

SHOULD interstellar space be considered as perfectly transparent outside the great clouds and the dark nebulae visible on photographs of the Milky Way? The idea that light of far-away stars could, in the course of its path, undergo appreciable absorption has had a somewhat chequered history. Time after time it was held, abandoned, then taken up again, and finally it was definitely adopted around 1930, but in the rather restricted form given it by R. J. Trumpler: *in the neighbourhood of the Galactic plane there exists a very thin absorbing layer*. Its thickness does not exceed some hundreds of parsecs, while, in the Galactic plane, it certainly extends to thousands and in some directions perhaps, to tens of thousands of parsecs. This hypothesis has been suggested by a whole number of observations of quite different kinds.

Various methods have permitted the evaluation of the *absorption co-efficient* for radiation of determined wave-length in the interior of the layer, i.e. absorption measured in magnitudes per unit length. Others have furnished, with greater certainty, the *difference in the absorption co-efficient* of two chosen radiations (the *differential absorption co-efficient*). Finally others have led to the evaluation of the *transverse opacity* of the absorbing layer, and from this, of its thickness in the direction perpendicular to the Galactic plane.

The altogether remarkable agreement of the results obtained by entirely different procedures does not leave any doubt as to the real existence of the Trumpler layer. But we shall see later that it is not a homogeneous layer. It consists of an aggregate of clouds whose opacity is often too weak for each one of them to be individually distinguished.

Determination of the photographic absorption co-efficient based on the spatial distribution of stars

In 1847, Wilhelm Struve had remarked that the number of stars per unit volume seemed to diminish in all directions as we receded from the

Sun. This fact could be interpreted in two ways: either the Sun was at the very centre of a true stellar condensation, or the effect was only an apparent one due to absorption.

On the latter hypothesis, the evaluation of distance was adversely affected by absorption—only the *apparent density* of the stars was obtained. By taking the *true* stellar density to be constant, instead of diminishing, the apparent distance r_1' is changed into the apparent greater distance r_2', and we obtain an equation which contains the absorption co-efficient as an unknown.[1] The first tentative attempts of Kapteyn (1904) [161] and of Seeliger (1911) [270] have given respectively $a=1.6$ magnitude per 1,000 pc. (or, by taking the kiloparsec as unit of length 1.6 magn./kpc.) and $a=0.3$ magn./kpc. But other workers subsequently obtained values which were clearly larger or much smaller.

Owing to an important advance in the study of the spatial distribution of stars it has nowadays been possible to determine the absolute magnitude of distant stars, according to their spectrum. Thus Schalen (1929) [250] has evaluated the apparent stellar density of stars of the classes B and A in numerous regions of the Milky Way (Cygnus, Cepheus, Cassiopeia and Auriga) and he obtained, by the method which we have just indicated, very different values of the absorption co-efficient, particularly in Cygnus and in Auriga where, as we have mentioned, large, dark nebulae exist. He concluded that the absorption is not uniform and that it varies with distance and with the direction considered. *On the average* he found the absorption co-efficient to be approximately 0.5 magn./kpc. But it would be larger near the Sun (0.8 magn./kpc. for distances ranging between 1.0 and 2.5 kpc.) and smaller in the further regions (0.2 magn./kpc. between 2.5 and 4 kpc.).

Basing himself on other papers and using different methods of calculation (see page 134), Bok (1931) [39] came to similar conclusions. He considered the whole Galactic zone between 0° and 20°, with the exception of regions covered by dark clouds (Taurus, Ophiuchus). If we suppose space to be transparent, stellar density will diminish constantly as distance increases, up to more than 10 kpc., a result which is altogether improbable. If, on the other hand, we suppose that stellar density is invariable, we find with Schalén, that the absorption co-efficient which is very large near the Sun (larger than 2 magn./kpc.) begins to diminish rapidly at 400 pc. and tends towards 0.5 magn./kpc. He therefore had to assume that the Sun is in the middle of an absorbing cloud, an hypothesis which Bok judged to be inacceptable. Bok finally concluded that the diminution of stellar density, in a radius of 400 to 500 pc. around the Sun, is real. The Sun, therefore, is part of a limited stellar condensation—"the local cluster"—whose existence has been suggested by other workers as well.

But it is probable that from there on the average stellar density becomes constant. The absorption co-efficient would, therefore, be in the neighbourhood of 0.4 magn./kpc.

This work has contributed to demonstrating the existence of a certain absorption in the neighbourhood of the Galactic plane. But it can never give us more than the order of magnitude of the absorption co-efficient. The hypothesis of uniform distribution of stars in space is obviously far too idealized, and the problem contains two unknowns: the stellar density and the absorption co-efficient. Today we prefer to determine the latter by another method, and from it directly to evaluate the density of stars at various distances from the Sun. It is in this way for instance, that Bok and his collaborators have proceeded in their most recent work.

Determination of the photographic absorption co-efficient based on open Galactic clusters

At first sight we distinguish two types of stellar clusters: *globular clusters* and *open Galactic clusters,* or simply, *Galactic clusters.*

The first, of more or less spheroidal shape, contain tens of thousands of stars, with a very marked concentration towards the centre. Plate XVII shows a typical example of them: the clusters of Hercules (M 13). Barely a hundred globular clusters exist, all far away and often situated at the borders of the Galaxy, of which in some way they are satellites.

Galactic clusters are assemblies of stars which are much less compact and of various forms, containing some tens, some hundreds or at most a thousand stars only. Amongst the best known are the Pleiades, Praesaepe, and the double cluster in Perseus. Plate XII shows one of them. We know some hundreds of open clusters, almost all situated in the immediate neighbourhood of the Galactic plane, but at various distances ranging between a hundred and some thousands of parsecs. It is these Galactic clusters that R. J. Trumpler has had the ingenious idea of investigating in order to bring the absorbing stratum into evidence (1930).

Their apparent distance can be evaluated in a sufficiently accurate manner if we suppose space to be perfectly transparent. To this effect we determine the photographic apparent magnitude m and the spectral type of a sufficiently large number of stars of the cluster. We assume that each one of them possesses absolute mean magnitude M (known) corresponding to its spectral type, by clearly distinguishing the giants from the dwarfs starting with the class F. The apparent distance r' of the star is then given by the classic relation

$$5 \log r' = m - M + 5. \qquad (0.9)$$

Each star provides an independent determination and, since all those which belong to the cluster are practically at the same distance from the

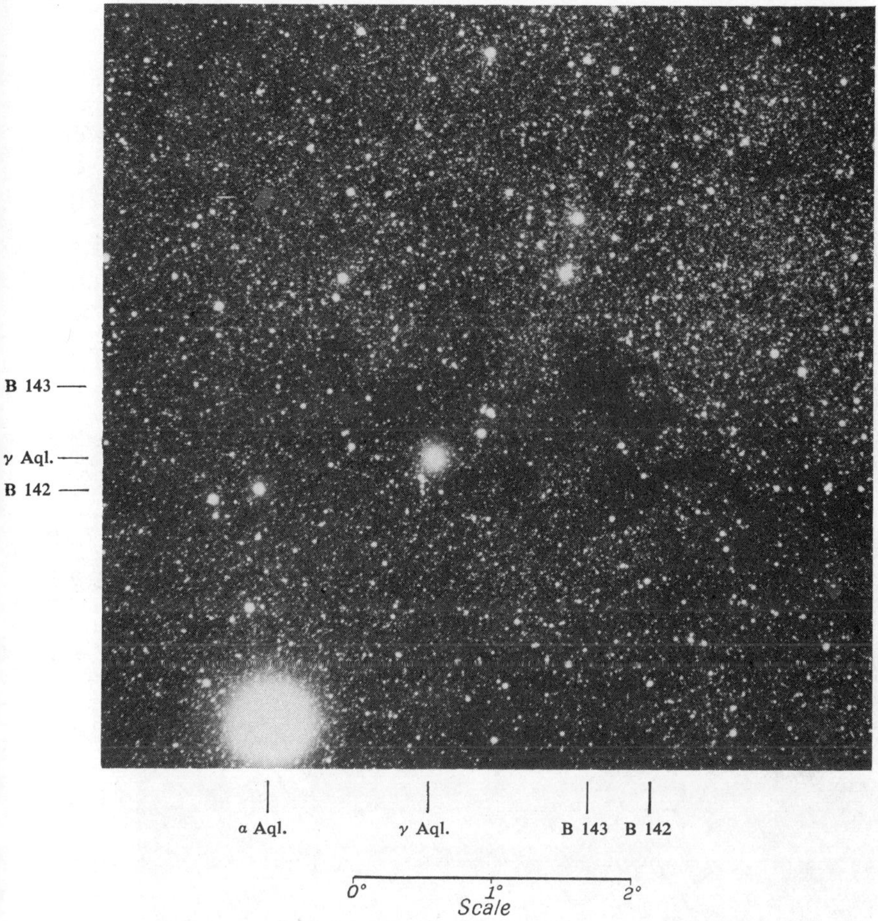

PLATE XV.—Dark nebulae of Barnard 142 and 143, near γ Aquilae. 17 cm. Zeiss objective (M. de Kérolyr).

PLATE XVI.—NGC 5146 in Cygnus. Combination of bright and dark nebulae. The marked absence of stars around the bright region is due to absorbing clouds. 193 cm. telescope of the Haute-Provence Observatory.

Sun, we can take the differences $m-M$ as the average of all of them. This compensates for the errors resulting from the fact that the absolute magnitudes of the individual stars differ more or less from the average of the type. The apparent distance of the cluster is much better determined than it would be in the case of an isolated star.

On the other hand, we estimate the *apparent diameter* of each cluster, i.e. the angular diameter of the circle in which one could approximately inscribe the majority of its stars. The *linear diameter* of a cluster is evidently proportional to the apparent diameter α and the distance r. If we express α in radians we have

$$D'=\alpha r'. \qquad (0.4)$$

Trumpler [335] evaluated the apparent distances, and then the linear diameters of 100 galactic clusters in this way. These objects, varying so much in appearance as they do, are most certainly not all constructed according to the same model. Also, it is rather surprising to find very different diameters (varying from 1 pc. to some 20 pc.). But, in considering the degree of condensation of the clusters and their richness in stars, Trumpler managed to classify them into twelve categories. Each one of the groups thus constituted is now sufficiently homogeneous and it appears to be legitimate to suppose that clusters of the same type have similar diameters.

Now, if we arrange them according to increasing distance, we will find that their linear diameters D' increase with their distances r'. The diameter of the furthest clusters would be twice as large as that of the nearest. Such a systematic variation as a function of their *distance from the Sun* is altogether inconceivable. We are, therefore, obliged to conclude that the distances r' obtained from equation (0.9) increase too rapidly. Naturally this is also the case with the diameters D' proportional to r'.

The hypothesis of the transparency of space ought, therefore, to be abandoned and equation (0.9) ought to be completed by introducing absorption into it. If we assume that this is proportional to distance, the equation becomes

$$5 \log r=m-M+5-ar. \qquad \text{(VII.8)}$$

We fix the absorption co-efficient so as to cancel the systematic variation of linear diameter with distance, for all categories of clusters. This is done by taking $a=0.79$ magn./kpc. However, after discussion, Trumpler, in 1930, preferred a somewhat smaller value (0.67 magn./kpc.). In a more recent work [337], entirely new measurements of ninety-seven clusters have given him $a=0.70$ magn./kpc., but because of systematic errors which could affect our determinations of apparent diameters, it is possible that this number is a little too small, and Trumpler proposes to retain

$$a=0.8 \text{ magn./kpc.}$$

Since then, Sanford [248] has determined the *true* distance of thirty-one clusters, by means of the intensity of interstellar lines. The values obtained differ very notably from those of Trumpler and are smaller on the whole. The mean distance is changed from 1,600 to 1,200 pc. and is thus reduced in the ratio of 4:3. Therefore, the mean absorption co-efficient ought to be multiplied by $\frac{4}{3}$, which gives

$$a = 0{\cdot}8 \times \frac{4}{3} = 1{\cdot}07 \text{ magn./kpc.}$$

Here are some details of Trumpler's method of calculation.

Let $\overline{\log D'}$ be the average of the logarithms of the diameters calculated by (0.4) for a certain type of cluster. For each cluster of this type the difference from the average is calculated:

$$\epsilon = \log D' - \overline{\log D'}. \qquad \text{(VII.9)}$$

$\overline{\log D'}$ naturally varies from one category of cluster to the next, but we are entitled to treat the residuals ϵ of the various groups together. This is more expedient, since the number of clusters contained in each group is small.

The clusters of all the types are arranged according to increasing distances and if we consider the mean residuals $\bar{\epsilon}$ belonging to successive intervals of apparent distance ($r' < 500$ pc., $500 < r' < 1{,}000$ pc., etc.), a graph will show that the residuals increase with distance. The nearest clusters have diameters which are smaller than the average ($\epsilon < 0$); the furthest have larger diameters ($\epsilon > 0$).

Now, the known relation (VII.5) between the apparent distances and the true distances can be written

$$\log r' = \log r + 0.2\, ar \qquad \text{(VII.5'')}$$

and since, according to (0.4), $\frac{D'}{D} = \frac{r'}{r}$, we also have

$$\log D' = \log D + 0.2\, ar. \qquad \text{(VII.10)}$$

Let us suppose that all the clusters of the same type have the same true diameter D. The relation (VII.10), applied to the mean values $\log \overline{D'}$ and $\bar{r}'$ is written:

$$\overline{\log D'} = \log D + 0.2\, a\bar{r}. \qquad \text{(VII.11)}$$

By inserting in (VII.9) the values of $\log D'$ and of $\overline{\log D'}$ we obtain

$$\epsilon = 0.2\, ar - 0.2\, a\bar{r} = 0.2\, ar - b. \qquad \text{(VII. 12)}$$

where we denote the constant $0.2\, a\bar{r}$ by b.

The residuals ϵ ought to increase linearly with r, and the slope of the straight line gives the absorption co-efficient a. But, since the true distances are not known independently, we must, by successive approximation, determine a value best satisfying relations (VII.5'') and (VII.12). Fig. 18

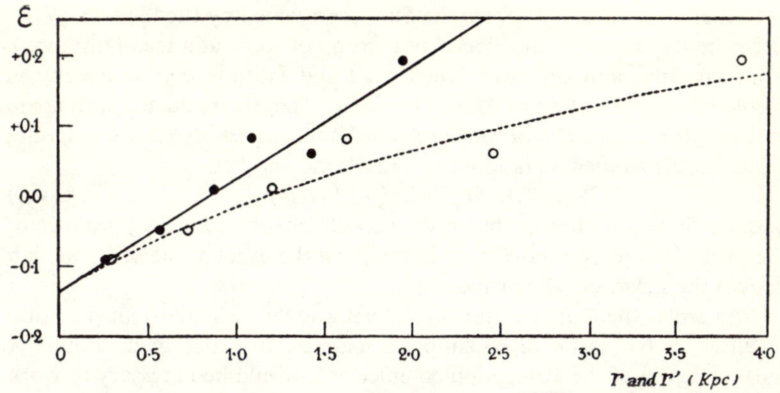

Fig. 18.—R. J. Trumpler's method. The residuals $\epsilon = \log D' - \overline{\log D'}$ plotted as a function of the apparent distance r' (open circles and dotted line) and of the "true" distance r (dots and thick line).
The equation of the straight line is $\epsilon = -0{\cdot}134 + 0{\cdot}158\ r$, giving an absorption co-efficient $a = 0{\cdot}79$ magn./kpc.

shows the increase of the residuals as a function of the apparent distance r' and of the true distance r, if we take $a = 0{\cdot}79$ magn./kpc.

If we add that Trumpler's first paper also contained a good determination of the differential absorption co-efficient (to which we shall return later on), we shall not be astonished by the fact that its publication ushered in a new epoch in the history of interstellar absorption.

Interstellar absorption and Galactic rotation

In every case we can transform the measured absorption co-efficient into a comparison between the apparent distance and the true distance of the stars, since, in accordance with (VII.5″) we have

$$a = \frac{5}{r} \log \frac{r'}{r}.$$

The first is well determined if we have a correct value of absolute magnitude. But it is always much more difficult to obtain the second.

In the preceding methods we have evaluated the true distance simultaneously with the absorption co-efficient, by means of the somewhat uncertain hypotheses of a uniform distribution of stars in space, and of the approximate equality of the diameters of Galactic clusters of a given type. In principle, the rotation of the Milky Way provides the possibility of a determination without supplementary hypotheses and, in consequence, it enables us to calculate the absorption co-efficient with a minimum of unccrtainty.

We have already seen (page 116) how the differential rotation of the Milky Way introduced into the radial velocity of each star a term proportional to its distance r from the Sun, represented by the formula (V.1). Let $\bar{v}_r$ be the mean radial velocity of a group of stars, at a mean distance $\bar{r}$ from the Sun, and of mean longitude $\bar{l}$ and latitude $\bar{b}$ (the velocity is corrected for translation of the solar system). These stars do not participate in a group movement and they are sufficiently numerous for their special velocities distributed at random to cancel one another.

$$\bar{v}_r = A\bar{r} \sin 2 (l - l_0) \cos^2 \bar{b} \qquad \text{(V.1)}$$

permits the evaluation of the product $\bar{q} = A\bar{r}$ of the mean true distance of the group by the Oort constant. If we know the exact value of A, we can deduce the mean distance $\bar{r}$ from it.

In reality, the Oort constant is not yet known with sufficient precision to allow us to reason in so simple a manner. In order to determine A independently of the absorption co-efficient it would be necessary to work on the near stars (negligible absorption). But then the effect of Galactic

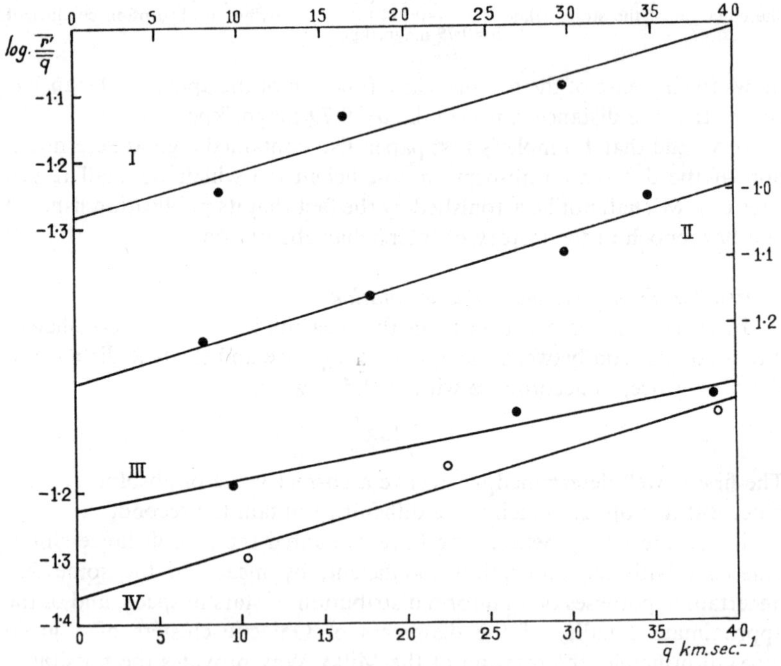

Fig. 19.—Simultaneous determination of the Oort constant A and of the absorption co-efficient a.

The straight lines $\log \frac{\bar{r}'}{q} = 0{\cdot}2 \frac{a\bar{q}}{A} - \log A$, are plotted for

I—interstellar lines,
II—interstellar lines, stars O, B, "c".
III—Cepheids—Joy grouping,
IV—Cepheids—Wilson grouping.

rotation on radial velocity is small and easily masked by the particular movements or the group movements. We are therefore forced to deal with distant stars and absorption necessarily intervenes. Thus in practice, we, still find that the evaluation of a is linked with that of A.

We are therefore forced to determine the values of the two unknown which best satisfy equations (V.1) and (VII.5″). Joy [158] has had the very good idea of making use of the Cepheids, supergiants whose brightness varies periodically and which are easily recognized by the form of their light curve. A simple law discovered by Miss Leavitt (page 34), relates the absolute magnitude to the period of variation in luminosity. Since this can be determined with extreme precision, the Cepheids are certainly, amongst the distant stars, those whose absolute magnitude is best known. This is why the *period-luminosity* law has played so important a role in fixing distances for the interior as well as the exterior of the Galaxy. Actually, the galactic distribution of the Cepheids also is very favourable, since they are closely concentrated in the lower latitudes, but are distributed sufficiently uniformly in longitude.

Joy used the radial velocity of 128 Cepheids, classified in four groups in accordance with their apparent distances (up to $r'=10$ kpc.). Each star gives an equation analogous to (V.1), in which we know l, b and v. We consider as unknown the quantities $q=\bar{r}\,A$ and l and we calculate them in each group by the method of least squares. The four values obtained for l agree well amongst themselves (between 323° and 331°). The four values of q would be proportional to the mean distances r' if space were transparent. We find that in reality, r' increases more rapidly than q, showing that apparent distances are affected by absorption. In order to re-establish the proportionality it is necessary that the absorption coefficient be

$$a=0.85 \text{ magn./kpc.}$$

and the Oort constant is then

$$A=20{\cdot}9 \text{ km. sec.}^{-1} \text{ kpc.}^{-1},$$

a value which is a little large, but still acceptable.

The analogous work of R. E. Wilson [381] has been carried out on very much more abundant material. In most of the Cepheids, Wilson in fact used the O and B stars and supergiants of all the spectral types, up to K5. He assumed that all the supergiants (not Cepheids) of the same type have the same absolute magnitude. He also used radial velocities measured on the interstellar lines by supposing a uniform distribution of the Na and Ca^{+} ions in space, such that their mean distance is half that of the star (page 118).

Wilson thinks that the longitude l_0 of the centre of rotation is well known ($l_0=325°$) and at first he divides the stars into groups according to longitude. He finds that the mean radial velocity is a sinusoidal function of 2 ($l-325°$). By then taking groups of increasing apparent distances, he agrees with Joy in finding that q increases less quickly than $\bar{r}$ and he arrives at the values

$$a=0{\cdot}65 \text{ magn./kpc.}, \ A=19{\cdot}2 \text{ km. sec}^{-1} \text{ kpc}^{-1}.$$

Nevertheless, this solution is not completely satisfactory because, with $a=0{\cdot}65$ magn./kpc., he obtains, for A, numbers which diminish when the distance increases. It is much more reasonable to think that A is a constant, as the Oort theory requires it to be, and that a can vary with distance. Wilson thus takes $A=17{\cdot}7$ km. sec.$^{-1}$ kpc.$^{-1}$ value which seems preferable to all the others. This leads to rather small variations in a at least up to 2 kpc. from the Sun, with a mean value of the order of 0·4 and 0·5 magn./kpc.[2]

Another point still needs elucidation. Both Wilson and Joy used the *photographic* magnitudes of the Cepheids, but the *visual* magnitudes of the other stars. Now, the two series of data lead to almost the same value for the absorption co-efficient. This result is quite incompatible with the numerous concordant determinations of the differential absorption co-efficient which we shall examine in the following chapter. An evaluation of the absorption co-efficient based on Galactic rotation, therefore, leaves much to be desired. It is, no doubt, a method of the future, which ought to be re-examined at a time when we shall possess more numerous and more certain data on radial velocities, apparent magnitudes, and absolute magnitudes.

The most probable value of the mean absorption co-efficient

All the values of the photographic absorption co-efficients cited up till now are between 0·4 and 0·85 magn./kpc. It is the same with those obtained by other methods, e.g. that of L. Berman [36], based on the study of planetary nebulae and which makes use both of their photographic apparent magnitudes, their diameters and the effect of Galactic rotation on their radial velocities, Berman found $a=0{\cdot}55$ magn./kpc., but, because of the great intensity of the 5007, 4959 and 4861 A lines in the nebulae, this value corresponds to a mean wave-length near 4900 A, which is higher than that of the usual photographic measurements.

The agreement can be considered as satisfactory, seeing how diverse were the methods—and sometimes even the hypotheses—employed. By giving the greatest weight to those measurements based on Galactic rotation and on open clusters, we are led to prefer a value of the absorption co-efficient in the neighbourhood of 0·7 to 0·8 magn./kpc.

Van Rhijn [351], however, has observed that even the best determinations can be affected by a *selection effect* which leads to an underestimation

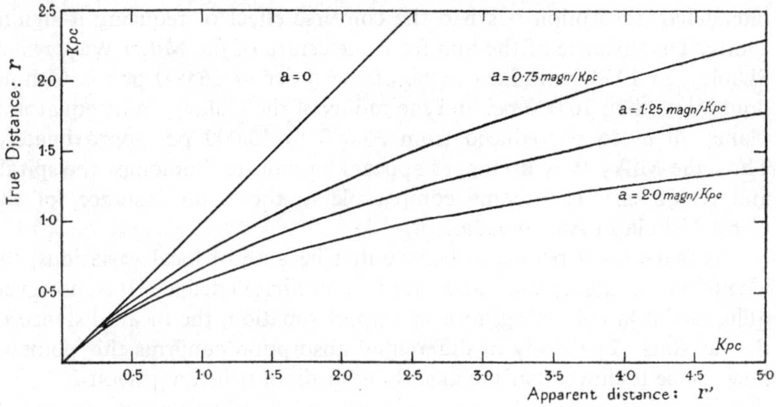

Fig. 20.—Relation between the true distance r and the apparent distance r' in kpc.

of interstellar absorption. We can in fact for the moment, take it that it is not really uniform and varies from one direction to the other of the galactic plane. If this is the case, the weakest Cepheids and the most distant clusters will preferably be observed in the least dark regions. Van Rhijn therefore thinks that in order to obtain a truly "mean" absorption co-efficient valid for the total of the lower Galactic latitudes, it would be necessary to increase the values of Trumpler and Joy appreciably. He first proposed to take a=1·10 magn./kpc. instead of a=0·85 magn./kpc. (Joy's value), and later [459] a—2·0 magn./kpc. Oort, Van de Hulst and other authors also think that in the Galactic plane the absorption co-efficient is of the order of 2 to 2·5 magn./kpc.

Influence of absorption on the evaluation of distances

Fig. 20 shows the variations of the apparent distance r' with the true distance r for various values of the absorption co-efficient. The ratio $\frac{r'}{r}$ increases more and more rapidly with the distance. *Thus the distance of far stars becomes considerably overestimated if absorption is neglected.*

Thus, with the minimum absorption co-efficient a=0.75 magn./kpc., the true distance is less than half the apparent distance if the latter exceeds 4020 pc. and less than a quarter if it exceeds 8040 pc. With a=1.25 magn./kpc., the most probable value, the ratio drops to a half at r=2400 pc. and to a quarter at r=4800 pc. Finally, with a=2.0 magn./kpc., the apparent distances which become 1500, 3000 and 5000 pc. ought to be divided by 2, 4 and 10 respectively.

Although previously the progress of astronomy had led to a progressive increase in the estimated size of our Galactic system, the discovery of interstellar absorption has had the converse effect of reducing it significantly. The distance of the Sun from the centre of the Milky Way, which Shapley, in 1930, estimated to be of the order of 16,000 pc., is without doubt, less than 10,000 pc. and the radius of the Galaxy, in its equatorial plane, ought to be reduced from 35,000 to 15,000 pc. approximately. Thus, the Milky Way no longer appears "a monster" amongst the spirals and its dimensions become comparable to those, for instance, of the Great Nebula in Andromeda.

We have good reason to believe that because of local variations, the idea of a *mean absorption co-efficient* is somewhat ill defined. It is, however, quite useful in correcting, to a first approximation, the mean distance of the far stars. The study of differential absorption confirms this point of view, since it shows that the absorbing medium is heterogeneous.

NOTES TO CHAPTER VII

[1]Throughout this book we shall follow the common usage of expressing the absorption co-efficient a in magnitudes per kiloparsec. This is the loss in magnitude suffered by the star when its light traverses a thickness of 1.000 pc. of the absorbing medium (supposed homogeneous). It is important to see how this co-efficient a is connected with the absorption co-efficient k as defined in physics.

A beam of parallel rays produces an illumination l_0 on a plane normal to the rays, at the point of entry into the absorbing medium. The illumination is reduced to l after traversing a thickness r of the medium. The absorption co-efficient k is defined by

$$\frac{l}{l_0}=e^{-kr} \qquad \text{(VII.1)}$$

where e is the base of Napierian logarithms (e=2.71828). By taking common logarithms of both sides we obtain

$$\log\left(\frac{l}{l_0}\right)=-\log e\; kr=0{\cdot}4343\; kr.$$

Further, the fall in magnitude in a distance r is, by definition

$$\Delta m=m-m_0=-2.5\log\left(\frac{l}{l_0}\right)=2.5\times0{\cdot}4343\times k\times r \qquad \text{(VII.2)}$$
$$=1{\cdot}086\; kr.$$

If k is expressed in kiloparsec^{-1}, we obtain, by taking r=1,

$$a=1.086\; k. \qquad \text{(VII.3)}$$

If k is expressed in cm^{-1}, r becomes 10^3 pc.=$3{\cdot}08.10^{21}$ cm. whence

$$a=1.086\times3.08{\cdot}10^{21}\; k=3{\cdot}34.10^{21}\; k. \qquad \text{(VII.3')}$$

Here then is Schalén's method of calculation [250];

Let $D'(r')$ be the apparent measured stellar density. The number n of stars contained in the element of volume $dV'=\omega r'dr'$ (ω=solid angle) is

$$n=D'(r')\,\omega r'^2\, dr'.$$

But the apparent distances are too great, and in reality these stars are contained in the element of volume $dV=\omega r^2dr$, at a distance $r<r'$. This gives

$$D(r)\; r^2dr=D'(r')r'^2dr'. \qquad \text{(VII.4)}$$

In the case where absorption is proportional to distance, equation (VI.3) connecting apparent with true distance becomes

$$5\log r'=5\log r+ar, \qquad \text{(VII.5)}$$

whence

$$5\log\frac{r'}{r}=ar=2.5\; kr\log e \qquad \text{(VII.5')}$$

or, by using Napierian logarithms

$$\text{Log}\left(\frac{r'}{r}=\frac{kr,}{2}\right)$$
$$r'=r\; e^{kr/2}$$
$$dr'=\left(1+\frac{kr}{2}\right)\, e^{kr'2}\; dr.$$

By taking these values of r' and dr' in equation (VII.4), it becomes

$$D(r)=D'\;(r\; e^{kr/2})\; e^{3kr/2}\left(1+\frac{kr}{2}\right) \qquad \text{(VII.6)}$$

Observations give the curve of apparent densities $D'(r')$. By writing that the true density is the same at two chosen distances r'_1 and r'_2 (or at the true distances r_1 and r_2).

$$D'\ (r_1 . e^{kr/2})\ e^{3kr/2}\ \left(1+\frac{kr_1}{2}\right) = D'\ (r_2\, e^{kr_2/2})\ e^{3kr_2/2}\left(1+\frac{kr}{2}\right),$$

or by taking common logarithms

$$\log\frac{D'\,(r_1\, e^{kr_1/2})}{D'\,(r_2\, e^{kr_2/2})} + \log\frac{1+\dfrac{kr_1}{2}}{1+\dfrac{kr_2}{2}} - \frac{3}{2}\times\log e.\ k(r_2 - r_1) = 0 \qquad \text{(VII.7)}$$

From the curve we obtain the two values of D' corresponding to the two chosen distances ${}_1'r$ and r'_2. Taking a hypothetical value of k, r'_1 and r'_2 are calculated by means of (VII.5′) and these numbers are put into equation (VII.7), to see if it is satisfied. This process is repeated with different values of k until the desired result is obtained.

[2]It is due to a systematic error in interpretation that Wilson again found a=0.65 magn./kpc. On taking A=17·7 km. sec.$^{-1}$ kpc.$^{-1}$, all the data of Wilson and Joy lead to values of a lying between 0·4 and 0·5 magn./kpc.

By combining equations (V.1) and (VII.5″) we obtain the equation

$$\log\frac{r'}{q} = 0{\cdot}2\frac{a}{A}q - \log A,$$

which can be represented, as in Fig. 19, by a straight-line graph of $\log\left(\frac{r}{q}\right)$ against q. The ordinate at the origin gives $-\log A$, and the slope gives a. This method applied to the data of Wilson and of Joy, leads to the following values [73];

Interstellar lines	A=19.2 km. sec.$^{-1}$ kpc.$^{-1}$	a=0.67 m^{pg}/kpc.
Interstellar lines and super giants (not Cepheids)	A=19.4 km. sec.$^{-1}$ kpc.$^{-1}$	a=0.72 m^{pg}/kpc.
Cepheids (Joy grouping)	A=22.8 km. sec.$^{-1}$ kpc.$^{-1}$	a=0.84 m^{pg}/kpc.
Cepheids (Wilson grouping)	A=16.8 km. sec.$^{-1}$ kpc.$^{-1}$	a=0.40 m^{pg}/kpc.

The discrepancy in the values obtained using Cepheids according to whether one adopts Joy's or Wilson's grouping, shows the great uncertainty.

In his work on open clusters in 1940, Trumpler determined A from the radial velocities of thirty-nine clusters. With a=0.79 magn./kpc., he obtained A=15 km. sec.$^{-1}$ kpc.$^{-1}$.

CHAPTER VIII

DIFFERENTIAL ABSORPTION IN THE NEIGHBOURHOOD OF THE GALACTIC PLANE AND THE HETEROGENEITY OF THE ABSORBING MEDIUM

General method

The study of differential absorption in absorbing Galactic layers is based, as in the case of the dark nebulae (Chapter VI, page 129), on the measurement of the *colour indices* of stars.

If we neglect absorption, we can assume that all stars belonging to the same spectral type (and to the same class of luminosity) should have the same colour index C_0. But stars of one type situated at various distances inside the absorbing layer are found to have indices C larger than C_0, varying with their distance r and perhaps with the direction considered as well. On the hypothesis of uniform absorption, the colour index increases proportionately with distance. Denoting by d the increase of index per unit distance, we can write:

$$C=C_0+dr;$$

d is the *differential absorption co-efficient* corresponding to the two monochromatic radiations or to the two spectral regions used. Since the colour index is the difference in the magnitudes of the star measured for the two radiations, d is the difference of the corresponding absorption co-efficients a.

We can construct a graph of colour indices of the stars against distance. If the absorption is truly uniform the points should lie on a straight line and the ordinate at the origin gives the "normal" index C_0 and the slope the differential absorption co-efficient d.

Separate graphs ought to be plotted for each spectral type (and perhaps for each class of luminosity), since the index C_0 differs from one to the other. Also it is more advantageous in each spectral type to consider the colour excess

$$E=C-C_0.$$

Since the effect of absorption is naturally independent of the spectrum,[1] we can treat all stars of different types by constructing the straight line

$$E=dr, \qquad \text{(VIII.1)}$$

which ought to pass through the origin. The method requires a preliminary and very accurate determination of the normal indices C_0 for each spectral type (and perhaps for each class of luminosity). It is necessary to make use of the very near stars situated in high Galactic latitudes, so that absorption should become practically negligible.

Thus the determination of the differential absorption co-efficient is very simple in principle. It does not use any arbitrary hypotheses and, in consequence, it is generally much more certain than that of the photographic absorption co-efficient. It suffices to measure the colour indices accurately (this is necessary since the colour excesses are often small) and to know the distances of the stars exactly.

The principal difficulty is the correct evaluation of the distances of the far stars. Their parallaxes cannot be determined by the "trigonometric" method. We can sometimes make use of the distances deduced from *group movements* and, since we are trying to fix the mean distance of a group of stars, of distances derived from *shifts in parallax* or from the rotation of the Milky Way. For individual stars the absolute spectroscopic magnitudes remain the principal source of information. But they only give the *apparent* distances, systematically too large, and it is necessary to correct them by means of a provisional value of the visual or the photographic absorption co-efficient. When the distances are not too large, the correction is fortunately rather small so that we obtain a near enough value with a rough approximation to the absorption co-efficient.

Only stars of great luminous intensity can be observed at all distances. But giants of classes F, G, K and M are frequently eliminated since their normal colour index C_0 varies, not only with spectral type, but also in a very appreciable manner with absolute magnitude. On the other hand, today (especially since the work of Wilhelm Becker) we know that this *absolute magnitude effect* is negligible in the B stars. Measurements are, therefore, essentially made on the stars of classes O, B and A. Strictly speaking, we should only consider the latter since the absolute spectroscopic magnitudes of the O and B stars are still uncertain.

Results derived from the usual colour indices

In the study of differential absorption, the usual colour indices of the stars are often used, that is to say, the difference between their photographic magnitude and their visual, or rather their photovisual magnitude.

In 1909, with the aid of photographic and visual magnitudes measured at Harvard, for stars of known spectra, whose distances were deduced from their proper motions, Kapteyn [162] obtained the differential absorption co-efficient

$$d=(+0{\cdot}30\pm0{\cdot}06) \text{ magn./kpc.},$$

which agrees surprisingly well with recent determinations. Some years later, Spencer Jones (1914) [157], by means of photographic and photovisual magnitudes measured at the Yerkes Observatory, obtained a value which was slightly larger, but of the same order of magnitude (0·47 magn./kpc.) and Van Rhijn (1916) [350], by observing stars of high galactic latitudes, obtained half the value (0·15 magn./kpc.).

The stars used in these first investigations were, at most, 350 pc. from the Sun, such that their colour excess hardly reached 0·1 magn. Thus, the existence of differential absorption could still have been doubted, particularly since other observations tended to support the idea of the almost complete transparency of space. Shapley [277] showed that in fact *blue* stars existed (with negative colour indices) in the globular cluster M 13, some thousands of parsecs away from the Sun. Their colour excesses could reach up to 0·1 magn.

It was not until the work of Trumpler (1930) [335] that the existence of differential absorption was definitely established. His hypothesis of a *thin* absorbing layer, localized in the neighbourhood of the Galactic plane led to the interpretation of the observations of Shapley: the cluster M 13 is in fact at 40° of Galactic latitude and the light which comes from it traverses a thin section of the absorbing layer.

For seven clusters whose distances had been determined by Trumpler (between 610 and 2,170 pc.), the colour indices of numerous stars of known spectra had been measured by different authors. Trumpler calculated the colour excess of the individual stars and then the mean colour excess of each cluster. He came to the conclusion that the latter increased proportionately with distance, with the slope

$$d=(0{\cdot}32\pm0{\cdot}03) \text{ magn./kpc.}$$

This time the reddening was as much as 0·7 magn. for the furthest cluster and its existence became incontestable.

The publication of Trumpler's paper immediately led to many investigations of a similar kind. Some were still made on Galactic clusters (Zug [391]), others on individual stars or on groups of stars taken outside of the cluster (Van de Kamp [348]; Miss Slocum [283], etc.). Fig. 21, taken from one of Van de Kamp's papers, shows that the colour excess of a certain number of groups of stars of types B0 and A5 increases directly with the distance, with

$$d=(0{\cdot}33\pm0{\cdot}02) \text{ magn./kpc.}$$

The most distant group is at 2,600 pc. from the Sun and has a colour excess of almost 0·9 magn. Schalén [252] used the most distant cluster investigated by Trumpler (NGC 663, in Cassiopeia) and nearby stars. He found them to be at very varied distances whose apparent closeness was only the effect of perspective. He obtained $d=0{\cdot}30$ magn./kpc.

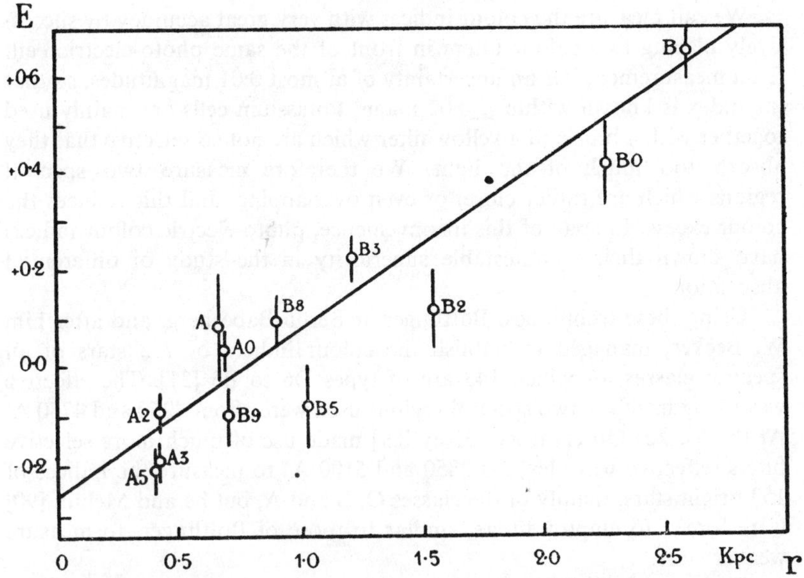

Fig. 21.—The colour excess *E* of *B* and *A* stars as a function of the distance *r* according to Van de Kamp. The vertical lines represent the probable errors in *E*, the slope of the line corresponds to a differential absorption co-efficient $d=0.33$ magn./kpc.

It would be of little interest to review all the other results because they are not directly comparable with one another. By using various filters and plates of various spectral sentivities, the pairs of wave-lengths used to form the colour indices very often differ from each other. We shall return to this in a later chapter (Chapter IX, page 168). For the moment it suffices to recall that this entire research has clearly confirmed Trumpler's mean value of the differential absorption co-efficient relative for photographic and photovisual magnitudes ($d \backsim 0{\cdot}3$ magn./kpc.). Further, absorption is strongly localized in a thin layer, for Van de Kamp found an appreciable decrease in *d* as one deviates from the Galactic plane by some degrees. Finally, inside this thin layer, absorption appears to vary appreciably with longitude (Zug).

Use of photo-electric colour indices

The photographic and visual (or photovisual) magnitudes used in the preceding investigations are never measured to more than a hundredth or more often, a tenth of a magnitude. The error in estimating colour indices must be at least 0·1 magn. From this there follows a dispersion of the points on graphs, and we must restrict ourselves to using only those stars which are rather far away, and often we must take them in groups to reduce the importance of the errors of observation.

We can measure the colour indices with very great accuracy by successively placing two colour filters in front of the same photo-electric cell. Each measurement has an uncertainty of at most 0·01 magnitudes, so that the index is known within ±0·02 magn. Potassium cells are mainly used together with a blue and a yellow filter which are not so selective that they absorb too much of the light. We therefore measure two spectral regions which are rather close, or even overlapping, and this reduces the colour excess. In spite of this inconvenience, photo-electric colour indices have shown their incontestable superiority in the study of differential absorption.

Using these techniques, Bottlinger at Berlin-Babesberg, and after him W. Becker, managed to publish the colour indices of 773 stars of all spectral classes, of which 143 are of types O6 to B5 [31]. The effective wave-lengths of the two spectral regions used were about 4250 and 4750 A. At the Yerkes Observatory, Elvey [89] made use of much more selective filters (effective wave-lengths 3850 and 5100 A) to measure the indices of 153 bright stars, mainly of the classes O, B and A, but he and Mehlin [90] were forced to employ filters, similar to those of Bottlinger, to measure weaker stars.

With a cell of caesium on oxidized silver, J. S. Hall [119] and Bennet [33] measured the colour indices between the red and the near infra-red. But by far the most important work was that of Stebbins, Huffer and Whitford at Madison [296] and at Mount Wilson, for their last catalogue contains colour indices of 1300 B stars (1939) [297]. In the course of their work they replaced the electrometer and amplified the photo-electric current before detection. This permitted the measurement of even the weakest stars up to the ninth magnitude, situated more than 2,000 pc. from the Sun. The effective wave-lengths chosen for their filters naturally varied with the temperature of the stars. They are in the neighbourhood of 4260 and 4760 A for stars of the solar type (6,000°K), from 4215 and 4740 A for A stars, and from 4170 to 4710 A for B stars (20,000°K). A discussion of these numerous excellent measurements has led Stebbins and his collaborators to most important results.

The measurements of Bottlinger and Becker, as also those of Elvey, had already been interpreted by different authors (Van de Kamp [348]; Becker [31]; Miss Williams [379]; Dufay and Liau [75]; Miss Westgate [371], etc.). Later Hunter and Martin [153] discussed their measurements of the gradient, and also all the photo-electric measurements of Bottlinger and Becker, Elvey, Stebbins and Huffer. As far as the mean value of the differential absorption co-efficient is concerned, the indices determined by Bottlinger and Becker had led to values between 0·10 and 0·16 magn./kpc., at a time when the absolute spectroscopic magnitudes of the B stars were still rather uncertain. The discussion has shown that the dispersion of the colour indices of the B stars was, in all cases, due to differential absorption and not to an "absolute magnitude effect". The yellow colour of certain B stars comparable to that of the F or G stars, results from a strong absorption, generally produced by a dark cloud. It is also because of absorption, that the mean colour indices of Bottlinger and Becker, instead of increasing all along the spectral series, diminish first from the O6 to the B3 stars, and pass through a minimum for that type before beginning their regular increase. The mean distances of the stars are, in effect, larger for the first type and the anomaly disappears if one places all the stars at the same distance by taking into account the differential absorption.

Stebbins and his collaborators determined the *apparent* distances of the B stars by adopting an absolute and well-determined magnitude for each spectral type. They then corrected these distances by assuming that the "visual" absorption is seven times the colour excess measured with their filters (nine times the "photographic" absorption). We will see further on to what extent this hypothesis is justified. All their measurements [298] have led to the conclusion that the mean value of the differential absorption co-efficient, between effective wave-length of about 4215 and 4740 A, is between

$$0{\cdot}11 \text{ and } 0{\cdot}15 \text{ magn./kpc.}$$

Heterogeneity of the absorbing medium; regional studies of absorption

But, above all, photo-electric measurements have shown very clearly that the absorbing medium is heterogeneous. Stebbins, Huffer and Whitford determined the differential absorption co-efficient in thirteen separate Galactic sectors each of 20° longitude [298]. The results differ from one sector to the other in the ratio of 1 to 3 or 4 (from 0·05 to almost 0·20 magn./kpc.). In some of the sectors the dispersion of the points is much too large to be attributed simply to the dispersion of absolute magnitudes. It certainly demonstrates the existence of great local variations in differential absorption.

Already in 1929 Otto Struve and Wilhelm Becker [31] had stated, in plotting colour excesses on a celestial chart, that the strongly reddened B stars were frequently grouped next to each other and that this was also the case with those stars which had remained white. W. Becker thus distinguished, on the chart, "red domains" with strong colour excess from "white domains" without notable colour excess and also "mixed domains" where the white and red stars mingled.

It therefore became necessary to measure differential absorption in limited galactic areas. The *regional studies* to this effect during the last fifteen years are already too numerous for us to describe in detail. Many were made with a view to determining the structure of the Milky Way: for this one must first know the absorption in order to be able to evaluate stellar density at various distances. We shall content ourselves with giving some examples only, and with stating the general conclusions drawn from this work.

Clifford Smith [284] measured the colour indices ($m_{pg} - m_{pv}$) of the weak stars of various known spectral types in Aquila, where the Milky Way appears to be divided into two branches. Differential absorption is much stronger in the dark region which separates them, but there remains, in the bright areas, a thinner veil, more opaque in the north branch than in the south branch.

The study of this region was resumed by Mrs. Shain [276] and, more recently, by Weaver [367], who, with the Schmidt telescope of 45 cm. aperture at Mount Palomar, determined photographic and photo-visual magnitudes, as well as the spectral types of 388 stars (up to $m_{pv}=12{\cdot}0$), in an area of 15 square degrees of the darkened zone and, among others, enumerated stars up to the fourteenth magnitude. The graph representing the variation of the colour excess as a function of the modulus of apparent distance $m-M$, shows a plateau marking a region of weak absorption between $m-M=8$ and 9. A deep discussion led Weaver to conclude that the ratio χ between photographic absorption and colour excess (on the international scale) is $\chi=4{\cdot}3\pm0{\cdot}2$. This value permits the transformation of apparent distances into true distances. Weaver thus found that the cloud starts abruptly at 110 pc. from the Sun (Mrs. Shain found 130 pc.) and, until about 260 pc., is of considerable density, with an absorption co-efficient of 7·4 magn./kpc. The most transparent region ($a_{pg}=0{\cdot}75$ magn./kpc.) extends from 260 to 350 pc. and from there on the absorption increases again until 1,000 pc. at least, with an absorption co-efficient $a_{pg}=3{\cdot}2$ magn./kpc. At 1,000 pc. from the Sun, the total absorption already reaches 3·7 magn.

In certain areas investigated by Smith, W. Becker [32], limiting his studies to B, A and F stars, found that the colour excess increases by successive jumps, as if it were due to the crossing of discrete absorbing clouds. He obtained similar results in other stellar regions: two deep clouds, at apparent distance of 200 and 1,200 pc. in the area chosen by Kapteyn (S.A. No. 26 of Auriga), two clouds of thickness 50 pc. or more at 200 or 1,000 pc. in S.A. No. 42 of Lacerta. On the other hand, in S.A. No. 40, near α Cygni, the differential absorption seems to increase proportionately with the apparent distance, up to $r=2{,}000$ pc.; it is less in the Scutum Sobiescii, one of the brightest areas of the Milky Way.

We might also mention the work of R. H. Baker and Elaine Nantkes [17,215] on regions of Cassiopeia and Cepheus, that of Calvert [51] on a very large region of Aquila, of Hercules and of Lyra (650 square degrees in extent) and that of Heerschen [128] on a very large area of Perseus (1,200 square degrees). They are based essentially on the count of stars by magnitude up to $m_{pg}=15$ and on the colour excess of the stars, and also on counts of extra galactic nebulae. In Cassiopeia, Baker and Nantkes found two systems of clouds. The first, at 500 pc., is mainly responsible for the varied appearance of this part of the Milky Way; the second, at 800 pc., is much more uniform and absorbs about one photographic magnitude. Calvert also interprets his results by the presence of two clouds, of variable opacity, situated at 500 and 1,250 pc. respectively, of which the second is more opaque. It is necessary to assume the existence of supplementary absorption in two places, at only 200 pc., notably in the dark band which separates the Milky Way into two branches. In this band, at a longitude of 24°, total absorption may attain five magnitudes (approaching the results of Smith and Weaver). Finally, Heerschen also places two extended clouds in Perseus, the first at about 300 pc., of very irregular opacity, reaching 2·5 magn. in places, the second, much more uniform, at 800 pc., covering almost the entire region studied ($\Delta m \backsim 0{\cdot}8$ magn.).

The extensive investigations of Bok and his collaborators on the spatial distribution of the stars has also given us important information on the local variations in differential absorption. At Harvard, the colour index is taken as the difference in magnitudes determined by photography in the blue and in the red and, it is assumed, that under these conditions the photographic absorption is three times the colour excess. It is thus that Bok obtains the "true" distance of the stars, necessary for the study of their distribution in space. In numerous areas of Monoceros [44], the colour excess of the B8 to A7 stars increases directly with the distance, up to 1,250 or 2,000 pc. from the Sun, with differential absorption co-efficients between 0.18 and 0.33 magn./kpc. At the junction of Gemini, of Monoceros and of Orion [43], the photographic absorption co-efficient varies, from one area to the next, from 0·3 to 1·4 magn./kpc. In other regions there is evidence of absorbing clouds. In Vela, the differential absorption co-efficient is rather small (0·17 magn./kpc.) and in S.A. No. 193 of Centaurus, there seems to be no absorption up to 1,600 pc. from the Sun [45].

By determining magnitudes in the blue and in the red, and also the spectral types of stars and their classes of luminosity by means of a prism objective, Nassau and MacRae [216] studied a bright region of Cygnus measuring 11 square degrees. Here, the colour excess was almost proportional to the distance up to 3,500 pc. and the photographic absorption co-efficient (obtained by multiplying the differential absorption co-efficient by 2·6) is small ($a_{pg}=0{\cdot}53$ magn./kpc.). But, over an area of 1 square degree, an absorbing cloud situated at 900 pc. exercises a supplementary absorption of 0·9 photographic magnitudes.

By the same method and with the aim of investigating local variations of the distribution function of absolute magnitudes, McCuskey and Seyfert, and later McCuskey alone, studied a certain number of very transparent Galactic regions each of approximately 15 square degrees. In Aquila ($l=12°$, $b=-7°$), photographic absorption increases rather slowly up to 1,250 pc., then more quickly up to 2,500 pc., where it reaches 1.7 magn.; in Cygnus ($l=33°$, $b=+3°$), it is only 1 magnitude from 2,000 pc. onwards [420]. The same is the case at the junction of Cepheus and Lacerta ($l=70°$, $b=—3°$) [418]. The absorption is negligible up to 500 pc. in Orion, near Monoceros ($l=114°$, $b=+4°$) [416], and then increases rather quickly from 500 to 1,500 pc., then more slowly up to 2,500 pc., where it reaches 2·8 magn. In Auriga, on the other hand, it is proportional to the distance ($l=133°$, $b=+2°$), and begins at 200 pc. from the Sun, with a coefficient of about 1·2 magn./kpc. [417]. Finally, in Camelopardus ($l=114°$, $b=+4°$) [419], an absorbing cloud extends from 200 to 500 pc., and absorbs 2 magn. From 500 to 2,500 pc., absorption increases more slowly, up to 3 magn.

At Stockholm, Tord Elvius [93] very carefully investigated twelve selected areas of Galactic longitude ranging between 55° and 90° at different latitudes (between $-13°$ and $+49°$). In these areas he determined the photographic magnitudes of 2,550 stars and also their spectral types. Great care was taken in the determination of the "normal" colour indices of non-reddened stars. All spectral types were used with the exception of the first B, A5, F0 types and the M stars, whose absolute magnitudes do not appear to be known with sufficient exactitude. For each one of the selected areas Elvius plotted a graph of the colour excess E against *the modulus of apparent distance*

$$m-M=5 \log r'-5.$$

The graphs differ greatly from area to area. In areas having low latitudes, the colour index increases much more with the modulus of distance (S.A. 18, $b=+7°$; S.A. 41, $b=-7°$ and especially S.A. 19, $b=-1°$; S.A. 40, $b=0°$) than in areas having high latitude. The increase is almost inappreciable in S.A. No. 15 at 49° latitude.

Apparent distances may be converted into true distances by assuming the total absorption to be three to four times the colour excess (see page 156). If we now plot new graphs with the true distances as abscissae, they will show that the colour index generally increases in rather sudden jumps. Therefore, the absorption seems to be produced in separate clouds whose distance and whose co-ordinate z above or below the Galactic plane can be determined.

If we try to derive a mean value of the differential absorption co-efficient from these measurements, we find, for latitudes between 10° and 20°

$$d=0{\cdot}35 \text{ magn./kpc.},$$

while in the immediate neighbourhood of the Galactic plane ($z<100$ pc.), we find practically double this absorption ($d=0{\cdot}70$ magn./kpc.). By multiplying these numbers by 3·5, the corresponding photographic absorption co-efficients would be respectively

$$a=1{\cdot}26 \text{ and } a=2{\cdot}5 \text{ magn./kpc.}$$

The first of these two values is close to that found by Van Rhijn, the second to that found by Oort and Van de Hulst (Chapter VII, page 142).

By selecting 675 stars situated at less than 50 pc. above or below the Galactic plane, Kukarkin [173] had already obtained the differential absorption co-efficient on the Stebbins scale:

$$d_1=0{\cdot}26 \text{ magn./kpc.}$$

As we shall soon see (page 179), the total photographic absorption ought to be about 8·2 times larger, which gives us 2·1 magn./pc.

Stebbins, Huffer and Whitford [298] found, towards the direction of the centre of the Milky Way, photo-electric colour excesses E_1 greater than 0.6 magn., in stars at only a distance of 360 to 480 pc. The total corresponding photographic absorption would be almost 5 magn., thus reducing the apparent brightness of the stars in the ratio of 100 to 1. This explains the fact that we cannot observe temporary stars or *novae* near the dark clouds of Ophiuchus.

This region of the Galactic centre has been studied recently ir greater detail by Bok and van Wijk [398] who measured the photo-electric colour indices of forty-six B stars in an area of 20 square degrees whose centre is at eliptic longitude of 328° and latitude −3°. Absorption there is most irregular. While it is of the order of 1.6 magn. at a distance of 1,600 pc. in the large bright cloud of Sagittarius, and of 2.0 magn. at the west and south-west of it, in places it reaches very much larger values, up to 6.8 magn. at only 200 pc. from the Sun. Here the relatively close absorbing clouds of the group of Ophiuchus have a great effect. In other points, from at least 800 pc. onwards, the absorbing masses are probably elongated. At the east of the great cloud, absorption becomes very much smaller and does not reach more than 0.5 magn. at 1,000 pc. (see Chapter IX for a study of the centre of the Milky Way by means of near infra-red radiation).

Along the Galactic plane, from longitude 252° to 261°, in Carina, Crux and Centaurus, Bok and van Wijk found the absorption to be much weaker, roughly varying from 0.6 to 1.5 magn. at 1,500 pc., with a maximum of 2.0 magn. at 1,100 pc. to the north of the nebula of η Carenae.

In the direction opposite to that of the Galactic centre, absorption seems to increase almost proportionately with distance in the first 500 or 1,000 parsecs, and then to remain constant (Stebbins, Huffer and Whitford).

Particular interest is attached to the region of the north celestial pole, because it contains stars of the *international polar sequence*, used as standards of photographic and photovisual magnitudes. The work of Seares (1936), Shapley and Jones (1937), W. Becker (1939), and Stebbins and Whitford [299] has shown that the pole, although situated at a rather high Galactic latitude ($\simeq +28°$), is covered with an absorbing veil of more than 0·5 photographic magnitudes. The colour excesses E_1 of the B8 and A2 stars attain 0·08 magn. and, on the hypothesis of uniform absorption, correspond to $d_1=0{\cdot}21$ magn./kpc. But this is probably a cloud of 10 to 20 pc. thickness only, situated approximately 200 pc. from the Sun.

Owing to the accuracy of photo-electric measurements Stebbins and his collaborators could show colour excesses of only 0·03 magn. in A stars situated near the two Galactic poles, whose distance varies from 50 to 250 pc. from the Sun. The differential absorption co-efficient in the direction of the poles is probably $d_1=(0{\cdot}11 \pm 0{\cdot}05)$ magn./kpc.

Conclusion

From all this work we may clearly conclude that differential absorption—and surely total absorption—varies within wide limits with longitude and with distance, in the neighbourhood of the Galactic plane. Thus, not far from the centre of the Milky Way, the presence, near the strongly reddened stars, of stars not exhibiting colour excesses, shows that there exist, in places, veritable *holes* in the absorbing medium (for instance towards $l=335°$, $b=-2°$ according to Stebbins). This discontinuous structure, this, so to say, "patchy" appearance is very similar to the jagged contours of the *zone of absence*, in which we observe no external galaxies (see page 184). In most cases, absorption certainly results from the juxtaposition and the superposition of distinct clouds which are more or less thick and opaque. It is permissible to think that even though absorption seems to be proportional to distance, it is produced by rather a large number of clouds of small thicknesses, distributed almost uniformly along the line of vision. Therefore, the "*absorbing layer*" *of Trumpler would be essentially heterogeneous.*[2]

In these conditions, even the appearance of the Milky Way ought to be governed by absorption phenomena. Barnard's dark nebulae and the large clouds studied in Chapter VI only represent condensations of matter which are easier to distinguish either because of their strong opacity, or because of their great extension. On the other hand, the appearance of bright clouds of stars in the Milky Way does not necessarily always correspond to a true stellar condensation, but sometimes to relatively transparent regions of space. In general, these remain covered by a lighter but nevertheless appreciable veil, absorbing, for instance, approximately one photographic magnitude of the same bright cloud of Sagittarius and of that of Cygnus, thus reducing their brightness in the ratio of 2·5 to 1 (Stebbins and Whitford).

In consequence, the mean differential absorption co-efficient

$$d = 0{\cdot}30 \text{ to } 0{\cdot}35 \text{ magn./kpc.}$$

in the cases of photographic and photovisual magnitudes, and

$$d_1 = 0{\cdot}13 \text{ magn./kpc.}$$

in the cases of Stebbins's photo-electric measurements, cannot be used except with the greatest precautions, and only for statistical applications. Of course, this is also the case with Van Rhijn's first mean photographic absorption co-efficient

$$a_{pg} = 1{\cdot}10 \text{ magn./kpc.}$$

The values are probably double near the Galactic plane.

Some workers have tried to represent absorption at different Galactic latitudes and longitudes by practical formulae, particularly in order to correct the distance of the stars for absorption. Such are, for instance, the attempts of Guintini [118], Wilkens [377] and G. Alter [6]. By far the best solution was given in 1945 by Parenago [233], and, in 1949, by Van Rhijn [459].

It is natural to suppose that the density of absorbing matter diminishes exponentially with the distance z above (or below) the Galactic plane. Parenago adopts, according to Miss Williams [379], the simple formula

$$\frac{D\,(z)}{D\,(0)} = e^{-z/\beta} \qquad \text{(VIII.2)}$$

It is easily found that the absorption $A\,(r,\, b)$, at distance $r = \dfrac{z}{\sin b}$ and at latitude b, is given by the expression

$$A\,(r,\, b) = \frac{a_0 \beta}{\sin b}(1 - e^{-r \sin b/\beta}) \qquad \text{(VIII.3)}$$

in which a_0 is the photographic absorption co-efficient in the same Galactic plane.[3] In analysing the colour excesses of 2878 objects obtained mainly from the photo-electric measurements of Stebbins, Becker, etc., and by evaluating the total corresponding photographic absorptions (see Chapter IX), Parenago found that the constant β is practically the same for all the regions of the sky and is $0{\cdot}100 \pm 0{\cdot}004$ kpc., in excellent agreement with the determinations of Van de Kamp (0·105 kpc.). On the other hand, he finds that the mean value of a_0, in the Galactic plane at all longitudes, is $3{\cdot}5 \pm 0{\cdot}1$ magn./kpc.

Using these values, I have calculated the mean values a_{pg} of the photographic absorption co-efficient for various Galactic latitudes b, by applying equation (VIII.3). They are given in Table IX.

TABLE IX

Average Values of the Absorption Co-efficient a_{pg} for Various Galactic Latitudes

b	a_{pg} (magn./kpc.)	b	a_{pg} (magn./kpc.)
± 0°	3·50	±25°	0·82
5°	2·34	30°	0·70
10°	1·66	45°	0·49
15°	1·27	60°	0·40
20°	1·01	90°	0·35

The reader will note that in the neighbourhood of $b=5°$, the absorption co-efficient found by Elvius or by Oort and Van de Hulst reappears (page 162) for the stars situated near the Galactic plane, and that the first co-efficient of Van Rhijn (1·10 magn./kpc.) corresponds to a latitude between 15° and 20° and that of Trumpler to $b=25°$.

But the values of a_0 determined by Parenago vary enormously with longitude (from 0·7 to 9·0 magn./kpc.). Parenago plotted the most convenient values of a_0 to adopt for the different regions of the sky on a celestial planisphere. The distribution of absorbing matter is mainly characterized by the existence of two regions of very strong absorption, situated in two opposite directions, the one towards the centre of the Milky Way, the other corresponding to the clouds of Taurus.

It therefore becomes possible to determine corrections to the moduli of apparent distances by successive approximations, and so to evaluate the true distances of the stars. With this aim, Parenago constructed monograms giving the modulus of true distance and the true distance, as a function of the apparent modulus, of the latitude b and of the values of a_0.

Van Rhijn [459], on his part, represents the colour excess E_1 (on the Stebbins scale) for a star situated at Galactic latitude b, and at a distance r parsecs, by

$$E_1\,(r, b)=\frac{0.031}{\sin\, b}(1-e^{-0{\cdot}0081\, r \sin b})$$

and then constructs a convenient double table giving the colour excess E_1 as a function of b and r.

In the Galactic plane itself, Van Rhijn takes as differential absorption co-efficient $d_1 = 0.25$ magn./kpc. We shall see later on that the photographic absorption co-efficient is 8.16 d_1, from which we obtain $a_{pg} = 2.04$ magn./kpc., given above.

NOTES TO CHAPTER VIII

[1]This is only strictly true for monochromatic light. When working with a large spectral band the *effective wave-length* varies with spectral type. This is also the case with absorption.

[2]The recent photo-electric measurements of Walraven and Fokker [437] made at the Haute Provence Observatory in four different spectral regions on 462 stars situated in fifty-four of Kapteyn's selected areas, further supports this conclusion.

[3]Let k_z be the absorption co-efficient (as used in physics) at $z = r \sin b$. The optical thickness of an element of length dr, at the co-ordinate z, is by definition $d\tau = k_z dr$, and the total optical thickness $\tau = \int k_z dr$. Since from (VIII.2) $\frac{k_z}{k_o} = \frac{D(z)}{D(0)} = e^{-r \sin b/\beta}$, τ becomes

$$\tau = \frac{k_o \beta}{\sin b}(1 - e^{r \sin b/\beta})$$

We obtain the absorption in magnitude from the absorption co-efficients and the optical thicknesses by multiplying both sides by 1.086 (page 152). From this equation (VIII.3) follows.

CHAPTER IX

THE ABSORPTION CURVE

1. Plotting the absorption curve

Problem and general method

Measurements made on various pairs of wave-lengths suggest that, in the neighbourhood of the Galactic plane, "general" absorption increases regularly from the red to the violet, as does that of the true dark nebulae (page 130). An examination of spectrograms confirms this impression qualitatively. Fig. 22, for instance, shows the microphotometer records of the spectra of two stars of the B1 type. The first, β Cephei, is very blue; the second, 26 Cephei, however, is strongly red. It is clear that their difference in magnitude increases considerably and *continuously* from the red to the ultra-violet.

As in the case of the dark nebulae, it is therefore legitimate to suppose that absorption is essentially due to a scattering phenomenon. The colour excesses for the pair of wave-lengths studied are again capable of giving some information on the nature and the probable dimension of the particles scattering the light. But, in order to identify these particles in space more definitely, we must plot that curve representing, for the whole length of the spectrum, the variation of absorption co-efficient with wave-length.

With this in view, we compare the energy distribution in the spectrum of a reddened B star to that of a normal B star. This can be done, either by photographic spectrophotometry, or by means of a photo-electric cell placed behind a mono chromator, or—if this is not available—by means of a set of coloured filters each having a narrow transparent band. In each case, we finally obtain, for a certain number of wave-lengths, the difference in monochromatic magnitude of the two stars. The results are represented on a graph on which Δm is plotted against the wave number $\tilde{\nu}=\frac{1}{\lambda}$ (cm.$^{-1}$ or μ^{-1}).

If the energy curve of each of the two stars can be considered as that of a black body, Δm is a linear function of $\tilde{\nu}$ and we can measure the

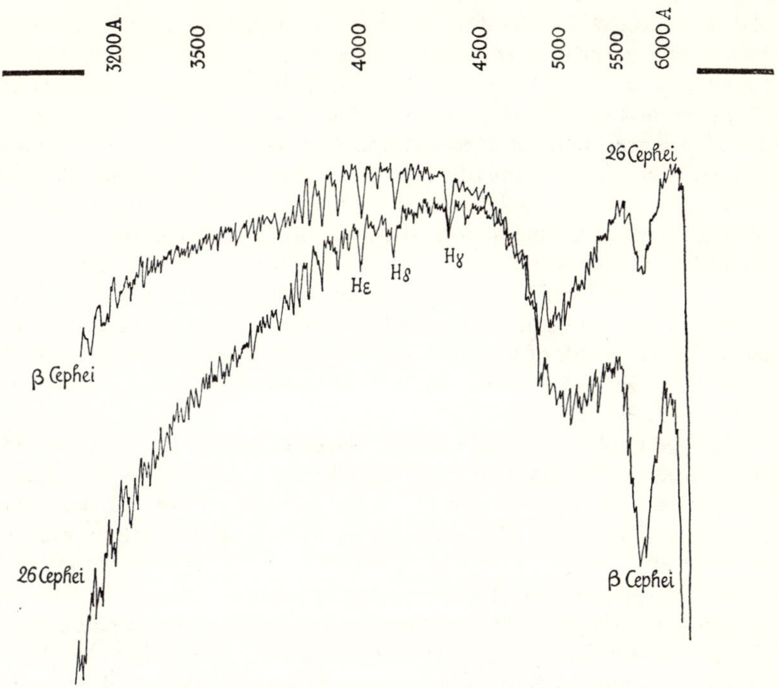

Fig. 22.—Microphotometer records of the spectra of two B1 stars: β Cephei (very blue) and 26 Cephei (strongly reddened). Quartz spectrograph mounted on the 80-cm. telescope of the Haute Provence Observatory (D. Chalonge and L. Divan).

relative gradient of the two stars which is associated with their colour temperatures (page 27). When the points do not fall along a straight line, we must conclude that, in the spectral regions studied, the energy curve of at least one of the stars differs from that of a black body. This is precisely what must occur when the radiation of one of the stars is modified by interstellar absorption. The *curve* on the graph will, therefore, lead us to the law of absorption.

The above reasoning can only be wrong in one single case: in that where absorption (in magnitude) is proportional to wave number $\tilde{\nu}$ (or inversely proportional to the wave-length λ). It is evident that we shall then obtain a straight line: the only effect of absorption will be to change its slope.

Measurement by photographic spectrophotometry

Most of the measurements up to now have been made by photographic photometry, by means of prism-objectives. The first of them have rightly shown that the difference of monochromatic magnitude Δm between a red B star and a normal B star is almost a linear function of $\tilde{\nu}$. In 1928 this was found independently by Kienle and Gottlingen [168] and by Greaves, Davidson and Martin at Greenwich [104], who compared the two stars ζ and ϵ Persei, both of type B1, and of which the first is abnormally yellow. Nine (α) Camelopardis (B0) and 9 Cephei (B2ρ), both reddened, gave Greaves and his collaborators similar results. We can interpret them in two ways: either the abnormal colour of the red B stars is due to a really low colour temperature, or else it is the result of an absorption which is proportional to the wave number $\tilde{\nu}$. It was natural, at the beginning of these investigations, to accept the first explanation. Thus, according to the measurements of Kienle, if we attribute to ϵ Persei a colour temperature of 25,000 to 30,000°K, that of ζ Persei ought to be only between 11,600 to 12,000°K.[1]

This point of view would appear to be confirmed by the measurements of Gerasimovic at Harvard (1929) [100], which were made on a larger number of stars. Gerasimovic has stated that the yellowest B stars were generally supergiants. We might ask if their abnormal colour was the result of absorption at a very large distance, or if it was an intrinsic property of those stars which were of great luminous intensity. Gerasimovic argued that if the light coming from certain of these stars, such as o and ζ Persei, f_1 Cygni, had in effect, crossed dark nebulae, this did not happen in general. 9 Cephei, 9 Sagittae are not in darkened regions and 55 Cygni, one of the reddest B stars, is found in a bright stellar cloud. Gerasimovic finally noted that the radiation from B stars, reddened or not, was more intense in the ultra-violet region (not used for determining colour temperature), which he thought was a contradiction of the hypothesis that absorption increased continuously from the red to the ultra-violet. He therefore concludes in favour of an *absolute magnitude effect*: the yellow colour of certain B stars ought to be associated with their great luminous intensity. This he has tried to explain by means of super-excitation phenomena in the atmospheres of these stars, due to an intense flux of radiation of short wave-lengths. Their *ionization temperature*, associated with the absorption lines defining the spectral type, ought to be higher than their *colour temperature*, measured according to the continuous spectrum.

Today we know that the great intensity of ultra-violet radiation in the case of the stars of the first B types, compared to the A0 stars, for instance, is in fact due to the presence, in the latter, of a continuous absorption spectrum of hydrogen which extends the Balmer Series, and depresses the energy curve in the ultra-violet. All the statistical work based on photoelectric colour indices has further shown that the absolute magnitude effect is weak or non-existent in the case of B stars (W. Becker, Greenstein,

Dufay and Liau), so that the ideas of Gerasimovic have had to be abandoned.

We must therefore consider a hypothesis in which absorption will be proportional to the wave number $\tilde{\nu}$. This is what was done by Trumpler (1930) [336], who made the first spectro-photometric measurements with a view to investigating the law of absorption as a function of wave-length. He compared the spectra of stars situated in the distant Galactic clusters (more than 2,000 pc.) with those of much nearer stars of the same type (350 to 500 pc.). In the whole spectral range studied (from 3420 to 6320 A), all the graphs representing variations of Δm as a function of $\tilde{\nu}$ are almost linear (Fig. 23).

At the Yerkes Observatory, Struve, Kennan and Hynek [327] obtained similar results by studying the yellow stars 55 Cygni (B2) and 13 Cephei ($B9_p$), as did also Mrs. Rudnick [245], by comparing 55 Cygni to 22 Cygni, from 4000 to 6300 A. Nevertheless, in the case of the classical pair ζ—ϵ Persei, studied over a longer spectral range (0400 to 7500 A), Mrs. Rudnick has shown that the distribution of the points would suggest a slight curving which could already have been suspected from Trumpler's measurements: Δm seems to increase less and less as $\tilde{\nu}$ becomes bigger. Although he worked in a much narrower spectral range (4000 to 5000 A), Melnikov [194] found, at Poulkovo, a similar curve by comparing the B and A supergiants contained in the cluster χ and h Persei (at a distance of almost 2,500 pc., found from the

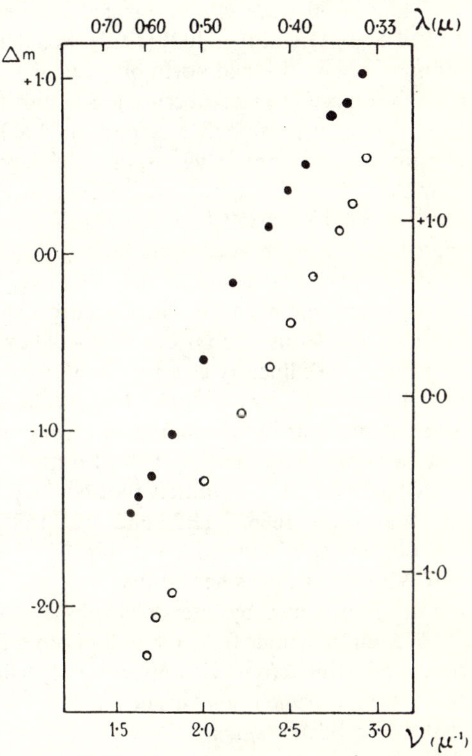

Fig. 23.—Variation of Absorption with $\tilde{\nu}=\frac{1}{\lambda}$ according to Trumpler.
Ordinates: Differences in monochromatic magnitude between the stars of the clusters and near stars used for comparison.
Circles; NGC 6910 (1 star). Scale at right.
Dots; NGC 6913 (3 stars). Scale at left.

effect of Galactic rotation on their radial velocities) with much nearer stars (150 to 170 pc.).

The important work of Greenstein (1938) [107] and of Strohmeier (1939) [310] has established the general validity of this result. The observation of thirty-eight red B stars (compared with normal B stars) led Greenstein to a slope systematically greater in the visible regions (4950–6250 A) than in the violet (3510–4950 A). Strohmeier, whose work extended from 3470 to 6350 A, and sometimes into the far red, investigated thirty-eight supergiants of various types and twelve weaker blue stars, and also found that the points seem to fall along two distinct straight lines, on either side of the wave-length 4800 A, with a greater slope in the red. Only the star χ^2 Orionis gives a differing result, with a greater slope in the violet.

These results are also confirmed by the recent work of Schalén [426] and of Miss Divan [68, 402] (1952). Schalén photographed the spectra of numerous stars situated north of γ Cygni and in a region of Cepheus, with a prism-objective placed before a Schmidt telescope of short focal length (F=68 cm. dispersion 525 A/mm. at 4000 A, 2000 A/mm. at 6000 A). For thirty-two stars of Cygnus, of which twenty were reddened, and thirty-four stars of Cepheus, he made measurements on twenty-five wave-lengths between 3860 and 6350 A, and the author constructed a graph representing the difference in monochromatic magnitude as a function of $\tilde{\nu}$, between reddened and non-reddened stars. The very careful observations of Miss Divan were made with a much more dispersive spectrograph on stars situated in various regions of the sky. They show that interstellar absorption diminishes linearly as a function of $\tilde{\nu}$ from 3·19 to 2·32 μ^{-1} (λ=0·313 to 0·43μ), then more rapidly from 2·32 to 1·45μ^{-1} (λ=0·43 to 0·65μ). The energy distribution in the spectrum of a reddened star thus differs systematically from that of stars of a more advanced spectral type.

Miss Divan also accurately plotted the parallel absorption curves of the two stars HD 168625 (B5) and BD 168607 (B8), by comparing each one of them to a non-reddened star. Their colour excess, on the Stebbins and Whitford scale, is near +0·8; they are, therefore, amongst stars which are most weakened by interstellar absorption.

We thus conclude from the best photographic observations that absorption proportional to wave number is only a first and very rough approximation; *the absorption co-efficient increases less quickly with $\tilde{\nu}$ in the violet than in the visible region.*

Photo-electric measurements

The use of photo-electric cells made of caesium on oxidized silver permits the extension of the investigations into wave-lengths which are higher than 1μ, provided that the cathode is refrigerated to obviate thermionic emission, which is appreciable at ordinary temperatures. By this means, J. S. Hall [121] resumed the study of the pair ζ and ϵ Persei from 4800 to 10300 A, by forming, in the focal plane of a long-focus objective, spectra of small dispersion made with a grid of fine wires acting as a diffraction grating. The graph $\Delta m = f(\tilde{\nu})$ again shows a certain curvature.

With an electron multiplier of similar spectral sensitivity mounted on the 2·5 m. telescope at Mount Wilson, Stebbins and Whitford [302] isolated, by means of filters, six fairly broad spectral regions whose equivalent wave-lengths, in the case of uniform energy distribution, would be:

$$0{\cdot}350,\ 0{\cdot}422,\ 0{\cdot}488,\ 0{\cdot}570,\ 0{\cdot}719 \text{ and } 1{\cdot}030\mu.$$

The filters were chosen so that, in each spectral region, the cell would give almost the same response for a star of the Solar type. These six regions overlapped considerably and, since each time filters with very large transmission bands were being used, the equivalent wave-lengths varied appreciably with the temperature of the source studied. In spite of this inconvenience, the measurements of Stebbins and Whitford can be considered as the best at our disposal because of the extent of the spectral regions studied, and also because of their accuracy of measurement which is much superior to that of photographic spectrophotometry.

For the six spectral regions considered we obtain the difference in magnitude between a reddened B star and a normal B star, chosen from the same spectral type. Each pair of stars thus gives a particular absorption curve and the curves of the different pairs cannot be superimposed, since the rise in absorption varies from one case to the other. To make the results comparable with one another, Stebbins and Whitford assumed that, as the absorption varies, all the monochromatic magnitude differences vary proportionally from one to the other and they arbitrarily fix at 1 magnitude the differential absorption for two extreme spectral bands. Let Δm (2·83) and Δm (0·97) be the difference of magnitude for the wave-lengths $0{\cdot}350\mu$ ($\tilde{\nu} = 2{\cdot}83\mu^{-1}$) and $1{\cdot}03\mu$ ($\tilde{\nu} = 0{\cdot}97\mu^{-1}$). Forming the difference $d = \Delta m\,(2{\cdot}83) - \Delta m\,(0{\cdot}97)$, we divide the differences in magnitude for the four intermediate spectral bands by d, in order to put the results on the desired scale.

The procedure implicitly supposes firstly that the two stars forming a pair have the same colour temperature and secondly, that the absorption law is the same for all reddened stars. These two conditions seem to be satisfied on the whole, since the curves thus brought to a uniform scale can almost be superimposed.

These measurements have confirmed the fact that absorption is not proportional to wave number. Curve I of Fig. 24 represents the deviations from the straight line obtained by joining the extreme points of the graph $\Delta m = f(\tilde{\nu})$ (average of the measurements of thirty pairs of stars).

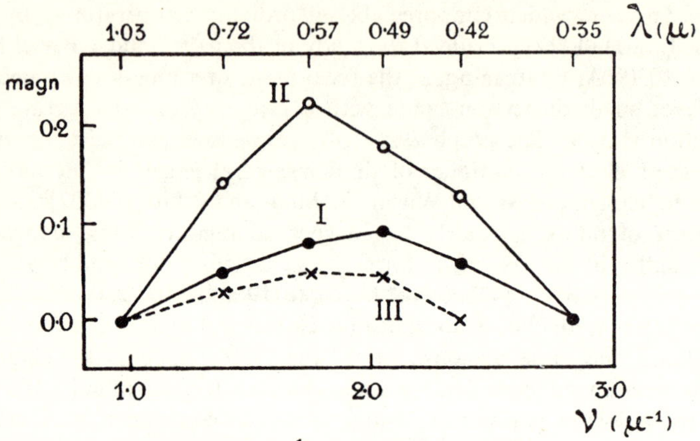

Fig. 24.—Deviations from the $\frac{1}{\lambda}$ law, according to Stebbins and Whitford.
I—Collection of reddened stars.
II—θ_1 Orionis.
III—Reddened stars, reduced by Greenstein.

Only the four components of the star θ_1 Orionis [303] (average spectrum B0) behave in a different manner: the deviations from the straight lines are almost twice as great as for other reddened stars (curve II of Fig. 24). Thus, as Baade and Minkowski [15] had already mentioned, the law of absorption in the region of the trapezium of the Nebula in Orion differs from the general law applicable to all interstellar space.[1]

Greenstein [109] rediscussed the measurements of Stebbins and Whitford. He prefers to eliminate the point corresponding to the ultra-violet (0·350μ) situated near the limit of the Balmer Series and which might be affected by absorption or emission of stellar hydrogen. He reduces the differential absorption between 0·422μ ($\tilde{\nu}=2{\cdot}37\mu^{-1}$) and 1·03$\mu$ ($\tilde{\nu}=0{\cdot}97\mu^{-1}$) to 1 magnitude. The deviations from the straight line joining the points for the violet radiation and for that of the near infra-red retain the same appearance, but are naturally smaller (curve III of Fig. 24).

Although Miss Divan's spectrophotometric measurements [68, 402] seem to confirm the fact that the absorption law is practically the same for various regions of the sky, Schalén [426] found appreciable differences between certain regions of Cygnus and Cepheus: the deviations from the $\tilde{\nu}=\frac{1}{\lambda}$ law are, nevertheless, similar to those found by Stebbins and Whitford, although in the opposite sense. This result would indicate the existence of local variations in the composition of interstellar matter, but before being able to accept this conclusion, it would be necessary to study a larger number of the regions of the Milky Way systematically.

For the identification of the scattering particles, it is particularly important to extend the absorption curve into infra-red. Very important progress in this direction was made by Whitford [375] (1948), by means of a photo-conducting lead-sulphide cell, whose senstivity, many hundreds of time greater than that of a thermo-couple, extended beyond 3 microns. Two filters permitted him to isolate the spectral bands centred on wavelengths $0{\cdot}90\mu$ ($\tilde{\nu}=1{\cdot}11\mu^{-1}$, and $2{\cdot}1\mu$ ($\tilde{\nu}=0{\cdot}48\mu^{-1}$). He observed four pairs of B stars and, even though the measurements were somewhat inaccurate, they sufficed to show that, in the infra-red, the curve is concave downwards and not upwards as in the ultra-violet. The curve, therefore, has a very characteristic S shape. Fig. 25 shows, according to Whitford, the absorption curve of the pair ζ—ϵ Persei, extended up to $2{\cdot}1\mu$ into the infra-red, and up to $0{\cdot}32\mu$ ($\tilde{\nu}=3{\cdot}13\mu^{-1}$) into the ultra-violet, by means of a measurement of Stebbins made through a silver filter.

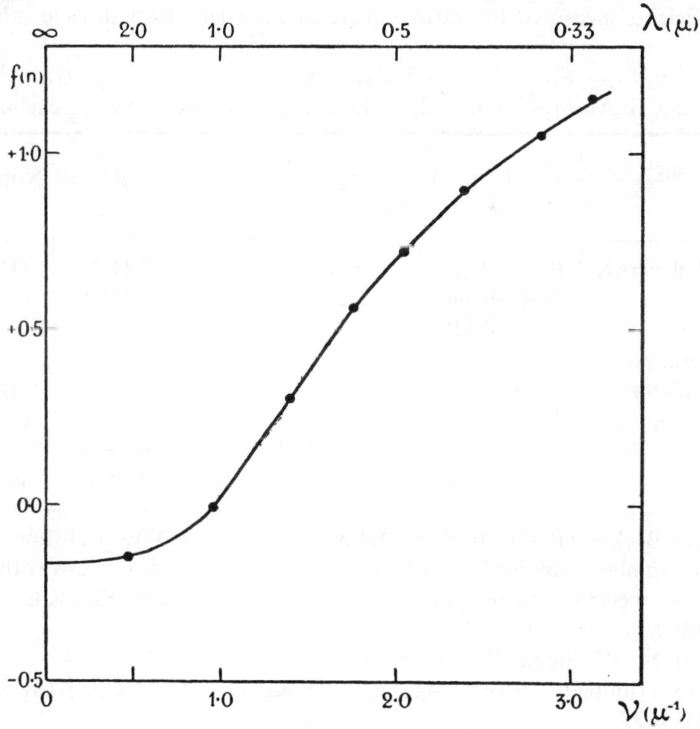

Fig. 25.—Absorption curve as a function of $\tilde{\nu}=\frac{1}{\lambda}$ due to Stebbins and Whitford. Revised values given by Van de Hulst (1949).

This curve was plotted by fixing $f(\tilde{\nu})=0$ arbitrarily for $\tilde{\nu}=09{\cdot}7\mu^{-1}$. The colour excess between 0·4215 and 0·4740μ is 0·130 magn. This is the mean value adopted as the differential absorption co-efficient between these two wave-lengths (page 159). We can thus say that, apart from a constant, the curve represents the mean absorption co-efficient in magnitudes per kiloparsec.

2. Applications of the absorption curve

Comparison of differential absorption co-efficients

Apart from the fundamental problem of the identification of scattering particles, which will be examined later (Chapter XII), the knowledge of the absorption curve f($\tilde{\nu}$) valid at least as a first approximation for the entire Galaxy, apart from the trapezium or Orion and perhaps some other special regions, immediately leads to interesting applications.

In the first place it will help us to compare the differential absorption co-efficients measured for various pairs of wave-lengths with each other.

TABLE X

Differential Absorption Co-efficients reduced to the Scale of Stebbins

	λ_1 (μ) λ_2 (μ)	d_1 magn./kpc.	F	d magn./kpc.	Notes
Babelsberg .	0·425–0·475	0·11	1·04	0·115	(1)
Uppsala . .	0·395–0·440	0·10	1·15	0·115	(2)
Yerkes . . .	0·385–0·510	0·36	0·44	0·16	(3)
International Indices . .	0·426–0·543	0·33	0·46	0·15	(4)
Lick.	0·430–0·620	0·40	0·31	0·125	(5)

Mean: 0.13 magn./kpc.

(1) Becker (1932): 0·095; Dufay and Liau (1933): 0·11 (mean of the numbers obtained or taking as the criterion of distance: (a) the absolute spectroscopic magnitude, (b) the intensity of interstellar K line).

(2) Schalen (1930, 1931).

(3) Miss Williams (1932): 0·38; Miss Westgate (1933): 0·34.

(4) Trumpler (1930): 0·32; Van de Kamp (1930): 0·33; Elvius (1951): 0·35.

(5) Zug (1933): 0·36; Miss Slocum (1931): 0·44.

Plate XVII.—The globular cluster in Hercules, M 13 (NGC 6205). 120 cm. telescope (J. Texereau).

The co-efficient $d(\tilde{\nu}_2, \tilde{\nu}_1)$ for wave-lengths of wave numbers $\tilde{\nu}_2$ and $\tilde{\nu}_1$, is proportional to $f(\tilde{\nu}_2)-f(\tilde{\nu}_1)$. If then we consider a second pair of wave numbers $\tilde{\nu}_4$ and $\tilde{\nu}_3$, we can write:

$$\frac{d(\tilde{\nu}_2, \tilde{\nu}_3)}{d(\tilde{\nu}_4, \tilde{\nu}_4)}=\frac{f(\tilde{\nu}_4)-f(\tilde{\nu}_3)}{f(\tilde{\nu}_2)-f(\tilde{\nu}_1)}=F.$$

The curve gives the factor F by which the first co-efficient $d(\tilde{\nu}_2, \tilde{\nu}_1)$ must be multiplied to obtain the second $d(\tilde{\nu}_4, \tilde{\nu}_3)$.

In Table X the differential absorption co-efficients d measured for different pairs of wave-lengths have been reduced to the scale d_1 of Stebbins. Their main uncertainty is due to the uncertainty in the equivalent wave-lengths contained in the second column. We have taken, in the case of Stebbins's measurements, 4215 and 4740 A, values given by the authors for the A stars and which seem to fit the reddened B stars equally well. As indices of the international scale, the wave-lengths indicated by Seares and Joyner (1943) have been taken. The third column gives the differential absorption co-efficient d measured in each case, the fourth the factor F by which we must multiply the co-efficients d to obtain the corresponding co-efficients d_1 in the Stebbins scale (fifth column).

Let us recall that the discussion of Stebbins's measurements has led to the conclusion that the mean value of d_1 ought to be between 0·11 and 0·15 magn./kpc. (page 159). Now, all the values drawn from measurements made in other spectral regions are between 0·115 and 0·16 magn./kpc. These results show that Stebbins and Whitford's absorption curve on the one hand, and the various measurements of the differential absorption co-efficient on the other form a sufficiently coherent picture.[2]

Does neutral absorption exist?

Relative spectrophotometric measurements only acquaint us with an absorption in magnitude apart from a constant. We know that the curve $f(\tilde{\nu})$ of Stebbins and Whitford (reproduced in Fig. 25) can be considered as representing the variation of the average absorption co-efficient, in magn./kpc., as a function of the wave number $\tilde{\nu}$. The absorption co-efficient is thus

$$a(\tilde{\nu})=f(\tilde{\nu})+\gamma$$

where γ is an unknown constant.

Let us first imagine that the absorption tends to zero with $\tilde{\nu}$ (i.e. for an infinite wave-length). In this hypothesis we can try to determine γ by extrapolating the curve $f(\tilde{\nu})$ to $\tilde{\nu}=0$. It is so flat in the infra-red that there is a little uncertainty about the extrapolation. The ordinate at the origin is certainly near $-0{\cdot}17$. Then, if $a(\tilde{\nu})$ is zero for $\tilde{\nu}=0$, the constant γ is almost $+0{\cdot}17$ magn.[3]

But let us now suppose that a *neutral* absorption, i.e. independent of the wave-length in the entire spectral region explored, is superimposed on the absorption varying with λ. It would completely escape relative spectrophotometric measurements and also differential measurements on any pair of wave-lengths whatever. The constant γ can then take any value whatsoever higher than 0·17 magn.

But this neutral absorption would necessarily affect the measurements of total absorption. We can investigate its existence if we know both the absorption curve $f(\tilde{\nu})$ and the ratio

$$\chi = \frac{A(\tilde{\nu})}{A(\tilde{\nu}') - A(\tilde{\nu}'')}$$

between the total absorption for the wave number $\tilde{\nu}$ and the differential absorption corresponding to the same Galactic section between the wave numbers $\tilde{\nu}'$ and $\tilde{\nu}''$ (one of which can be identical with $\tilde{\nu}$). This ratio can in effect be written

$$\chi = \frac{f(\tilde{\nu}) + \gamma}{f(\tilde{\nu}') - f(\tilde{\nu}'')}$$

The values of $f(\tilde{\nu})$, $f(\tilde{\nu}')$, $f(\tilde{\nu}'')$ are read off from Stebbins and Whitford's curve and the preceding relation permits the calculation of the constant γ. We could only conclude as to the existence of a neutral absorption if we obtained a number greater than +0·17 magn. [74].

The best determinations of the ratio χ are, without doubt, those which result from the work done at Uppsala on the absorbing clouds of Auriga, of Cygnus and of Cepheus. The measurements of monochromatic magnitudes determined on spectrograms at 3950 and 4400 A leave no uncertainty in the wave numbers $\tilde{\nu}'$ and $\tilde{\nu}'' = \tilde{\nu}$. According to Table VIII (page 140), we have, in this case $\chi = 9{\cdot}2$ or $9{\cdot}3$, numbers which lead respectively to:

$$\gamma = +0{\cdot}19 \text{ and } \gamma = +0{\cdot}18 \text{ magn.}$$

Thus, all this takes place as if the clouds of Auriga, of Cygnus and of Cepheus produced no neutral absorption.

I have further assembled thirteen different determinations of the ratio χ which, though of smaller weight, are of interest in that they use various wave-lengths (sometimes a little uncertain). Four of them give obviously wrong values.[4] The other nine give for γ respectively:

0·00, +0·06, +0·11, +0·11, +0·15, +0·18, +0·22, +0·31, +0·44,

fairly well grouped around the general average +0·18. By eliminating the two extreme numbers, that average becomes +0·16 magn.

Instead of considering particular absorbing clouds, it is interesting to look at all the B stars in the vicinity of the Galactic plane. Stebbins' measurement of the photo-electric colour indices and those of his collaborators Henyey and Greenstein [114] have led, by a very ingenious method too complicated however, to be discussed here, to the *ratio X* of the absorption co-efficients for two spectral regions isolated by filters. They found $X=1{\cdot}142$. Now we can write this ratio

$$X=\frac{a(\tilde{\nu}_1)}{a(\tilde{\nu}_2)}=\frac{f(\tilde{\nu}_1)+\gamma}{f(\tilde{\nu}_2)+\gamma}=1{\cdot}142$$

and again obtain the constant γ from it. If, with Henyey and Greenstein, we take as equivalent wave-lengths $\lambda=4215$ A and $\lambda_2=4740$ A, we find $\gamma=+0{\cdot}15$ magn. With $\lambda_1=4170$ and $\lambda_2=+0{\cdot}17$ magn.

The almost constant ratio of these various determinations of the constant γ shows that *there is no place in the Galaxy where there is neutral absorption.*[5]

Return to total absorption

The colour excesses are generally much better known than the total absorption, the determination of which meets with great difficulties. It is, therefore, advantageous to know as accurately as possible the ratio χ between total photographic absorption and differential absorption, so that the first may be simply deduced from the second. As we have seen in the preceding chapter, this is what a number of authors are working on at the moment. But the evaluation of the ratio χ by direct observation is also extremely difficult.

Knowing that there is no neutral absorption, we can make use of the Stebbins and Whitford curve to calculate χ by means of the relation

$$\chi=\frac{f(\tilde{\nu})+0{\cdot}17}{f(\tilde{\nu}_1)-f(\tilde{\nu}_2)}.$$

The most frequently measured colour excesses are those of the international scale, those of the Stebbins scale and of the Harvard "blue-red" system. We find, in the three cases, the ratios

$$\frac{a\,(0{\cdot}426)}{a\,(0{\cdot}426)-a\,(0{\cdot}543)}=3{\cdot}77\ (\sim 3{\cdot}8),$$

$$\frac{a\,(0{\cdot}426)}{a\,(0{\cdot}4215)-a\,(0{\cdot}4740)}=8{\cdot}16\ (\sim 8{\cdot}2),\text{[6]}$$

$$\frac{a\,(0{\cdot}426)}{a\,(0{\cdot}426)-a\,(0{\cdot}620)}=2{\cdot}53\ (\sim 2{\cdot}5).\text{[7]}$$

Between photovisual absorption and colour excesses, we would also have

$$\frac{a\,(0{\cdot}543)}{a\,(0{\cdot}426)-a\,(0{\cdot}543)}=2{\cdot}78\ (\sim 2{\cdot}8)\ \text{and}\ \frac{a\,(0{\cdot}543)}{a\,(0{\cdot}4215)-a(0{\cdot}4740)}=6{\cdot}00.$$

If we take as mean values of the differential absorption co-efficient $d=0{\cdot}33$ magn./kpc. in the international system and $d_1=0{\cdot}13$ magn./kpc. in the Stebbins system, we find for the photographic absorption co-efficient respectively

$$a_{pg}=3{\cdot}77\times 0{\cdot}33=1{\cdot}24\ \text{magn./kpc.}$$

and

$$a_{pg}=8{\cdot}16\times0{\cdot}13=1{\cdot}06 \text{ magn./kpc.};$$

in good agreement with Van Rhijn's value (1.10 magn./kpc.).

The mean photovisual absorption co-efficient is

$$a_{pv}=2{\cdot}78\times0{\cdot}33=0{\cdot}92 \text{ magn./kpc.}$$

or

$$a_{pv}=6{\cdot}00\times0{\cdot}13=0{\cdot}78 \text{ magn./kpc.}$$

Observation of the centre of the Milky Way

The absorption curve shows that the Galaxy is much more transparent to radiation of long wave-lengths than to violet radiation. Just as it permits us to photograph distant landscapes through some terrestrial fogs, infra-red radiation pierces Galactic dust more easily and favours the observation of distant celestial objects through absorbing clouds.

Stebbins and Whitford [304] were the first to try to prove the existence of a maximum in the infra-red emission in the direction of the centre of the Milky Way, a region poor in stars on ordinary photographs (1947). Their observations were made by means of the 2·5 m. telescope at Mount Wilson, with the electron multiplier and with the filter which they generally used for isolating a fairly broad spectral band around 1·03μ (page 173). Since the telescope was fixed, the rotation of the Earth brought an arc of constant declination across the cathode. A series of observations, at different declinations permitted the exploration of the region between longitudes 321° and 331°, up to 2° on either side of the Galactic plane. In the neighbourhood of this, photometric records showed an increase in brightness of the order of 20 per cent. The bright zone in the infra-red seems roughly to occupy an elliptical area extending over 8° of longitude and 4° to 5° of latitude, with its centre at 326·5° approximately. The increase is hardly appreciable if one uses a filter whose pass band is centred on 0·719 μ.

A little later (1949), Kaliniak, Krassovsky and Nikonov [160] managed to photograph this luminous region by means of an *electronic image converter*, working on a mean wave-length of 0·98μ. The principle of this apparatus, often rather inappropriately called an "electronic telescope", is quite simple. With an objective and a suitable filter an infra-red image of the source is formed on a semi-transparent cathode of caesium on oxidized silver, whose sensitivity extends up to 1·2μ. The electrons released from the cathode are localized by an electrostatic lens either directly on to a photographic plate (A. Lallemand), or on to a fluorescent screen which can be photographed. On the screen or on the plate a true replica of the infra-red image is obtained. In this way Russian astrophysicists have photographed a large bright cloud of irregular contours whose isophotes (contours of equal intensity) they have traced and which, in spite of certain differences in detail, correspond sufficiently well with Stebbins and Whitford's emission maximum.

It was at once realized that the cloud could be photographed directly without having to use electronic apparatus. The emulsions actually used (Eastmann I N) have a good sensitivity up to approximately 0·89μ. At 0·8μ, interstellar absorption is stronger than at 1μ, but the emission from the upper atmosphere—in the near infra-red essentially due to bands of the OH molecule and, less so, to those of O_2—is much weaker there, such that the contrast between the bright stellar cloud and the background of the sky can be as great at 0·8μ as at 1μ. Very rough calculations based on Stebbins and Whitford's absorption curve, the spectral sensitivity of the receptors and the relative intensities of various emission bands of the night sky, have convinced me that the experiment can be tried with success. In fact, during the spring and summer of 1952, we tried, at the Lyon and Haute Provence Observatories, to photograph directly, in the near infra-red, the stellar cloud situated in the direction of the Galactic centre (J. Dufay, J. Bigay, P. Berthier and J. Texereau) [403, 404, 405, 436].

With a small and wide open objective (F/2), of short focal length (16 cm.), we traced isophotes of the blue, the red and the near infra-red regions. The latter are very similar to those traced by the Russian astronomers for the wave-length 0·98μ. The cloud can hardly be distinguished in the red and it is practically invisible in the blue.

Apart from its simplicity, direct photography has the advantage of giving much sharper images than do current electronic devices. With the Schmidt telescope of the Haute Provence Observatory (see page 112) we have been able to resolve the cloud almost completely into individual stars.[8] Thousands of invisible blue stars have been counted on the infra-red negatives. Some of them appear very bright: they are probably red giants, further reddened by absorption. The enlargements reproduced on Plates XVIII and XIX show one of the densest parts immediately to the east of 45 Ophiuchi (the brightest star to the right, almost at the edge of the plates).

If the Galaxy were a spiral, similar to that of Andromeda (type Sb), it ought to possess, as does the latter, a large and very bright nucleus but which might be hidden by an absorbing cloud. Stebbins and Whitford undertook their observation in order to try to show the existence of such a nucleus; the object which they did discover in fact appeared to them to represent the nucleus looked for. Kaliniak, Krassovsky and Nikonov share this opinion and, according to Kholopov [167], who has discussed the two series of observations, the large cloud of Sagittarius, situated a little to the south of the Galactic plane and clearly visible in the blue, would belong to the nucleus as a new infra-red cloud.

Now, using the magnitudes of the brightest stars of the globular cluster NGC 6522, inside the great cloud, Baade [394] estimates the distance of the latter as approximately 9,000 pc., taking the absorption into account. This would be approximately 2·6 photographic magnitudes, according to the colour excess of the cluster measured by Stebbins and Whitford. On the hypothesis of Kholopov, supposing amongst other things that the "blue cloud" and the "infra-red cloud" separated from each other by a very opaque absorbing region, have the same proper brightness on the average, our measurements permit the evaluation of the total absorption before the "infra-red cloud" for the three spectral bands studied. We find less than two magnitudes at 8300 A, almost three magnitudes at 6560 A, and more than five magnitudes at 4260 A. The aggregate of the two clouds constituting or surrounding the nucleus have a diameter of approximately 1,600 pc.[9]

These absorptions are truly very small for a distance of 9,000 pc. and, in this respect, the most surprising figure remains the absorption of 2·6 photographic magnitudes before the NGC 6522 cluster. On the certainly false hypothesis of uniform absorption, it would correspond to an absorption co-efficient a_{pg} of hardly 0·3 magn./kpc. As the observations of Bok and Van Wijk [398] have confirmed the existence, in that region, of absorbing clouds only some hundreds of parsecs from the Sun, it would be necessary that, behind these, space should be extremely transparent.

In order to establish these conclusions definitely, one would have to determine the distance of the infra-red cloud directly. With this in view, one would have to determine the apparent infra-red magnitudes of the stars which compose it and to try to fix their absolute magnitudes. Here, research on the variables of clusters or of the Cepheids would be particularly interesting.

NOTES TO CHAPTER IX

[1]L. Divan (1953, [402]) has shown that the absorption law in the trapezium of Orion is the same as in other galactic regions. The observed anomaly appears to be due to the fact that the three stars exciting the nebula (H D 37061, θ_1C and θ_2 Orionis) have, in the ultra-violet, a colour temperature higher than that of other stars of the same type.

[2]In order to compare the differential co-efficients of absorption among themselves, use is often made of empirical relations between the colour indices of different systems. These relations have been established by considering the stars of all the spectral classes not, in general, affected by absorption. Their approximately linear form is due to the fact that the stars radiate almost like black bodies. The differences in monochromatic magnitude are then a linear function of the wave number $\bar{\nu}$ (page 29). Assuming that the same relations hold for reddened stars, one may then suppose that the absorption in magnitudes is also proportional to $\bar{\nu}$. This procedure is, in fact, quite incorrect.

[3]This value depends essentially on the accuracy of the point at $\bar{\nu}=0.48\ \mu^{-1}$, which only depends on a few observations.

[4]The measurements of Müller and Hufnagel on the nebula North America and those of Müller on another region of Cygnus give $\gamma\sim2.0$ magn. If this were really the case (but the results are doubtful), it would lead to the existence of a strong neutral absorption in Cygnus. On the other hand, two other sets of measurements lead to negative values (−0·28 and −0·13 magn). These were made by W. Becker on S.A. No. 4 and by Madame Lehman-Balanowskaja on the region of ξ Persei.

[5]It is possible that while this conclusion is on the whole correct, it does not apply to certain limited regions particularly of the Milky Way. This is contrary to the idea of G. Alter [6] who, by a detailed analysis of the colour excesses found by Stebbins and Whitford, believes he has demonstrated variations of the ratio χ as a function of the galactic longitude. Alter supposes that the total absorption and differential absorption are proportional to the distance. Combining the equations (VII.5′) and (VIII.1) one can write

$$\log\left(\frac{E}{r'}\right) = \log d - 0{\cdot}2\chi\, E.$$

Each star gives an equation of this form in which one knows the apparent distance r and the colour excess E. The solution of the system of equations by the method of least squares fixes the two unknown log d and χ. For all the stars situated at least 15° from the plane of the Galaxy, Alter obtained $a_{pg}=0{\cdot}83$ magn./kpc. and, on the international colour index system $\chi=3{\cdot}5$. But along the galactic longitudes χ varied from 2·5 to 4·7 and at the same time reached the value 7·7 between $l=140°$ and $l=190°$. It seems doubtful that these variations are real.

[6]The average of all the experimental determinations discussed by Van Rhijn [352] is 8·2; by a very different method Greenstein and Henyey [114] have found $\chi=8{\cdot}1$.

[7]Nassau and MacRae [216] take, in this case $\chi=2{\cdot}6$, a value which seems preferable to that adopted by Bok ($\chi=3{\cdot}0$).

[8]Complete resolution was obtained in 1953 with the 81 cm. aperture telescope, and has led to the discovery of a new globular cluster, invisible in the blue [460].

[9]1,300 pc., only if one changes the distance of the centre of the Milky Way to 7,500 pc., as proposed by Parenago [424].

CHAPTER X

TRANSVERSE OPACITY AND THICKNESS OF THE "ABSORBING LAYER"

IT has been known for some time that spiral and elliptic nebulae, which are abundant near the poles of the Milky Way, become scarce and then disappear near the Galactic plane. On either side of the latter there extends a "zone of absence" in which it is impossible to photograph the spiral nebulae.

Conclusions from extragalactic nebulae

Since these nebulae are external to our stellar universe, and their spatial distribution can in no way be related to our Galactic co-ordinates, this appearance certainly results from the absorption of light in the Milky Way. Historically, it was the study of the distribution of spirals in the celestial vault that first proved the existence of Galactic absorption.

The extension of the *zone of absence*, at varying longitudes, is on the average from 10° to 20°, but with extensions which attain +35° of latitude. Its very irregular contours follow those of the large absorbing clouds very well: the two principal extensions correspond to the dark nebulae of Ophiuchus and of Taurus, and another to that of Cepheus.

To draw some quantitative conclusions on absorption from these observations, we shall have recourse to a very rough picture (Fig. 26). Let us suppose that the absorbing medium is homogeneous, bounded by two planes P and P′, parallel to the Galactic plane G, and h kpc. apart. Let S be the Sun, supposed to be in the Galactic plane. The light ray N S, coming from the nebula N, at Galactic latitude b traverses a distance $L=$ MS through the absorbing layer. The *brightness* of the nebula is the less the longer the path, i.e. the smaller the latitude.

Let μ be the magnitude per unit solid angle corresponding to the measured brightness ($\mu=-2{\cdot}5 \log B$), μ_0 the value corresponding to the original brightness of the nebula ($\mu_0=-2{\cdot}5 \log B_0$), above the absorbing layer. If we again represent by a the absorption co-efficient in magnitude per kiloparsec, the fundamental law of absorption is:

$$\mu=\mu_0+aL.$$

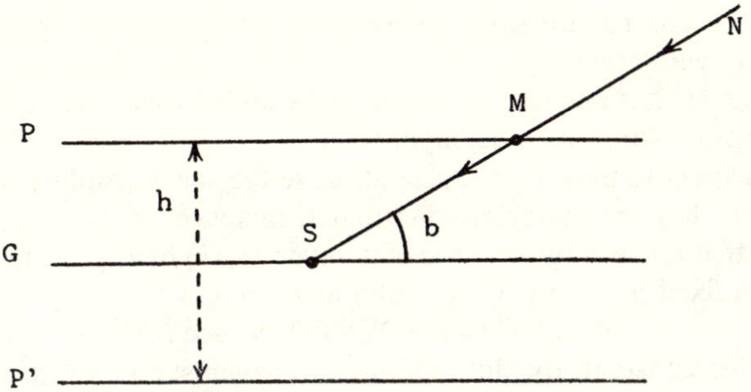

Fig. 26.—Diagram of the Galaxy.

Further, if we suppose that the two planes P and P' are equidistant from the Galactic plane G, the figure gives directly

$$L=\frac{h}{2\sin b}=\frac{h}{2}\operatorname{cosec} b$$

and the preceding equation becomes

$$\mu=\mu_0+0{\cdot}5ah \operatorname{cosec} \mathrm{b}. \qquad \text{(X.1)}$$

Let us now measure the brightness of a sufficiently large number of nebulae belonging to the same type (so that they always have the same actual brightness B_0, and thus the same μ_0), but situated at different latitudes. The points obtained by plotting μ against cosec b ought to lie on a straight line whose slope gives the product $t=ah$, i.e. the absorption in magnitude which a light ray would undergo if it crossed the entire absorbing stratum, perpendicular to the Galactic plane. For purposes of abbreviation, I shall call this quantity the *transverse absorption* of the Milky Way. If we further know the absorption co-efficient a we can derive the total thickness h of the absorbing layer.[1]

The method is very similar to that of Bouguer for the study of atmospheric absorption. It has not as yet led to really satisfactory results, because few measurements of brightness of the nebulae have yet been made and often they involve systematic errors. Qualitatively, the discussion of Holeschek's estimations and of Wirtz's visual measurements is evidence for a progressive decrease in average brightness with Galactic latitude (Carpenter [53]). But the value obtained for the transverse opacity of the Milky Way from Wirtz's measurements [383] (0·14 magn., Dufay [70]) is altogether uncertain. In fact, after having tried to correct systematic errors, 0·04±0·10 magn. was found for the northern hemisphere, and 0·10 ±0·10 for the southern hemisphere. Re-examining the entire problem, Eigenson [88] concluded that the photometric data known in 1939 could only have led to a qualitative result. More precise measurements on the brightness of the nebulae (or on their total magnitude and apparent diameter) would be necessary to evaluate the transverse opacity of the Milky Way with certainty.

Fortunately it is possible to determine t simply by *counts* of the nebulae, together with a supplementary hypothesis.

Let us suppose that the nebulae all have the same absolute magnitude M and that they are uniformly distributed in space. A simple calculation will show that the average number of nebulae $N_m(b)$ of apparent magnitude less than a fixed limit m per unit solid angle ought to satisfy the relation

$$\log N_m(b) = \log N_o - 0{\cdot}3\ t \operatorname{cosec} b, \qquad \text{(X.2)}$$

where N_o is a constant. By plotting $\log N_m(b)$ against cosec b we will obtain a straight line of slope $-0{\cdot}3\ t$.[2]

The photographic counts of the nebulae of Shapley-Ames (Harvard), of Mayall (Lick) and mainly of Hubble (Mount Wilson) may be used. The latter involve approximately 44,000 nebulae. Fig. 27 shows the straight line drawn by Hubble [150] for both Galactic hemispheres together. Each of them was divided into nine zones of approximately 10° of latitude, and Hubble took the average of $\log N_m$ in the two hemispheres for each zone. The straight line has a slope of -0.15, whence $t = \dfrac{0{\cdot}15}{0{\cdot}3} = 0{\cdot}50$ magn. Nevertheless, a new discussion of the same data led Oort to a straight line of slope $-0{\cdot}17$, whence $t = 0{\cdot}57$ magn. [221]. The results obtained by other authors are grouped in Table XI.[3]

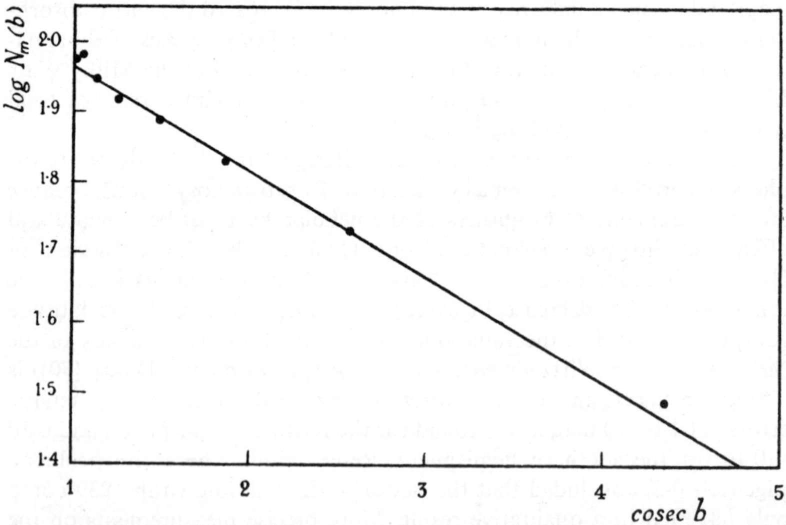

Fig. 27.—Determination of the transverse opacity t of the Milky Way by counts of external galaxies (Hubble). The straight line of slope $-0{\cdot}15$ corresponds to $t = 0{\cdot}50$ magn.

Conclusions from globular clusters (*Van de Kamp*, 1932)

The globular clusters situated in the most extreme regions of the Milky Way (see page 144), are also outside the "absorbing layer". At a latitude b, they therefore undergo absorption 0·5 t cosec b (in magnitude). But their number is much too small for the application of the same statistical methods used in the case of nebulae.

In Lindblad's theory of Galactic rotation, the system of globular clusters can be thought of as lying on an ellipsoid of revolution about the axis of the Milky Way. Knowing the longitude and the Galactic latitude of a cluster, and also its *apparent* distance from the Sun, we can calculate its co-ordinates in a system of three rectangular axes having as origin the centre of the Milky Way and two of its axes (Ox, Oy) in the Galactic plane (for convenience the axis Ox is directed towards the Sun). The ellipsoidal distribution requires that, for the total of the clusters, the mean values of the co-ordinates x and y should, to all intents and purposes, be equal ($\bar{x}=\bar{y}$). We find that this is not the case.

In reality, the *apparent* distances r' are affected by absorption. Van de Kamp [349] calculated the "*true*" distance r, giving successively various values of the transverse opacity t

$$5 \log r = 5 \log r' - 0{\cdot}5\, t \operatorname{cosec} b$$

and from them he deduced new rectangular co-ordinates of the clusters. He found that the condition $\bar{x}=\bar{y}$ was practically satisfied for all values of $t > 0{\cdot}40$ magn. It is sufficient that the transversal absorption attain this value, for ellipsoidal symmetry to be re-established.[4]

Conclusions from star counts

Counts of stars by photographic magnitudes give the number $n(m)$ of stars of magnitude between $m-\frac{1}{2}$ and $m+\frac{1}{2}$, which, on the average, are present in 1 square degree of the sky, at a particular Galactic latitude. Let us take as unit illumination that which is produced by a star of photographic magnitude 1·0. By definition, a star of magnitude m then produces an illumination $10^{-0.4\,(m-1)}$ and all the stars of magnitudes between $m-\frac{1}{2}$ and $m+\frac{1}{2}$ produce the illumination

$$e(m) = n(m).\ 10^{-0.4\,(m-1)}$$

The weaker the stars, the greater becomes their number, but each one of them illuminates less. The first factor $n(m)$ of the product thus increases, while the second ($10^{-0.4\,(m-1)}$ diminishes. The curves representing the variations of $e(m)$ are very similar to those of Fig. 28 [77]; they show a maximum for some value of the magnitude between nine and thirteen, according to the Galactic latitude. This shows that, near the poles, for instance, stars of the ninth magnitude are those which give us most light; near the Galactic plane, those of the thirteenth. The area between the

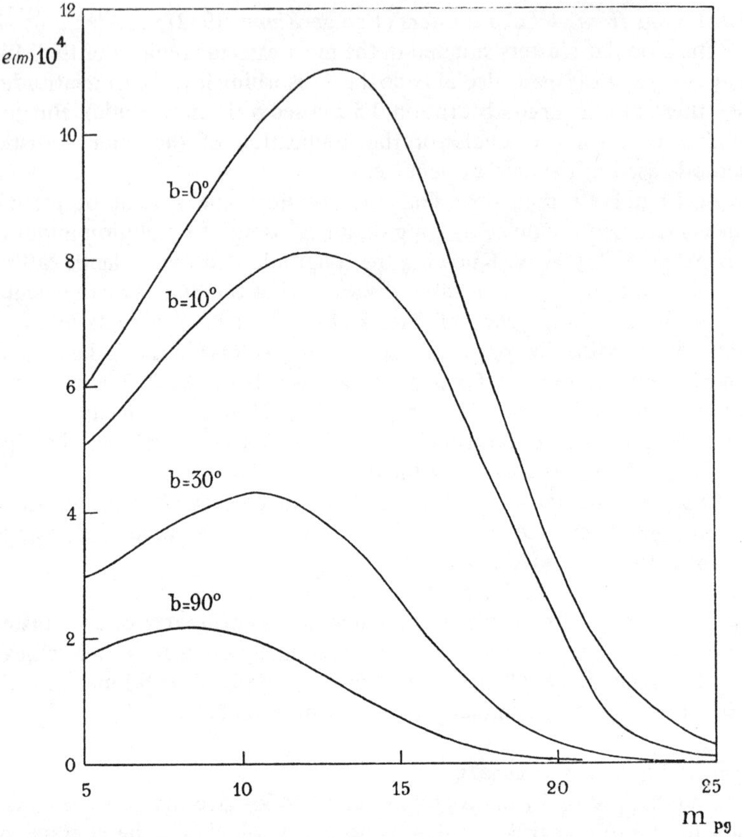

Fig. 28.—Curves for various Galactic latitudes showing, as a function of the photographic magnitude m_{pg}, the variations of the illumination $e(m)$ produced by stars of magnitude m_{pg} contained on the average per square degree. $e(m)$ is evaluated for stars of magnitude 1.00, according to counts of Seares, Van Rhijn, Joyner and Richmond.

curve and the abscissa measures the total illumination produced by all the stars contained in 1° of the sky, which we can call the *brightness of stellar origin* B_s. This is sufficiently well determined although the counts of stars are naturally limited by a limiting magnitude (m_{pg}=17 or 18 for instance), because the weakest of the observable stars do not produce anything but quite insignificant illumination in spite of their large number.

Thus, by constructing the curves $e(m)$ for different Galactic latitudes, we can evaluate the corresponding brightnesses of stellar origin. The points of Fig. 29 represent the brightnesses taken from the counts of stars of Seares, Van Rhijn, Joyner and Richmond. We find that in the Galactic plane B_s is hardly seven times larger than in the neighbourhood of the poles. This result is altogether surprising. Let us suppose that as a first approximation, the stars are uniformly distributed in the limited space which the Galaxy occupies: if there were no absorption, the extent of the Milky Way in its equatorial plane, would only be three and a half times larger than its total thickness *h* from one pole to the other. This conclusion completely contradicts everything we know about the structure of the Milky Way.

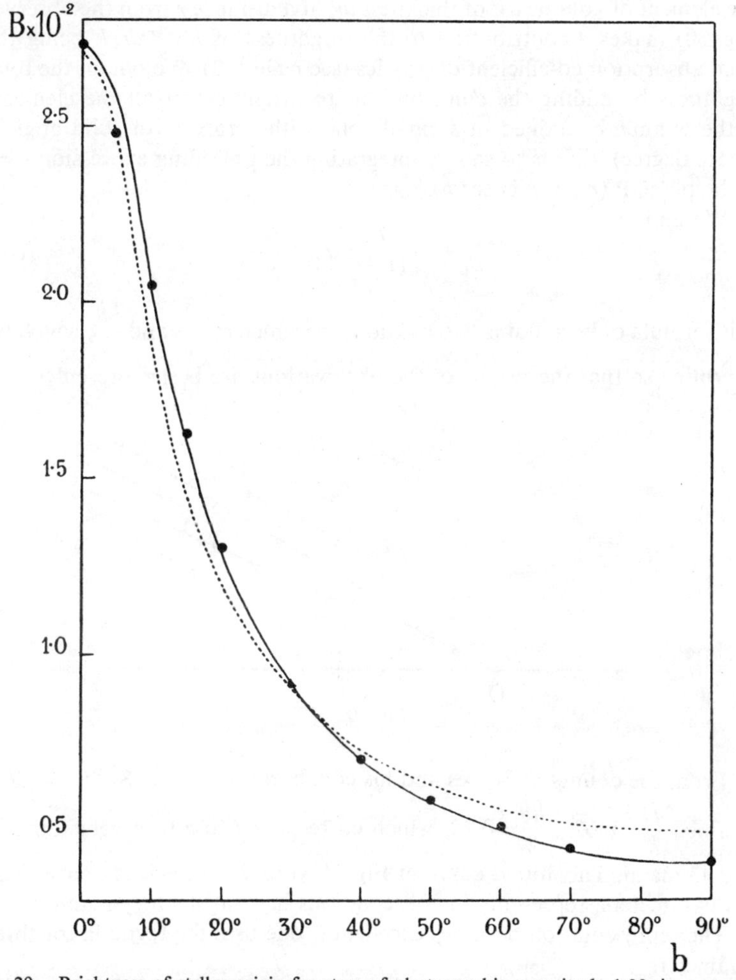

Fig. 29.—Brightness of stellar origin for stars of photographic magnitude 1·00, in a square degree as a function of Galactic latitude b. The points represent the values derived from the curve e(m), the broken line corresponds to the equation (X.3) containing two parameters. The full line corresponds to the equation (X.4) containing three parameters.

Thus, the hypothesis of absorption is forced upon us once more. Let us return to Fig. 26 and to simplify it as much as possible, let us suppose that the stars are uniformly distributed between the two planes P and P′ which bound the absorbing layer. (There are certainly stars above *P* and below *P′* but we do not take them into account.) It is easy to calculate what the brightness of stellar origin would then be. Let us designate by J_0 the luminous intensity of unit volume (constant on our hypothesis). The element of volume dV of thickness dr, at a distance r from the observer (Fig. 30) makes a contribution to the brightness of $J_0 e^{-kr}\,dr$, k being the usual absorption co-efficient of physics (see page 152). We obtain the total brightness by adding the contributions to brightness of all the elements of the volume contained in a small cone with vertex S (of solid angle 1 square degree), that is to say by integrating the preceding expression $r=0$ to the plane P ($r=\frac{1}{2}.\ h$ cosec b).

We find

$$B_S=\frac{J_0}{k}\left(1-e^{-\frac{kh}{2}\operatorname{cosec} b}\right) \tag{X.4}$$

This formula only contains two unknown parameters: $\frac{J_0}{k}$ and $\frac{kh}{2}$, which we determine so that the results of the observations are best represented.

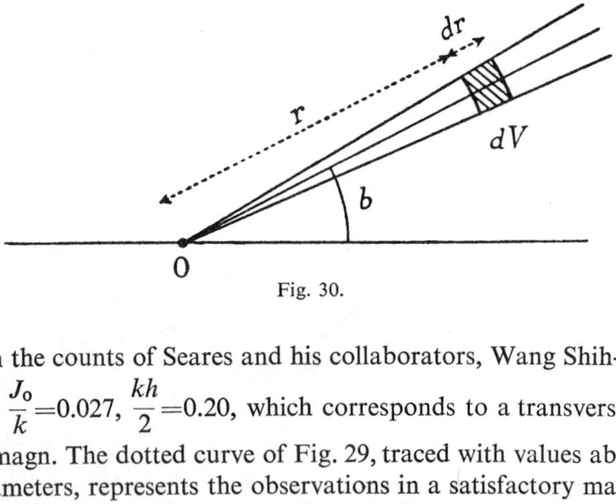

Fig. 30.

From the counts of Seares and his collaborators, Wang Shih-Ky [365] obtained $\frac{J_0}{k}=0.027$, $\frac{kh}{2}=0.20$, which corresponds to a transverse opacity $t=0.43$ magn. The dotted curve of Fig. 29, traced with values above these two parameters, represents the observations in a satisfactory manner.

The representation becomes excellent if one uses the formula for three parameters

$$B_S=\frac{J_0}{k}\left(H-e^{-\frac{kh}{2}\operatorname{cosec} b}\right). \tag{X.5}$$

Thus the full curve of Fig. 29 corresponds to

$$\frac{J_0}{k}=0{\cdot}030,\ H=0{\cdot}914,\ \frac{kh}{2}=0{\cdot}25,$$

whence $t=0{\cdot}54$ magn. (Dufay and Smoukovich [76]). The introduction of the parameter H, a little below unity may become necessary either because of imperfections in observation or more often, because of the inadequacy of the simple picture adopted for the Galaxy.

Transverse differential absorption

By measuring the colour excesses of the stars outside the absorbing layer at various Galactic latitudes, we can also apply the scheme of Fig. 26 to the determination of *transverse differential absorption* (difference of transverse absorptions corresponding to two groups of wavelengths). From globular clusters, Van de Kamp [348] obtained a difference of the order of 0·07 to 0·09 magn. between photographic and visual transverse absorptions, Vissotsky and Miss Williams [357] a difference of 0·11 magn., whereas Stebbins and Whitford [300] found, between 4340 and 4670 A, transverse differential absorption of 0·047 magn. In Table XI these various values have been reduced to the customary scale of Stebbins by means of the absorption curve. More recently, Stebbins and Whitford (1937) obtained, on this scale (colour excess E_1), 0.034 magn. from extragalactic nebulae and 0·026 magn. from globular clusters [301].[5]

We can consider the stars whose distance z from the Galactic plane is very large, as being outside the absorbing layer. Thus, Stebbins and Whitford [301] found a very weak differential absorption (0·014) magn. in the B stars with z above 100 pc., but the discussion of the same data led Oort [221] to a value which was much larger (0·064 magn.), while Mme Martel [186], with more numerous data, obtained an intermediate value (0·030 magn.). She also established that the differential transverse absorption varied with longitude of the stars and was particularly strong in the direction of the centre of the Milky Way. Here perhaps we have a partial explanation of certain discrepancies.[6] But their main cause must be looked for elsewhere; differential transverse absorption is very small and, in consequence, the least uncertainty in the "normal" colour indices which are used to evaluate the colour excesses, can seriously affect the results. Oort thought it necessary to increase the colour excesses of Stebbins by 0·045 magn. and he therefore finds a greater opacity.

Very different methods have given sufficiently compatible values for the photographic transverse opacity, which appears to be of the order of 0·4 to 0·5 magn. The differential opacity in the usual Stebbins scale ought to be 8·2 times smaller (see page 179), approximately 0·05 to 0·06 magn.; this latter value is sufficiently close to that found by Oort.

TABLE XI

Transverse Opacity of the Absorbing Layer

Transverse Photographic Opacity t_g			Transverse Differential Opacity t_o		
Method	Author	t_g	Method	Author	t_o
		magn.			magn.
Counts of Galaxies			*Globular clusters*	Van de Kamp (1930) [348]	0·032 to 0·041
Harvard	Van de Kamp (1932) [349]	0·72			
Lick	Mineur (1937) [203]	0·45		Vissotzky-Williams (1933) [357]	0·051
Mount Wilson	Hubble (1934) [150]	0·50		Stebbins-Whitford (1933) [300]	0·076
. . . .	Mineur (1937) [203]	0·51		. . . . (1937) [301]	0·036
. . . .	Oort (1937) [221]	0·57			
Distribution of the Globular Clusters	Van de Kamp (1932) [349]	0.40	*External Galaxy*	. . . . (1937)	0·028
Counts of stars			*B stars*	. . . . (1937)	0·014
Mount Wilson	Wang (1936) [365]	0·43		Oort (1937) [221]	0·064
. . . .	Dufay and Smoukovich (1939) [76]	0·54		Martel (1949) [186]	0·030
. . . .	Henyey and Greenstein (1940) [133]	0·33			

PLATE XVIII.—Region of the Galactic centre, east of 45 Ophiuchi, photographed in the near infra-red. 45 Ophiuchi is the brightest star (top right). 80 cm. telescope of the Haute-Provence Observatory (J. Dufay).

PLATE XIX.—Region of the Galactic center, east of 45 Ophiuchi, photographed in blue light. 45 Ophiuchi is the brightest star (top right). 80 cm. telescope of the Haute-Provence Observatory (J. Dufay).

ζ Ori.

ζ Ori.

Plate XX.—Bright and dark nebulae near ζ Orionis photographed in the light of the H α line. Schmidt telescope of the Haute-Provence Observatory (R. Mévolhon).

Thickness of the absorbing layer

By dividing the transverse opacity by the mean absorption coefficient, we obtain what we may call the *reduced thickness* of the absorbing layer. With t_g=0·4 or 0·5 magn. and a_g=1·1 magn./kpc., we obtain 365 to 455 pc.

The method used by Oort [221] permits the determination of both the differential transverse opacity and the thickness of the layer. Let E be the colour excess of a star situated inside the absorbing layer at distance z above the Galactic plane. Perpendicularly to the latter, the opacity, for a distance z, is evidently $E \sin b$. This idea leads to the construction of a graph of $E \sin b$ against $z=r \sin b$. Oort did this with the colour excesses of Stebbins, increased by 0·045 magn., by grouping the stars in accordance with their mean distance r, determined by means of the Galactic rotation. The curve obtained increases rapidly at first from the origin, then the product $\overline{E \sin b}$ increases less and less quickly and tends towards a limit (0·032 magn.) which, in practice, is reached at z=100 pc. The thickness of the absorbing layer, supposed to be homogeneous, would be of the order of 200 pc. In the same way, Mme Martel [187] obtained h=100 pc.[7]

Being some hundreds of parsecs in thickness only, the absorbing layer is really very thin, compared with its extension in the Galactic plane. But we ought not to loose sight of the fact that because the absorbing medium is certainly heterogeneous, the figures obtained by means of a vastly over-simplified scheme both for the transverse opacity and for the thickness, can only be considered as giving an extremely rough order of magnitude.

NOTES TO CHAPTER X

[1]If we suppose the Sun to be at a distance z above the Galactic plane, the nebulae of the Northern Galactic hemisphere give the relation

$$\mu=\mu_0+a\left(\frac{h}{2}-z\right)\operatorname{cosec} b$$

and those of the Southern Galactic hemisphere

$$\mu=\mu_0+a\left(\frac{h}{2}+z\right)\operatorname{cosec} b$$

Two straight lines of different slope are obtained. The sum of the slopes again measures the product ah, their difference the product $2az$, whence we obtain z, if a is known.

[2]The nebulae of magnitude less than m, at latitude b, are at a maximum distance r, so that

$$5 \log r=m+5-M-A\,(b),$$

where $A(b)$ is the absorption in magnitude, at latitude b. If there are n nebulae per unit volume (n=constant) a sphere of radius r would contain $\frac{4\pi r^3 n}{3}$ nebulae, and we should have, per steradian

$$N_m(b)=\frac{nr^3}{3} \text{ nebulae.}$$

Taking logarithms of this number, and replacing log r by the value found above

$$\log N_m(b)=\log n-\log 3+3\log r$$
$$=\log n+3-\log 3+0.6\,(m-M)-0.6\,A(b).$$

The limiting magnitude m being fixed, and the absolute magnitude M being constant, this relation becomes

$$\log N_m(b)=\log N_0-0.6\,A(b).$$

Using Fig. 26, $A(b)=0.5\,t \operatorname{cosec} b$, which leads to equation (X.2).

Conversely if, at a given Galactic latitude, we vary the limiting magnitude m (for instance by longer exposure) we have

$$\log N_{\mathrm{m}}(b) = \text{Constant} + 0.6\, m. \tag{X.3}$$

This provides a means of checking the supplementary hypothesis of uniform spatial distribution of nebulae.

[3]By considering each of the Galactic hemispheres separately, a somewhat stronger absorption has often been found in the Southern hemisphere, which would seem to place the Sun a little above the Galactic plane.

[4]The co-ordinates of each cluster in a system of axes parallel to the first but centred on the Sun, have been evaluated from the true distance r. From this the co-ordinates of the centre of the system are obtained, and then the co-ordinates of each cluster in the first system of axes.

[5]Since the extragalactic nebulae on the whole have the colour of the G stars, we should really take as equivalent wave-lengths 0.4260 and 0.4770μ, in place of 0.4215 and 0.4740μ. To pass to the usual wave-lengths it is necessary, as is shown by the absorption curve, to multiply the colour excesses by 1.07. This has been done in Table XI.

Eigenson [87] also tried to prove the existence of differential transverse absorption by comparing the total photographic magnitudes of the extragalactic nebulae of Shapley–Ames with visual magnitudes of Holeschek. But the data are very uncertain, and the value found ($\sim$0·2 magn.) seems too large.

[6]Of the sitxy-nine globular clusters used by Stebbins and Whitford in their first work, Mme. Martel noted fifty-seven were either in the direction of the Galactic centre, or in that of the bulges of the zone of absence, whereas the twenty-two clusters used in their second work were outside these bulges. The opacity found is thus greater in the first case.

[7]Van de Kamp (1930) had already found 300 pc. (the average of numbers between 210 and 500 pc.). Oort represented the curve $\overline{E \sin b} = f\,\overline{(r \sin b)}$ by supposing that the absorbing matter was concentrated in the Galactic plane in such a way that the absorption co-efficient was proportional to

$$e^{-8\cdot 28.10^{-5} z^2}.$$

Saturation is then reached at $z = 200$ pc.

CHAPTER XI

DIFFUSE NEBULAE AND THE GENERAL SCATTERING OF LIGHT IN THE MILKY WAY

1. Diffuse nebulae

Dark nebulae and diffuse nebulae

We have already seen that, if the dark nebulae seem to absorb the light of the stars, it is probably because they scatter it in all directions (Chapter VI, page 129). The illumination which they receive from all the stars must give them a certain luminosity and if they appear dark, it is only because they are far from all bright stars.[1]

If, however, the nebula is strongly illuminated by a very bright star which is sufficiently close, its brightness may become greater than that of the background of the sky. We then observe an *irregular* continuous-spectrum nebula (page 41). We know already that these nebulae are associated with stars of a spectrum more advanced than that of BO (page 57). In 1912, Slipher managed to photograph the spectra of some of them and he found the same dark lines in them as in the spectrum of the associated stars.

This proves that the light which we receive from nebulae comes originally from the stars. Each time that a sufficiently pure spectrum can be obtained, the presence of stellar absorption lines permits one to distinguish with certainty diffuse nebulae from those whose continuous spectrum is essentially of another origin (page 83).

Thus there is most probably no difference in the nature of dark nebulae and of diffuse nebulae. The proximity of a bright star is probably sufficient for the transformation of the one into the other—just as a beam from a projector changes a dark cloud into a luminous fog.

Are diffuse nebulae permanently and physically associated with the stars which illuminate them, or is their association rather a fortuitous and, in consequence, a temporary one? Such is the first question which poses itself in the case of bright-line nebulae. It was resolved in a most elegant manner by Ambarzumian and Gordeladse [8]. For a cloud to acquire sufficient brightness to become visible, the illumination which it receives must be larger than a certain threshold l_0. In this sense we can say that each star illuminates a sphere whose fixed radius R_0 is proportional to the square root of the luminous intensity J of the star $\left(R_0=\sqrt{\frac{J}{l_0}}\right)$. We have a fair knowledge of the average number of stars of each spectral class per unit volume; we also know approximately the law of distribution of luminous intensities ("luminosity function") for each spectral class. If we take a certain volume V, inside the Milky Way, it is possible to evaluate the fraction of volume V illuminated by all the stars of the B* class, or even all the A, F, G, K or M stars. On the hypothesis of a fortuitous association, this fraction represents the probability of a cloud being illuminated by a star of a particular class.

Now observations show that the number of diffuse nebulae illuminated by stars of different spectral classes is almost proportional to the probabilities calculated in this manner. We can, therefore, conclude that *the association of stars and diffuse nebulae is purely fortuitous.*

The statistics of Ambarzumian and Gordeladse have a further interesting consequence: all the stars of all the spectral classes illuminate, on the whole, a fraction of volume of the order of 1/2000. This is thus the probability of a dark cloud becoming transformed into a diffuse nebula due to the vicinity of a star.

The study of light scattered by nebulae is fortunately complementary to the study of transmitted light, and can contribute to the solution of the essential problem: the determination of the nature and dimensions of the scattering particles. The properties of the light which we receive from different points of the apparent surface of the nebula, however, depend to a large extent also on the shape of the nebula and on the position of the illuminating star with respect to it. These questions have been dealt with in great detail from a theoretical point of view, by Henyey [132] and by Schalén [258]. In some particular cases these authors managed to tell from observations whether the star was situated in front of, behind or even inside the nebular mass.

The main difficulty encountered in the observation is due to the small brightness of the diffuse nebulae. Their light is, amongst other things, distributed over a continuous spectrum instead of being concentrated in a limited number of bright lines. Photography of their spectra, therefore, requires very long exposure, even with spectrographs with relatively large apertures. Progress has been very much slower than in the case of emission-line nebulae. In spite of the interesting work of the last fifteen years, notably by Struve and his collaborators, the subject is still far from being solved.

*With the exception of the B_0 stars around which only bright-line nebulae can be observed.

Colour of scattered light

It will be particularly instructive to compare the energy distribution in the spectrum of the nebula and in that of the star which illuminates it. If we represent by $\mu(\tilde{\nu})$ the monochromatic magnitude of a particular area of the nebula (for instance 1 square minute), for radiation of wave number $\tilde{\nu}$, and by $m(\tilde{\nu})$ the magnitude of the star for the same wave-length, measurements of the energy distributions would lead to the graph

$$\mu(\tilde{\nu})-m(\tilde{\nu})=\phi(\tilde{\nu}).$$

Since we lack such spectrophotometric comparisons, which are difficult and still rarely carried out, we must content ourselves with the measurement of the difference in magnitude $\mu-m$ for two groups of wave-lengths, isolated by means of colour filters. The difference of the two differences in magnitude

$$E=(\mu_1-m_1)-(\mu_2 m_2)$$

represents the *colour excess* of the nebula relative to the star for the average wave-lengths λ_1 and λ_2 isolated by the filters ($\lambda_1<\lambda_2$). Naturally this is also the difference between the colour index $C'=\mu_1-\mu_2$ of the nebula and that of the star $C=m_1-m_2$:

$$E=C'-C.$$

Photographic and photovisual magnitudes, or even better, photo-electric colour indices can be employed. The brightness of the nebulae, however, is of the same order of magnitude and often smaller than the general brightness of the night sky which is superimposed on them. The "sky correction" is thus most important, and very often it is this which limits the accuracy of measurements.

All observers agree that, *in general, the colour of the nebulae differs little from that of the illuminating star*. A particular striking example is given by the nebula near the red giant Antares (α Scorpionis, spectrum cMO). Hardly visible in blue light, this nebulosity is very clear on photographs obtained in yellow light. Its colour index, of about $+2$, is almost the same as that of Antares (photographic magnitudes minus photovisual magnitudes). On the other hand, the nebulae associated with the B stars of the same region of the sky (σ, ν, 22 Scorpionis, B D$-24°$ 12684, ρ Ophiuchi) have, as do the B stars, a negative colour index (Struve, Elvey and Roach, 1936 [326]).

The nebulosities in the Pleiades have been especially studied. For those which surround Maïa, Struve, Elvey and Keenan [325] have found $E=-0.3$ magn. (±0.2 magn.) between $\lambda_1=3900$ and $\lambda_2=5000$ A by photographic spectrophotometry, and $E=0{\cdot}1$ magn., between 4250 and 4750 A by photo-electric photometry. The nebulosity associated with Merope, studied by Greenstein and Henyey [113] by spectrophotometry gave the somewhat uncertain value $E=-0{\cdot}1$ magn., whereas the photographic and photovisual measurements made by the method of Ch. Fabry led to the preferable value $E=-0{\cdot}35$ magn. ($\pm 0{\cdot}18$) (both these are expressed in the international colour index system). It is true that more recently Schalén (1948) found a nebula somewhat redder than Merope but, as the author himself suggests, it is possible that its colour excess is due to a systematic error.

Keenan [165] also found the nebula NGC 7023 a little redder than the illuminating star, but to the contrary, spectrophotometric measurements by Greenstein [111] showed that it is a little bluer. From Collins's [59] photographic and photovisual measurements of fifty nebulae, Greenstein [108] derived the mean value $E=-0{\cdot}19$ magn. ($\pm 0{\cdot}06$).

In general, thus, *scattered light is probably a little bluer than incident light*. But attention must be drawn to the fact that colour excess can vary from point to point in the nebula, according to the angular distance to the star. Keenan confirmed this in the case of NGC 7023, photographed in both blue light and red light. The colour excess seemed to be at a maximum at 1′ from the star. In the case of the nebula associated with Merope, Schalén [259] clearly established a progressive increase in colour excess, up to approximately 13′ from the star. (It is possible that this is the maximum.)

Nevertheless, we are aware of some exceptional cases where the scattered light seems to be much bluer than that of the star. Greenstein [111] measured the *relative gradients* of three nebulae, compared with their associated stars, all three of which were of the variable type T Tauri, i.e. dwarf stars whose spectra showed numerous emission lines (see page 246). One of these nebulae, NGC 6729, itself variable, already is clearly bluer than the associated star R Corona Australis ($G=-0{\cdot}44$), the two others are much bluer still. In fact, Greenstein found $G=-1{\cdot}2$ for the nebula NGC 2261, associated with R Monocerotis and $G=-1{\cdot}4$ for that which accompanies the star B D $-6°$ 1253 (NGC 1999).

Polarization of scattered light

The light scattered by small particles is always more or less polarized and the degree of polarization can lead to the identification of the scattering particles, since it varies with their nature and above all, with their dimensions.

Let us recall that light, considered as a transverse electro-magnetic vibration, is said to be linearly polarized when the vibrations take place in a fixed direction. We can produce this by means of an *analyser*, which only passes vibrations in a fixed direction. By turning this analyser about the axis of a light beam, the intensity of the transmitted beam is at a maximum for a certain orientation; the effective vibration is then parallel to the privileged direction of the analyser. For an orientation perpendicular to the preceding one, the light is completely extinguished. A conveniently inclined glass mirror, a Nicol prism or a polaroid plate can be used as analysers.

The rotation of the analyser cannot vary the intensity of so-called "natural" light, such as that from an incandescent filament, from the Sun, or from most stars. We imagine this "natural" light as constituted of linear transverse vibrations, whose orientation changes so irregularly at each instance, that all the possible orientations occur in a very short time. We can thus consider it as being made up of two vibrations, of equal amplitude at right angles to each other, but having no phase relationship.

Light which is only *partially linearly polarized* can also be thought of as two incoherent vibrations at right angles to each other, but of different intensities I and i $(I>i)$. The rotation of the analyser varies the intensity of the beam, without ever completely extinguishing it. The intensity is at a maximum when the privileged direction of the analyser is parallel to the vibration I, minimum when the privileged direction is parallel to i. We can characterize the degree of polarization by the ratio

$$\rho=\frac{i}{I} \tag{XI.1}$$

called the "*depolarization*". This is a number between 0 and 1. The first value corresponds to complete polarization, the second to natural light. It is often preferable to make use of the *proportion of polarized light*, defined by

$$p=\frac{I-i}{I+i}=\frac{1-\rho}{1+\rho}. \tag{XI.2}$$

This is zero in the case of natural light, and unity in the case of total polarization. We also express it as a "percentage" of polarized light P, by multiplying it by 100 $(P=100\,p)$.

Observations on the polarization of the light scattered by nebulae are still small in number. Most of them have been made by photographing the nebula through a polaroid plate placed in front of the photographic plate. Let us imagine that we could produce, on the same plate, a great number of exposures of equal length, but with different orientations of the polaroid. In a fixed point of the image of the nebula the blackening would be a maximum when the direction of the vibrations transmitted by the analyser coincides with that of the privileged vibration I, and a minimum for an orientation of 90° to the latter. If we plot a curve of the blackening of the plate against illumination, the ratio $\rho=\frac{i}{I}$ between minimum and maximum illuminations can be measured. For each point of the nebula, the direction of the privileged vibration and the proportion of polarized light can then be determined.

In this form, the method is evidently impracticable, because we know that photography of diffuse nebulae requires long exposures. From one exposure to the next, the height of the nebulae above the horizon will vary and atmospheric absorption would change. But it may be reasonably supposed that, because of symmetry, polarization can only be either *radial* (vibration *I* perpendicular to the radius vector joining the point studied to the illuminating star), or *tangential* (vibration *I* along the radius vector). The number of exposures necessary can thus be reduced to two, between which the polaroid is turned through 90°. It is in this way that Whitney and Weston [376] studied the polarization of the NGC 6729. Nevertheless, it may still be feared that atmospheric absorption (or emission by the night sky) could alter between the two exposures, and Henyey [131] devised a procedure which obviated this cause of error. This, however, only gives the average degree of polarization of two points of the nebulae equidistant from the star, and situated on two axes at right angles. Finally, Gliese and Walter [102] used negatives taken on successive nights with the polaroid having nine different orientations. Each plate carried a photometric scale, given by a sequence of stars of known magnitudes which surround the nebula. Variations of absorption affected both the images of the stars and that of the nebula equally, and were thus automatically corrected. A somewhat complicated method of reduction permitted them to evaluate the degree of polarization (supposed radial or tangential) of many points of the nebulae, by comparing these two at a time.

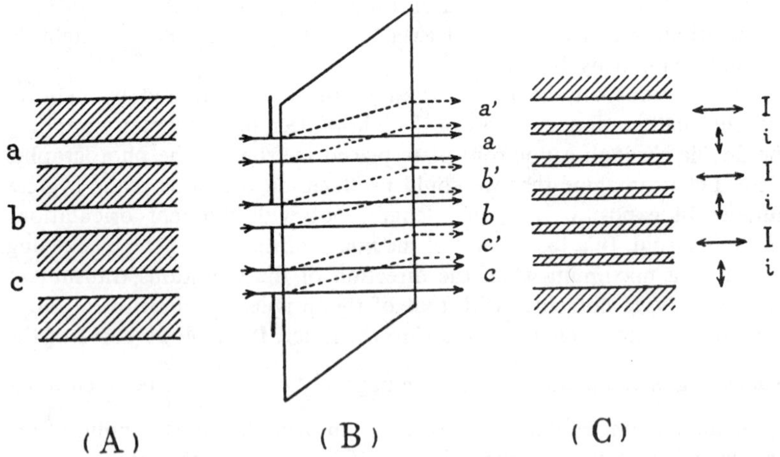

Fig. 31.—Diagram of the Öhman polarimeter. A—grid; B—path of the rays in the rhomb of spar; C—images on the plate.

Other devices can be used with advantage: e.g. that which Öhman [218] devised for the study of the polarization of comets and of the Nebula in Andromeda (see page 307), or the polariscope which Lyot and Dollfus [182] used in their work on light scattered by smoke. The first consists of a grille of identical opaque bars, separated by transparent intervals equal in size to the bars (Fig. 31 A)[2] and followed by a rhomb of birefringent spar. The *ordinary* rays, which traverse the crystal without deviation, are completely polarized. The *extraordinary* rays, displaced laterally, are also polarized, but at 90° to the former. The thickness of the rhomb is chosen so that the displacement is half the opening between the bars. It is, for instance, so arranged that the ordinary image contains the vibration i, and the extraordinary image the vibration I (Fig. 3 B and C). Each transparent interval, such as a, gives two contiguous images whose illuminations are respectively proportional to the intensities i and I. The measurement of the blackening of the images permits the evaluation of the ratio $\rho=\frac{i}{I}$ if one constructs a photographic scaling curve of the plate.

In the polariscope of Lyot and Dollfus, the opaque bars are replaced by half-wave cellophane plates at 45° to the optic axis. They turn the direction of a linearly polarized ray through 90°. In this manner the illumination is strongly increased; it would be doubled in the case of natural light. Mme Martel used these two types of polariscopes to study a great number of diffuse nebulae. The Öhman grid, or the cellophane grid, placed in the focal plane of a telescope is followed by a rhomb of spar and sometimes by a colour filter. Finally, an objective of short focus (50 mm.) gives an image of the nebula and of the grid on these plates.

Results of the observations

With a polaroid, Henyey [131] failed to discover appreciable polarization either in the case of the nebula near γ Cygnus,* in the Pleiades, or in the nebulosities attached to Maïa and Merope. A filament near Electra only showed a trace of polarization, as did also the nebula near ρ Ophiuchi. Walter [363] found 11 per cent of the light polarized in the nebula surrounding FU Orionis. In NGC 6729, Whitney and Weston [376] measured a polarization which varied from point to point and attained 17 per cent in places. Their results show that the diffuse nebula principally scatters the light of the star R Coronae Australis, and also appreciably so that of T Coronae.

NGC 7023, one of the brightest diffuse nebulae of very complex structure and having a spectrum which shows the emission line Hα, together with some forbidden lines of ionized iron, has been studied on

* It is really a bright-line nebula.

more than one occasion. Henyey [131] found an average polarization of 12 per cent in it. With a polariscope of the Lyot-Dollfus type, mounted on the 120-cm. telescope of the Haute Provence Observatory, Mme Martel [188] found that the polarization was radial. The proportion of polarized light varies rapidly from one point to the next and locally attains 25 per cent and 30 per cent. Gliese and Walter [102] also measured the polarization degree at many points by the method described above, and they found values which range between 0 and 56 per cent. Since they used a short-focus instrument (95 cm.), the images were small and their measurements were almost exclusively concerned with the most extreme regions of the nebula. On the other hand, those of Mme Martel, with a telescope of focal length 6·4 m., were only made on the inner region near the illuminating star. There are only three points common to the two series, but for these the numerical agreement is excellent.

However, Weston [438] stated definitely that he did not find the strong degrees of polarization indicated by Gliese and Walter in other points. Two exposures with the telescope of 2m. diameter of the McDonald Observatory, probably with a polaroid, have given proportions of polarized light which never exceeded 30 per cent, and are generally between 20 per cent and −5 per cent.[3] These results seem to agree with those of Mme Martel.

When the brightness of the nebulae is very small, the measurements must be corrected for the general light of the night sky superimposed on that of the nebula. These corrections are often very great and the interpretation of the results is very difficult. This, for instance, is the case with the group of nebulae IC 4601 a, IC 4601 b and BD −19° 4357, studied by Mme Martel [415] with a polariscope of the Öhman type mounted in the focus of the 80-cm. Cassegrain telescope of the Haute Provence Observatory. The proper brightness of these nebulae is hardly half that of the surrounding night sky. The proportions of polarized light found are, therefore, very uncertain, but in all cases less than those measured in NGC 7023.

2. General scattering of light in the Milky Way

All the dark clouds in the neighbourhood of the Galactic plane probably also scatter part of the light they receive from the stars. The "absorbing layer", in which the Sun is embedded, ought thus to have a certain brightness which contributes to the general illumination of the sky.

The "light of the night sky" is very complex. It contains lines and bright bands emitted in the upper terrestrial atmosphere and a continuous spectrum which shows the principal absorption lines of the Solar spectrum: Hβ, Hγ, Hδ, the lines H and K of Ca II, the band G, towards 4300 A

(Lord Rayleigh, 1923; J. Dufay, 1923). The origin of the continuous spectrum itself is not simple: the weak stars which are, in the main, of a spectral type similar to that of the Sun, naturally contribute in a very marked way. But the study of polarization reveals, on the other hand, that the light of the sky also contains light from the Sun scattered outside the terrestrial atmosphere (Dufay, 1925). This is an extension of *zodiacal light* over the entire sky, visible along the ecliptic up to more than 90° from the Sun. The total obtained by adding the light of zodiacal origin to that of the weak stars, nevertheless, appears too small to account for the continuous spectrum which is observed. Today it seems established that a part of the continuous spectrum is emitted in the upper terrestrial atmosphere (Barbier, Dufay and Williams, 1951). But the light scattered by the absorbing Galactic layer could also be invoked to account for this deficit.

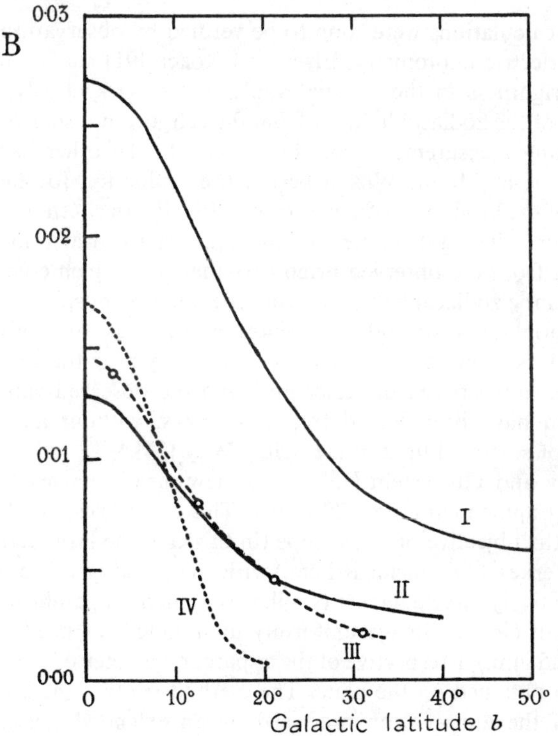

Fig. 32.—I—Brightness of stellar origin as a function of Galactic latitude (see Fig. 29). II—Brightness of the scattered light calculated by Wang. III—Brightness of the scattered light from the observations of Elvey and Roach. IV—From the observations of Henyey and Greenstein (average of Cygnus and Taurus).

A rough calculation made by Otto Struve [320] and completed by Dufay [71] for the case of multiple scattering (which is not negligible in a very extensive medium), has shown that the light scattered in the Galaxy can impart an almost measurable brightness to the sky. Wang Shih-Ky [364, 365] has given numerical solutions of the somewhat complicated integral equation occurring in the theory. The curve *I* of Fig. 32 represents, as a function of Galactic latitude, the distribution of brightness due to faint stars according to the star counts made at Mount Wilson (Seares, Van Rhijn, Joyner and Richmond, 1925). The brightnesses therein are expressed by the number of stars of magnitude 1·0 to which 1 square degree of the sky is equivalent. Below, the curve II represents the brightness which we ought to attribute to the scattering layer, in accordance with the calculations of Wang Shih-Ky; one can see that on the whole, it would be equivalent to a little more than 4/10 of the brightness coming from the stars.

These calculations were soon to be verified by observation. By means of photo-electric photometry, Elvey and Roach [91] studied the distribution of brightness in the celestial vault. Their work clearly showed the extension of the zodiacal light far from the ecliptic, and so gives the means of correcting measurements for this effect. On the other hand, the star counts made at Mount Wilson permit the evaluation for each Galactic latitude, of the brightness of stellar origin. Finally, the authors tried to take into account the light emitted in the upper terrestrial atmosphere. By separating from the observed brightness that part which comes from the stars, from the zodiacal light, and from the upper atmosphere, we obtain a residual brightness strongly concentrated towards the lower Galactic latitudes. It is shown by curve III in Fig. 32, very near the curve of Wang. The agreement between the calculated and the observed values is better than could have been hoped for, and shows, without any doubt, the existence of scattered light in the Milky Way [74].[4]

Henyey and Greenstein [133] made new measurements by means of the photographic method of Ch. Fabry. This consists essentially in photographing the objective of a telescope (in this case the large refracting telescope at Yerkes, of diameter 101 cm.) with an optical system of short focus, and of relatively large aperture (similar to a microscope-objective). In this way a small circular image uniformly illuminated, of fixed diameter (in this case 0.5 mm.), irrespective of the apparent diameter of the source being studied, is obtained on the plate. The method permits the accurate comparison of the illumination produced by an extended source and by a point source. In the first case, a diaphragm placed in the focal plane of the large objective limits the area of the source, and in the second case, a still smaller diaphragm is used to isolate the light of the comparison star.

In this way, Henyey and Greenstein measured the brightness of numerous small areas of diameter of only 126″ which did not contain stars brighter than the sixteenth magnitude. The importance of light of stellar origin is thus strongly reduced. They traced, point by point, two transverse sections of the Milky Way, one in Cygnus (at Galactic longitude $l=40°$), the other in Taurus and in Auriga (at $l=140°$). In Cygnus, the brightness in the Galactic plane due to scattering reaches that of 0·024 stars of magnitude 1.0 per square degree. It is strongly concentrated near this plane and appears to be almost negligible at 15° or 20° of latitude, north or south. In Taurus, the maximum is more than twice as small, very flattened, and clearly unsymmetrical. Scattering extends rather further to the south of the Galactic plane, where we know that there exist large absorbing clouds. In Fig. 32, curve IV represents the average of the results for Cygnus and for Taurus.

The curve is steeper than that of Wang Shih-Ky who, in order to make his calculations, had made two over-simplified assumptions: (1) the particles scatter the light falling on them without any true absorption; (2) the intensity of diffused light is independent of the angle between the incident ray and the scattered ray. Henyey and Greenstein took up these calculations in a more general and quite different manner, by introducing the *albedo*[5] of the scattering particles and by assuming that the intensity of the diffused light varies with the angle of scattering. They found that the curve deduced from the observations requires the particles to scatter much of the light *towards the front* (about the direction of propagation of the incident light) and that they have a very large albedo (at least 1/3). These indications could contribute to the identification of the scattering particles.[6]

NOTES TO CHAPTER XI

[1]Struve and Elvey [323] compared, by means of the photographic method of Ch. Fabry, the brightness of small areas where there are no stars brighter than the sixteenth magnitude chosen either inside certain dark nebulae of Barnard, or outside them but nearby. Sometimes positive, and sometimes negative, the differences are of the order of some hundredths of magnitudes.

[2]In practice the opaque bars of the Öhman polariscope are a little wider than the transparent intervals, as is shown in Fig. 31. The cellophane plates of the Lyot-Dollfuss polariscope must, on the other hand, have strictly equal width.

[3]The polarizations is considered to be negative when the privileged vibration I is radial instead of being perpendicular to the radius vector. Mme Martel found very weak negative polarization at certain points of NGC 7023.

[4]The light scattered in the distant regions of the Milky Way explains the poor contrast between the relatively near dark nebulae, and the background of the sky. The latter regions scatter almost as much light as the absorbing clouds. The scattered light also leads to a decrease in contrast between the bright Galactic regions and the large obscuring clouds.

[5]This is by definition the ratio between the total flux scattered by a particle and the incident flux when the particle is placed in a beam of parallel light.

[6]In Henyey and Greenstein's calculations, as also in those of Wang, use is made of the transverse opacity of the Milky Way determined by means of star counts. Wang has taken $t_g=0.43$ magn. from the counts of Seares; and Henyey and Greenstein have taken $t_g=0·33$ magn. from those of Van Rhijn (Table XI, page 192).

CHAPTER XII

NATURE AND DIMENSIONS OF THE SCATTERING PARTICLES

Statement of the problem

All the information which we have gathered on the properties of transmitted light and on those of scattered light now enables us to tackle the fundamental problem, the study of which has so far been deferred: *what are the scattering particles of space?* Observations would suggest that the dark clouds and the diffuse nebulae contain identical particles; let us suppose this to be the case, and let us reserve the right to discard this hypothesis later on if it proves untenable in practice.

Let us, therefore, very rapidly review the essentials of the known facts:

(1) Absorption increases from the infra-red to the ultra-violet, in accordance with Stebbins and Whitford's law represented as a function of the wave number $\tilde{\nu}=\frac{1}{\lambda}$, by an S curve (Fig. 25, page 175).

(2) By increasing the ordinates of this curve by about 0·17 magn. we obtain, for each wave-length, an adequate value of the mean absorption coefficient in the neighbourhood of the Galactic plane (page 175), in magnitudes per kiloparsecs.

(3) Scattered light is, in general, a little bluer than incident light (colour excesses $\simeq -0{\cdot}2$ to $-0{\cdot}3$ magn.) (page 198).

(4) It is partially polarized and the privileged vibrations are perpendicular to the plane of scattering (containing the incident and scattered rays) (page 199).

(5) It would finally appear that the particles have a rather large *albedo* and that they scatter the light "towards the front" (page 205).

We shall now compare these findings with the laws governing the scattering of light by various sorts of particles, beginning with the simplest cases.

1. Particles either smaller or larger than the wave-length

Very large particles are not involved

Particles which are very large in comparison with wave-length (such as the meteorites which streak across the Solar system), behave simply as so many opaque screens which present an obstacle to the passage of light and which diffract a part of it. Thus, they diminish the intensity of all radiation in the same ratio, and the colour of the transmitted light is, therefore, not modified. The same is the case with very large transparent particles.

The practically neutral absorption of very large particles is therefore incapable of producing the reddening of distant stars. Rather does the probable absence of all neutral absorption in space (page 179) show that *very large particles generally play a negligible part.*[1]

Let us now examine the polarization, colour, distribution and direction of scattered light. These depend very much on the optical properties of the particles. Polarization, probably very weak in the case of meteorites, could be total in certain directions for the case of transparent spheres. Opaque particles would give a maximum of scattering towards the rear, and not towards the front. The colour would be identical with that of the incident light for particles having no selective absorption. With meteorites, it would be appreciably redder, and this also is contrary to observation. The hypothesis of particles very large in comparison with wave-length must be definitely discarded.

Free electrons are not involved

Free electrons, under the action of an electro-magnetic wave, follow the oscillations of the electric field, with an amplitude which is independent of the frequency.[2] Scattered light and transmitted light thus have the identical colour of the incident light and *it is certain that free electrons cannot account for the observed phenomena.*

Nevertheless, it would be interesting to calculate the electron density which would be necessary to produce an average photographic absorption a=1·1 magn./kpc.

When a gas of electrons receives, in natural light, an illumination l, the luminous intensity $J(\theta)$, scattered by 1 c.c. of gas in the direction inclined at an angle θ to that of the incident light (Fig. 33), is given by the equation

$$\frac{J(\theta)}{l}=\frac{n_e\epsilon^4(1+\cos^2\theta)}{2m_e^2c^4}, \qquad \text{(XII.1)}$$

valid if the frequency is not too large (not in the case of X-rays). n_e is the number of electrons per c.c., ϵ is the charge of the electron ($4{\cdot}80.10^{-10}$ E.S.U.) and m_e is the mass ($9{\cdot}106.10^{-28}$ gm.) of each electron. As always, c represents the velocity of light. From (XII.1) we easily evaluate the luminous flux scattered throughout all space by 1 c.c. of the gas, and this immediately leads to the apparent absorption co-efficient, since the weakening of the light is only due to the scattering. We find:

$$k=\frac{16\pi J\left(\frac{\pi}{2}\right)}{3l}=\frac{8\pi n_e\epsilon^4}{3m_e^2c^4} \qquad \text{(XII.2)}$$

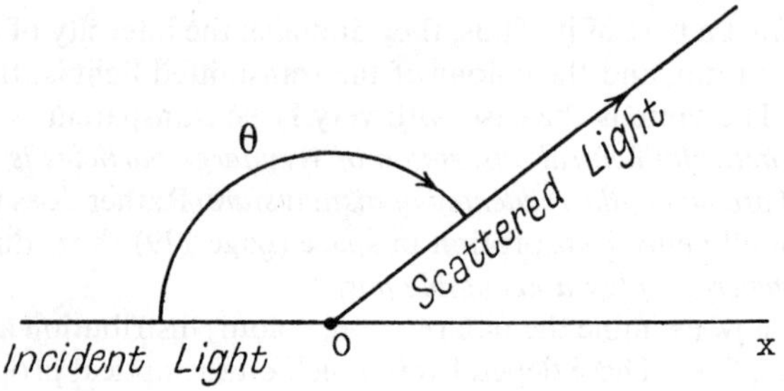

Fig. 33.—Definition of the angle of scattering θ.

This is naturally the usual absorption co-efficient of physics. If one expresses it in cm^{-1}, the absorption co-efficient of astronomy, in magn./kpc. (see page 152) is:

$$a=3{\cdot}34.10^{21}k. \qquad \text{(VII.3')}$$

For a to be 1·1 magn./kpc., the electron density must be

$$n_e=300\ cm^{-3}.$$

This is a density comparable to that which has been found in certain diffuse bright-line nebulae (page 82). According to Strömgren, inside the clouds of H I regions the density of the atoms is of the order of 12 per c.c. (page 82); it is perhaps a hundred times less on the outside. Short of supposing that space possesses an enormous negative electric charge (which seems extremely improbable), the density of the electrons ought to be almost the same as that of the positive ions, i.e. of the order of 1 electron per c.c. We can, therefore, take it as certain that there are not enough free electrons in space to account for the observed absorption.

Finally, let us examine the angular distribution and the polarization of light scattered by free electrons. By drawing about a point, in each direction, a line of length proportional to $J(\theta)$ (i.e. $1+\cos^2\theta$), we obtain a polar diagram of the scattering given by Fig. 34 A.

It is symmetrical about the incident direction; the intensity is a minimum at 90° to the incident beam and twice as large in the direction of propagation (either forwards or backwards).

From equation (XII.1), $J(\theta)$ is the sum of two terms: the first independent of θ corresponds to vibrations I, perpendicular to the plane of scattering, the second is a product of the first by $\cos^2\theta$, and corresponds to vibrations i in this plane. The depolarization is $\rho=\cos^2\theta$, the proportion of polarized light is $p=\dfrac{\sin^2\theta}{1+\cos^2\theta}$. At 90° to the incident rays, the light is completely polarized. It is totally depolarized in the forward and backward directions. For all other values of θ, polarization is only partial, the privileged vibration always being normal to the plane of scattering.

No atoms or molecules of gas are involved

The scattering of light by so-called "transparent" gases is very well known today. The theory of Lord Rayleigh, and later the work of Cabannes, which established the reality of the phenomenon, accounts for all laboratory observations. The Rayleigh theory also explains, quantitatively, the scattering and apparent absorption of Solar light in the upper atmosphere of the Earth, observed when the sky is clear.

In this case scattering and apparent absorption depend essentially on the wave-length of the incident light. For a monatomic gas on which there falls a beam of monochromatic natural light of illumination E, the luminous intensity $J(\theta)$ scattered by 1 c.c. of the gas is given by equation

$$\frac{J(\theta)}{l}=\frac{\pi^2\tilde{\nu}^4(\mu^2-1)(1+\cos^2\theta)}{2N}, \tag{XII.3}$$

in which μ represents the refractive index of the gas for radiation of wave number $\tilde{\nu}=\frac{1}{\lambda}$ and N the number of molecules (or atoms) per c.c.

The scattering indicatrix and the proportion of polarized light in each direction are exactly the same as in the case of free electrons.[3]

The apparent absorption co-efficient is:

$$k=\frac{16\pi J\left(\frac{\pi}{2}\right)}{l}=\frac{8\pi^3\tilde{\nu}^4(\mu^2-1)^2}{3N}. \tag{XII.4}$$

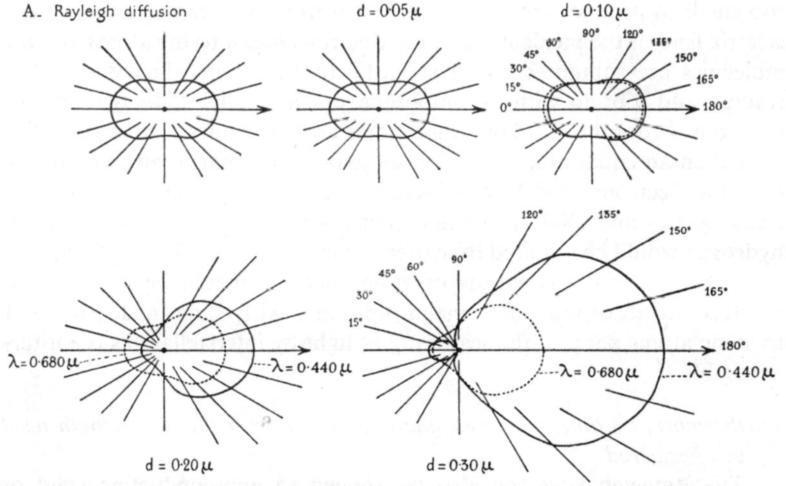

Fig. 34.—Scattering indicatrices of dielectric spheres very small compared with the wave-length (A), and for spheres of different diameters. The latter are calculated by the formulae of Schalén for wave-lengths 0.440μ and 0.680μ. The intensity of the light scattered at 90° has been made equal in the two cases.
Note the rapid increase in light scattered forward when the diameter of the particles increases from 0.10 to 0.30μ.

Equations (XII.3) and (XII.4) show that the *intensity of scattered light and the absorption co-efficient are very nearly strictly proportional to* $\tilde{\nu}^4$ $\left(\text{or } \frac{1}{\lambda^4}\right)$, since in the case of a transparent gas, the variation in refractive index μ with wave-length is very small. The wave-length $\lambda=0{\cdot}35\mu$ is scattered approximately sixteen times as much as the wave-length $\lambda=0{\cdot}70\mu$: the transmitted light is thus strongly reddened and the scattered light is much bluer than the incident light. Such is the origin of the blue colour in a clear sky, illuminated by the Sun.

It is almost impossible to explain the shape of the absorption curve of Stebbins and Whitford by means of atoms or molecules of a gas.[4] On the other hand the colour excesses of light scattered by a gas, compared with the light of the source which illuminates it, would be strongly negative. If we suppose that the photovisual and photographic measurements correspond respectively to wave-lengths 0·543 and 0·426 μ, the colour excess would be approximately

$$E=-2.5 \log \left(\frac{0{\cdot}543}{0{\cdot}426}\right)^4=-10 \log \left(\frac{0{\cdot}543}{0{\cdot}426}\right)=-1{\cdot}05,$$

whereas most of the diffuse nebulae give $l=-0{\cdot}2$ to $-0{\cdot}3$ magn.[5] *The scattering particles of the Galaxy are therefore neither atoms nor molecules of a gas.*

It is further easy to verify that the spatial density of the atoms is much too small to produce an appreciable absorption. Under the action of the electric field of the incident wave, the electrons *bound* to the atoms and the molecules are submitted to quasi-elastic forces (from which the $\tilde{\nu}^4$ law results) and vibrate with an amplitude much smaller than that of free electrons. From this it follows that atoms and molecules scatter much less light than an equal number of free electrons, and that in an ionized gas, only the electrons need be considered. For instance, in comparing equations (XII.2) and (XII.4), we find that approximately 700 molecules of hydrogen would be required to scatter as much light as a free electron, near $\lambda=0{\cdot}55\ \mu$. Tens of thousands of atoms per c.c. would be necessary to produce an absorption of 1 magn./kpc. and with a mean density of 1 to some atoms per c.c., the scattering of light by interstellar gas is entirely negligible.

Furthermore, no other particles small compared with the wave-length need be considered

The Rayleigh laws can also be applied to non-conducting solid or liquid particles (i.e. without true absorption), when they are small in comparison with wave-length.[6] The scattering indicatrix and the polarization are the same as discussed above.

In the case of metallic particles, that is to say having true absorption, the true absorption is added to the scattering and the absorption co-efficient is no longer proportional to $\tilde{\nu}^4$. For small perfectly conducting spheres, it would be *roughly* proportional to $\tilde{\nu}$ and the distribution of polarization would be very different. It is not always possible to reproduce the absorption curve observed in the Milky Way.

To conclude, we have been able to eliminate very large particles and free electrons producing respectively an approximately or strictly neutral absorption, and also atoms, molecules and small non-conducting particles which would give a law of reddening (proportional to $\tilde{\nu}^4$) incompatible with observation. Small perfectly conducting particles, to which we shall return a little later, do not seem to serve our purposes much better.

It is therefore necessary to examine the hypothesis of grains of matter of dimensions comparable to wave-lengths.

2. Particles whose size is of the order of the wave-length

General theory

The scattering of light by particles which are neither very large nor very small compared with wave-length, is a very complex phenomenon which cannot be represented by simple laws. The complete theory was developed by G. Mie (1908) [202] and by P. Debye (1909) [67] *for the case of spherical particles* and verified experimentally in colloidal solutions. It is based on the classical formulae of the electro-magnetic theory of light, but it is impossible here to give even an approximate idea of the truly laborious calculations involved. The theory can be equally well applied to metallic absorbing particles as to non-conducting particles without true absorption. In the case of metallic particles, the total absorption co-efficient $k\lambda$ is the sum of two terms: the first corresponds to *apparent absorption*, due to scattering, the second to *true absorption*. Very generally speaking it is:

$$k\lambda = \frac{N\lambda^2}{2} \text{ imaginary part of } \left[\sum_{j=1}^{j=\infty} (-1)^j (a_j - p_j)\right], \qquad \text{(XII.7)}$$

where N always represents the number of particles per c.c., and a_j and p_j are complicated functions of the wave-length λ, of the radius R of the particles and of their refractive index m or, more exactly, of the two parameters

$$x = \frac{2\pi R}{\lambda}, \qquad y = mx. \qquad \text{(XII.8)}$$

The refractive index m, real for dielectrics, is in the case of metals, a complex quantity

$$m = n - i\chi$$

χ being the extinction index or attenuation co-efficient (zero for dielectrics), $i = \sqrt{-1}$, and n is the real part of the refractive index.

The parameter x is a measure of the number of waves which can be established on the sphere of radius R.

The calculation of $k\lambda$ requires the summation of terms $a_1, a_2, \ldots a_j, \ldots p_1, p_2, \ldots p_j \ldots$, which become more complicated as their order j increases. But if x is sufficiently small, the importance of the successive terms diminishes rapidly, so that we obtain a satisfactory approximation if we limit ourselves to a restricted number of terms. As x increases, the convergence of the terms slows down and the numerical calculations become more and more difficult and, very soon, almost impossible.

Calculations for the case of non-conducting spheres have been made mainly by Blumer [38], Stratton and Houghton [309], Götz [103], Greenstein [105] and more recently by V. La Mer and by Van de Hulst [342, 343, 344]. Schalén [254, 255] has particularly concerned himself with the study of absorption and scattering by small metallic spheres, which have also been treated by Schoenberg and Jung [266].

Non-conducting particles

When x is very small, we can neglect all the terms involved in the calculation of the absorption co-efficient except a_1 and p_1. We then come back to the equation (XII.6) for very small spherical particles: the approximation remains good as long as x is less than 1 $\left(\text{i.e. for a diameter } 2R<\frac{\lambda}{3}\right)$.

For larger spheres we must take the subsequent terms into account and the scattering no longer obeys the Rayleigh law. It is convenient to plot the results as

$$E=\frac{k\lambda}{N\pi R^2} \tag{XII.9}$$

against x, or better still, against $x\,(m-1)$.

$N\pi R^2$ is the total cross section of all the particles contained in 1 c.c. If those particles only stopped those light rays which strike them, the absorption co-efficient would be precisely $N\pi R^2$. The factor $E\,(>1)$ thus measures the particular ability of these particles for scattering light. It is called the *efficiency factor*.

The curves of Fig. 35 (after Van de Hulst) show the variations of E as a function of $x\,(m-1)$, for spheres of refractive indices $m=1{\cdot}50$, $m=1{\cdot}33$ and m just a little over 1·0. Because of the choice of the variable representing the abscissae, their shapes are almost identical (Van de Hulst). The very accurately drawn curve, for $m=1{\cdot}50$, shows a "fine structure" characterized by indentations of amplitude $\pm 0{\cdot}15$. The curve for $m=1{\cdot}33$ has probably a similar structure, but with indentations which are only half as pronounced. For practical purposes these details can be neglected.

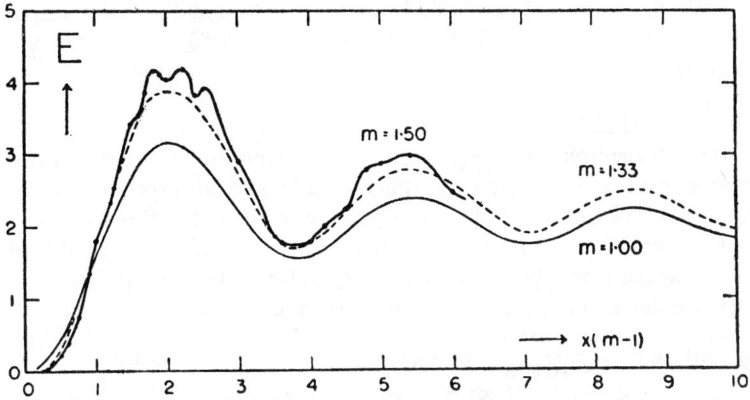

Fig. 35.—Efficiency factors E of dielectric particles as a function of the parameters $x(m-1)$ according to Van der Hulst, for refractive indices m=1.00, 1.33 and 1.50.

As $x\ (m-1)$ increases from 0 to 2, E increases from 0 to a maximum which is the higher, the larger the refractive index (approximately 4·2 for $m=1{\cdot}50$, 3·9 for $m=1{\cdot}33$, and near 3·2 when m tends to unity). Further on there are broad oscillations of diminishing amplitude and finally, for large values of x, that is to say when the particles are very large compared with wave-length, E tends to 2, for all values of m. So we again find the neutral scattering produced by large particles, but at first glance, it seems surprising that the limiting value of the efficiency factor should be 2 and not 1, as we should imagine it to be if we only took the occulation of light rays into account. The difference is fully explained by the diffraction.

For radiation of fixed wave-length, x is proportional to $2R$ and the curve represents the variation of absorption as a function of the diameter of the particles. Conversely, for spheres of a given radius, x is proportional to $\tilde{\nu}=\frac{1}{\lambda}$ and the curve represents the variation of absorption as a function of the wave numbers. Let us adopt the second point of view and let us give our special attention to the first part of the curve, between $x=0$ and the value corresponding to the first maximum. The efficiency factor starts from zero and first increases as $\tilde{\nu}^4$ (the Rayleigh law), up to $x=1$. We then observe an almost straight line where the absorption co-efficient is proportional to $\tilde{\nu}$. For larger or smaller wave numbers the curve bends and, on the whole, is of an S shape similar to that of Whitford. *It would thus appear possible for the first time to account for Galactic absorption by a convenient choice of diameter of the particles.* We may suppose that the quasi-linear portion is found on either side of $x\ (m-1)=\frac{4}{3}$ or $x=\frac{8}{3}$

if $m=1{\cdot}50$, and $x=4$ if $m=1{\cdot}33$. Taking the corresponding wave-length as approximately $0{\cdot}5\mu$, we find $2R{\sim}0{\cdot}4\mu$ for $m=1{\cdot}50$ and $2R{\sim}0{\cdot}6\mu$ for $m=1{\cdot}33$.

Metallic particles

The absorption curves of metallic spheres seem, at first sight, sufficiently similar to those of transparent spheres, at least if absorption is not very large. The maximum of the efficiency factor almost corresponds to the same value of x but it is lower than in the case of transparent grains. Nevertheless, since the metals have very varied absorbing properties, the curves differ in an appreciable fashion from one metal to the next.

By putting $\psi(R)=\dfrac{\lambda^2}{16\pi^3R^3}$ multiplied by the imaginary part of

$$\left[\sum_{j=1}^{j=\infty}(-1)^j(a_j-p_j)\right], \tag{XII.10}$$

the absorption co-efficient, given by equation (XII.7), becomes

$$k=8\pi^3NR^3\psi(R). \tag{XII.11}$$

or, introducing the volume V of the particle

$$k=6\pi NV\psi(R) \tag{XII.12}$$

and the efficiency factor becomes

$$E=8\pi R\psi(R). \tag{XII.13}$$

Using the complex refractive indices of metals (which vary with wave-length), Schalén constructed the function $\psi(R)$ for iron, nickel, zinc, copper, etc., for a certain number of wave-lengths between $0{\cdot}39$ and $0{\cdot}67\mu$. These curves permitted him immediately to determine the variation of the absorption co-efficient $k(\tilde{\nu})$ as a function of the wave number $\tilde{\nu}$. Fig. 36, for instance, represents the absorption curves for grains of iron and of nickel of 600, 800, 1000, 1200, 1400 and 1600 A in diameter.

When the radius R is very large compared with the wave-length, we again find neutral absorption as in the case of dielectrics. It is more difficult to see what takes place when R is extremely small compared with the wave-length. A rough calculation, neglecting in particular the variation of the refractive index with the wave-length, would lead to the result that the absorption co-efficient might be proportional to the wave-numbers. We shall soon see that the recent calculations of Güttler [408] do not confirm this point of view. In any case, Fig. 36 shows that the absorption co-efficient is appreciably proportional to $\tilde{\nu}$ for spheres of iron of diameter 600 A and for particles of nickel of 1000 A in the region of the visible spectrum.

It thus appears possible to reproduce an absorption curve similar to that of interstellar space, in the visible and in the near ultra-violet, by means of metallic spheres of convenient diameter. Is it possible to choose between dielectric particles and metallic particles?

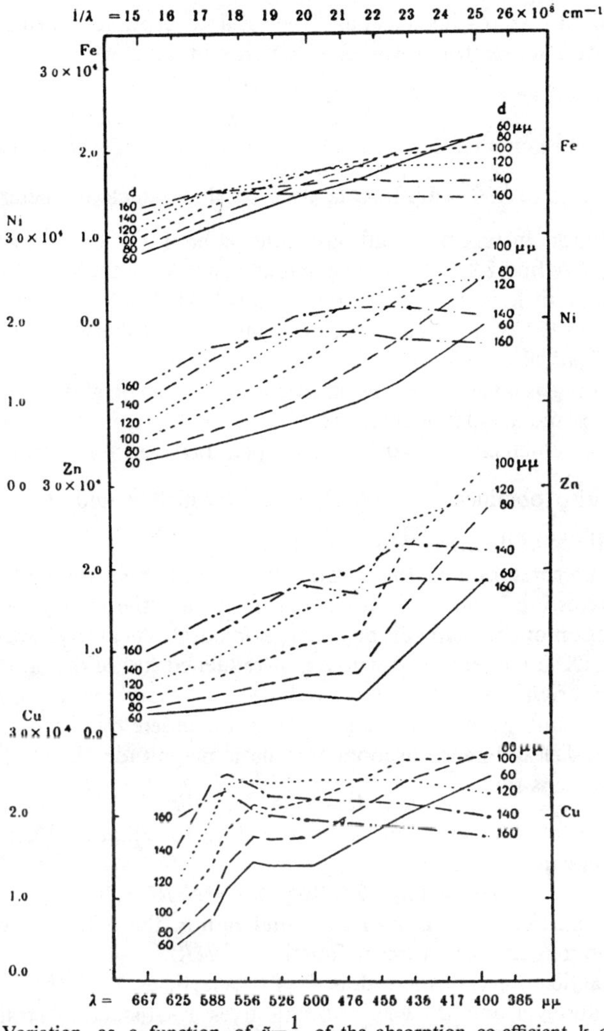

Fig. 36.—Variation, as a function of $\bar{\nu}=\frac{1}{\lambda}$, of the absorption co-efficient k of particles of iron, nickel, zinc and copper of different diameters (60 to 160 mμ, or 0.06 to 0.16μ) (after Schalén).

3. Comparison of theory and observation

The case of metallic particles

Schalén (in 1934–36) [254] gave an interpretation of some of the measurements on differential absorption and of total absorption in absorbing clouds. One knows the colour excess ΔC of the cloud (between

3950 and 4400 A for instance) and the total absorption Δm (for 4400 A). If there were no neutral absorption, we should have

$$\frac{k_{3950}}{k_{4400}}=1+\frac{\Delta C}{\Delta m}$$

Schalén's measurements on the clouds of Cygnus, Cepheus and Auriga (page 152) give $\frac{k_{4400}}{k_{3950}}=0{\cdot}91$. One tries to find which diameter of iron particles, for instance, would give the same ratio of absorption co-efficients. We find $2R=860$ A. The diameter would be the same for spheres of copper, a little larger in the case of nickel (1100 A$=0{\cdot}11\mu$) and of zinc (1210 A$=0{\cdot}121\mu$). The later measurements of Wernberg (page 137) give almost identical dimensions.

By taking as a mean absorption co-efficient in the neighbourhood of the Galactic plane $a=0{\cdot}9$ magn./kpc. for $\lambda=4400$ A, and as differential absorption co-efficient $d=0{\cdot}3$ magn./kpc., between 4400 and 5500 A, Schalén also obtained $\frac{k_{5500}}{k_{4400}}\simeq0{\cdot}67$, a ratio which would correspond to iron particles of diameter 500 A ($0{\cdot}05\mu$).

The interpretation of the spectrophotometric measurements is much more interesting because it shows to what extent the theory permits the reproduction of the shape of the absorption curve. As always, we compare (Chapter IX) a reddened star with a non-reddened star of the same spectral type. The brightness of the first is reduced by absorption in the ratio e^{-kr} and, as a consequence, its magnitude is increased by the quantity $2{\cdot}5$ $(\log e)\ kr$. The difference of monochromatic magnitude observed between the two stars is thus

$$\Delta m=\Delta m_0+2{\cdot}5\ (\log e)\ kr \qquad (\Delta m_0=\text{const.}).$$

By introducing the function $\psi(R)$ according to equation (XII.11), this expression can be written

$$\Delta m=\Delta m_0+2{\cdot}5\ (\log e).\ 8\pi^2NrR^3\psi(R)$$

Thus, for particles of a given nature and radius, the difference of monochromatic magnitude is a linear function of $\psi(R)$.

In practice, we vary the radius R of the particles, until the graph with $\psi(R)$ as abscissa and $\Delta m\lambda$ as ordinate gives a satisfactory straight line. The diameter of the particles is thus determined and the slope of the straight line gives the product Nr which measures the total number of particles contained in a cylinder of unit section having as length the thickness r of the absorbing medium.

Schalén (1936) [254] has shown that the spectrophotometric measurements of Trumpler (page 171) could be very well interpreted in terms of iron particles of diameter 750 A, and those of Struve, Kennan and Hynek (page 171) by particles of 800 A. Later (1939) [255], he interpreted the measurements of Strohmeier (page 172) on some B stars, and those of Baade and Minkowski on the stars of Orion (page 174). The curves of Strohmeier for the case of the stars H.D. 216411, H.D. 200775 and H.D. 2329 are very well explained by spheres of 700 A diameter for iron, of 1000 A for nickel and of 1400 A for zinc respectively, while we can account for Baade and Minkowski's measurement on the star H.D. 37061, for instance, either by grains of copper of 700 A, or by particles of zinc of 1000 A. In the case of θ_1(C) Orionis, the diameter of the particles of zinc is probably 1600 A.

In reality, it is improbable that the grains should all have the same diameter and that they should all be composed of the same metal. Schalén was therefore led to treating the problem much more generally as consisting of a mixture of spheres of different metals, whose diameter vary continuously in accordance with a given law. Let $N\phi(R)\,dR$, be the number of particles in 1 c.c. whose radius is between $R-\frac{dR}{2}$ and $R+\frac{dR}{2}$. Assuming a "normal" distribution law

$$\phi(R)=\frac{1}{\sigma\sqrt{2\pi}}e^{-(R-R_0)^2/2\sigma^2},$$

and a mixture mainly composed of iron, with the addition of varying proportions of nickel, zinc and copper. Schalén managed to reproduce the curves of Strohmeier, and of Baade and Minkowski extraordinarily well. He took, according to the case, an average diameter $2R$ between 800 and 1200 A, with *dispersion* $\sigma=150$ A (for the radius).

The total number Nr of particles contained in a cylinder of 1 square cm. section, stretching from the star to the Earth is, in general, of the order of 10^9 and their total mass is 10^{-6} to 10^{-5} gm. A law of distribution of the radii of the form $\phi(R)=KR^3$ (K=Constant) would still lead to satisfactory curves, but the total absorption would be increased in improbable proportions. $N\,r$ would then be of the order of 10^{10} to 10^{11} and the mass contained in the cylinder would reach 10^{-4} gm. Further, a uniform distribution of the radii (between 200 and 5000 or 10000 A for instance) would lead to results in complete contradiction with observation; they ought, therefore, to be eliminated.

It is not surprising that we manage to improve the representations by increasing the number of arbitrary parameters: chemical composition of the mixture, average diameter and dispersion. By choosing them in a convenient manner, we can reproduce all the absorption curves observed in the visible and in the ultra-violet (Schoenberg, Jung, Greenstein).[7] But this remark does not diminish at all the value of the results obtained by Schalén, since he had already managed to account for the observations with a single kind of particle of fixed diameter. Thus, due to his very important work, the balance stood for quite a long time in favour of metallic particles. But the extension of the interstellar curve into the infra-red has raised the question anew.

The case of non-conducting particles

In fact, it seems impossible to reproduce with metallic particles the flattening of the absorption curve in the infra-red, which is shown by Whitford measurements at $2{\cdot}1\mu$.

At great wave-lengths, the curve $f(\tilde{\nu})$ deviates enormously from proportionality to $\tilde{\nu}$; we must find out if it approaches proportionality to $\tilde{\nu}^4$, which was predicted for the case of non-conducting grains. With this in view, let us plot the log of the absorption co-efficient a, expressed in magn./kpc., against log $\tilde{\nu}$. When a is proportional to $\tilde{\nu}^p$, log a is represented as a function of log $\tilde{\nu}$, by a straight line of slope p. If, therefore, the $\tilde{\nu}^4$ law applies, the points ought to fall along a straight line of slope 4. We further know (page 178) that to obtain a in magn./kpc., it suffices to add the constant $+0{\cdot}16$ or $+0{\cdot}17$ magn. to the ordinates $f(\tilde{\nu})$ of the Whitford curve.

The curves obtained with the constants 0·16 and 0·17 magn. are shown in Fig. 37.

Both show, in the visible and in the near ultra-violet, a line of average slope of about 1·0 (corresponding to the $\tilde{\nu}$ law, but in the infra-red, the slope progressively approaches 4·0. The first curve (A) is particularly satisfactory, since its average slope between $\tilde{\nu}=0{\cdot}97\mu^{-1}$ ($\lambda=10{\cdot}3\mu$) and

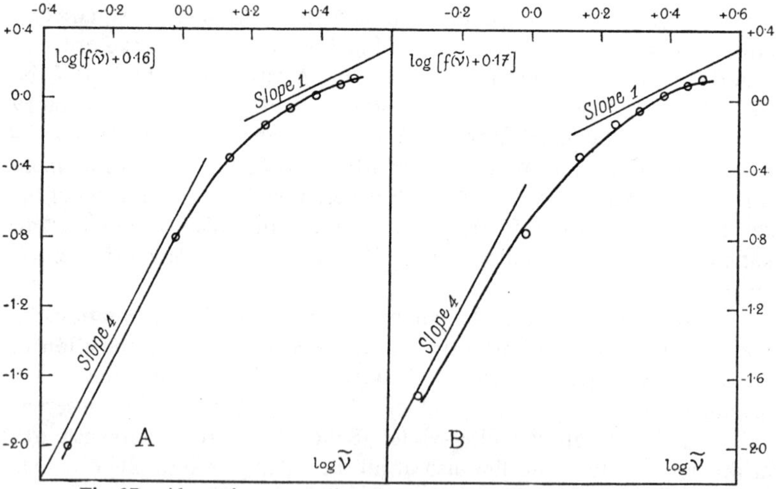

Fig. 37.—Absorption curves plotted logarithmically; log a against log $\tilde{\nu}$.
A—$a(\tilde{\nu})=f(\tilde{\nu})+0.16$ magn.,
B—$a(\tilde{\nu})=f(\tilde{\nu})+0.17$ magn.

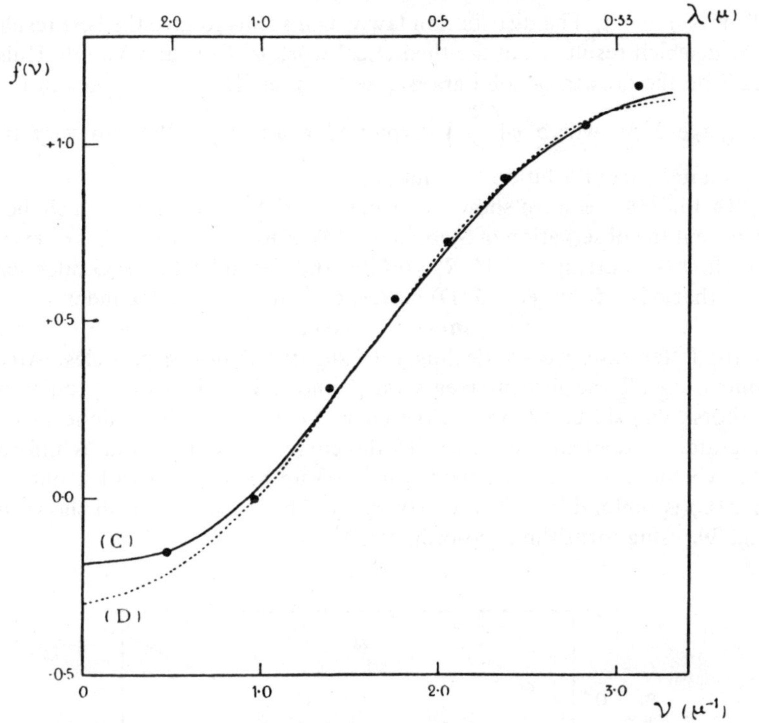

Fig. 38.—Theoretical absorption curves after Van de Hulst, compared with the observations of Stebbins and Whitford.
Solution C; $m=1.33$;
Solution D; $m=1.25-0.03$ i.

$\tilde{\nu}=0{\cdot}48\mu^{-1}$ ($\lambda=2{\cdot}1\mu$) is 3·93. Between the same points the average slope of the second curve (B) is 3·04. Neither of the two curves varies from each other by more than a minimal change of 0·01 magn. from the constant added to f($\tilde{\nu}$). In fact, the measurements of Whitford at 2·1μ are still too uncertain for the constant to be determined with such accuracy. The two solutions can, therefore, be taken as being practically equivalent.

The appearance of the graph clearly shows that the interstellar *absorption curve,* drawn in logarithmic *co-ordinates, is of the form predicted for the case of non-conducting particles.*

It remains to determine the refractive index of the particles and their dimensions, or more exactly, the law of distribution of their diameters. A priori, it is not at all certain that there exists a unique solution to this problem. Among the numerous curves traced by Van de Hulst (1949) [343], with various distribution laws and various values for the index, there are in fact, a number which represent the observations very well within the errors of measurement. The distribution law which seems to give the best results is that which results from the theoretical work of Oort and Van de Hulst [225] on the growth of solid grains (see Chapter XIV). It is shown in Fig. 41 (page 255), where $\phi\left(\frac{R}{R_1}\right)$ is plotted against $\frac{R}{R_1}$. We can vary the constant R_1 from solution to solution.

In Fig. 38 we have shown two curves of Van de Hulst which best represent the observation of Stebbins and Whitford of the pair ζ—ϵ Persei. The first (C) corresponds to $R_1 = 0{\cdot}410\mu$ and the index $m = 1{\cdot}33$, identical with that of ice; the second (D) corresponds to the complex index

$$m = 1{\cdot}25 - 0{\cdot}03\ i.$$

In the latter case, we are dealing with *slightly absorbing* particles. After comparing all the measurements on ζ and ϵ Persei made by different authors, Van de Hulst prefers the curve D. If, as we have done in the diagram, we compare this curve to the curve of Stebbins and Whitford, the two solutions are equivalent from 1·03 up to the ultra-violet, but the first (C) is preferable at $2{\cdot}1\mu$ ($\tilde{\nu} = 0{\cdot}48\mu^{-1}$). This solution is also shown in Fig. 39, using logarithmic co-ordinates.[8]

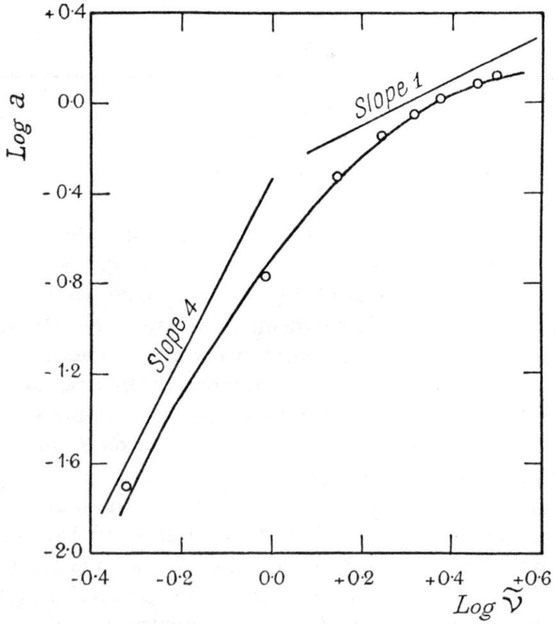

Fig. 39.—Theoretical absorption curve in logarithmic co-ordinates, and Stebbins and Whitford's observations. Solution C of Van de Hulst, with $a(\tilde{\nu}) = f(\tilde{\nu}) + 0.17$ magn.

The work of Güttler (1952)

After the work of Van de Hulst (1949), the problem of the nature of the scattering particles seemed definitely solved, for it seemed impossible to reproduce the interstellar absorption curve in the infra-red with metallic particles. But a recent work of Güttler [408] has shown that this conclusion was incorrect.

Güttler calculated the absorption by absorbing spheres, and stressed our almost complete ignorance of the variation of refractive and extinction indices of metals from the visible to the infra-red. For iron, even in the visible, the results of measurements differ enormously according to the state of division of the metal. Those of W. Meier (1910) on finely divided iron are usually taken ($m=1{\cdot}51$, $n\chi=1{\cdot}63$ for the D lines of sodium), but these values are still very uncertain. Güttler thus finds, as does Schalén, that the absorption curve $\Delta m=f(\tilde{\nu})$ for the ζ and ϵ Persei is well represented in the visible and the near ultra-violet by spheres of iron of diameter 750 A (0·075μ). In the infra-red, the two experimental points fall *below* the theoretical curve and not above, as was generally supposed. In other words, *the calculated curve is flatter in the infra-red than that of Stebbins and Whitford.*

The author then considers mixtures composed of particles of iron of two different arbitrary diameters, and he tries to find out for each of these mixtures, what proportion of the two kinds of spheres would lead to an absorption curve which would fall along the observed interstellar curve. He arrives, for instance, at satisfactory representations with:

6,750,000	grains of diameter	10 A	for 1 of diameter	1500 A
5,600	,,	100 A	,,	1500 A
360	,,	250 A	,,	1500 A
136	,,	750 A	,,	3000 A

In these four solutions the ratio χ between total absorption and differential absorption, in the Stebbins system, is 8·0, 8·9, 9·0 and 8·1 respectively. We know that the most probable value is 8·16 (page 179).

The problem posed has an infinite number of solutions and the conception of these double mixtures, formed of particles of two diameters only, is very artificial. It certainly does not correspond with reality, but it shows that it ought to be possible to find one (or more) continuous distribution functions of the diameters which represent the observed absorption curve faithfully. As in the distribution envisaged by Oort and Van de Hulst for the case of dielectric grains, the number of particles ought to diminish progressively with increase in diameter, such that the smallest particles would always be the most numerous.

Thus the hypothesis of metallic grains, which was rejected with too much haste, could account for the observations just as well as the hypothesis of non-conducting grains, and Güttler concludes that *a knowledge of the absorption curve, even if taken together with that of the total photographic absorption, can in reality give us no information on the nature of the scattering particles.*

The total mass of solid particles contained in 1 *c.c.*

Knowing the absorption co-efficient a in magn./kpc., the distribution function of the radius of the spheres, their refractive index and their specific mass s, we can evaluate the mean mass ρ_s of all the particles contained in 1 c.c. of interstellar space.

The calculation is immediate when all the grains have same radius R_1. If there are N particles per c.c., we obviously have

$$\rho_s = \frac{4}{3}\pi R_1^3 s N. \tag{XII.14}$$

Further, the total cross-section σ of these N particles is N times the section of one particle, for we can simply add the cross-sections of the individual grains, when they are sufficiently far apart

$$\sigma = \pi R_1^2 N.$$

The absorption co-efficient k is the product of σ by the *efficiency factor* E_1 of the particles

$$k = \sigma E_1 = \pi R_1^2 N E_1$$

and, in magn./kpc.,

$$a = 3{\cdot}34.10^{21}\, k = 3{\cdot}34.10^{21}\, \pi R_1^2 N E_1. \tag{XII.15}$$

By dividing (XII.14) by (XII.15), we obtain:

$$\frac{\rho_s}{a} = 4{\cdot}0.10^{-22} R_1 \frac{s}{E_1}$$

Dielectric grains

Let us suppose, for example, that the grains are of specific mass $s=1$ are similar to ice crystals, and have refractive index $m=1{\cdot}33$ and radius $R_1=4{\cdot}07.10^{-5}$ cm. (diameter $0{\cdot}814\mu$). The product $x\ (m-1)=2\pi R_1 (m-1)/\lambda$ is then 2 for $\lambda=4{\cdot}26.10^{-5}$ cm. ($0{\cdot}426\mu$) and corresponds to the maximum efficiency $E_1=3{\cdot}9$ (page 213). We then have

$$\frac{\rho_s}{a} = 4{\cdot}17.10^{-27}$$

and, with $a=1{\cdot}10$ magn./kpc. (for $\lambda=0{\cdot}426\mu$),

$$\rho_s = 4{\cdot}6.10^{-27} \text{ gm./c.c.}$$

With $a=2{\cdot}0$ magn./kpc., we shall have

$$\rho_s = 8{\cdot}3.10^{-27} \text{ gm./c.c.}$$

The calculation is naturally more complicated when we consider a mixture of grains of different diameters. With the distribution function shown in Fig. 41 and $R_1=4{\cdot}1.10^{-5}$ cm. for grains of refractive index

$m=1{\cdot}33$, and of specific mass $s=1{\cdot}0$ (solution C), Oort and Van de Hulst obtained

$$\frac{\rho_S}{a}=4{\cdot}3.10^{-27}$$

The same distribution function gives, for slightly absorbing grains, of specific mass $s=1{\cdot}1$ (solution D):

$$\frac{\rho_S}{a}=6{\cdot}9.10^{-27}.$$

With $a=1{\cdot}10$ magn./kpc., we get

$$\rho_S=4{\cdot}7.10^{-27} \text{ gm./c.c. (C) and } \rho_S=7{\cdot}6.10^{-27} \text{ gm./c.c. (D)};$$

while with $a=2{\cdot}0$ magn./kpc., we would have:

$$\rho_S=8{\cdot}6.10^{-27} \text{ gm./c.c. (C) and } \rho_S=1{\cdot}4.10^{-26} \text{ gm./c.c. (D)}.$$

Metallic grains

We could make similar calculations on the hypothesis of metallic grains. The specific mass of iron $s=7{\cdot}9$ gm./c.c. is much larger, but the spheres have much smaller radii, and for the various binary mixtures studied by Güttler, we find

$$\frac{\rho_S}{a}=8.10^{-27}$$

If we take $a=1{\cdot}10$ magn./kpc. or $a=2{\cdot}0$ magn./kpc., we obtain

$$\rho_S=7{\cdot}3.10^{-27} \text{ gm./c.c. or } \rho_S=1{\cdot}6.10^{-26} \text{ gm./c.c.}$$

The total mass of grains contained in 1 *c.c. is thus almost of the same order of magnitude as in the case of dielectric particles.*

We recall that the density of all the solid grains contained in space is hardly of the order of 10^{-26} gm./c.c. We are, therefore, dealing with a *mean* value, which ought to be considerably exceeded inside an absorbing cloud, but much reduced outside it.

Conclusion

The particles responsible for interstellar absorption are definitely tiny solid grains, but the study of absorption by itself is not capable of giving us precise information on their nature.

If they are dielectric grains similar to ice crystals, as Eddington suggested in 1937, their diameters are comparable with the wave-lengths. They might very well be particles of quite different dimensions (mainly much smaller ones) but those which scatter the light most have diameters of the order of 6.10^{-5} to 10^{-4} cm., or from 6000 to 10000 A (maximum of the efficiency factor E). It is, therefore, very difficult to determine the distribution function of the diameters by observation. Those used by Oort and Van de Hulst, drawn from their theory of the formation of grains (Chapter XIV), account well for the absorption curve. This is characterized by the abundance of very small particles. The number of grains diminishes constantly as their diameter increases and particularly quickly around 6000 A, so that there are practically no more particles of diameters greater than a few microns. This last point agrees perfectly with the absence of neutral absorption.

Basing his arguments on the cosmic abundance of elements and on the fact that atoms of hydrogen and helium probably escape condensation (Chapter XIV), Van de Hulst suggests the average composition below for solid grains:

for 100	molecules	H_2O,
30	,,	H_2,
20	,,	CH_4,
10	,,	NH_3,
5	,,	MgH, etc.

It must be well understood that the presence of heavier metals, such as iron, is not excluded, but that they only exist in small quantities.

If we were dealing with metallic particles, their diameters would be somewhat smaller, but the graph of the distribution function ought to be of similar appearance. In the two cases the mean mass of all the particles per c.c. remains of the same order ($\simeq 10^{-26}$ gm.).

Can the study of the colour and polarization of the scattered light, and also the diagram of the scattering indicatrix remove the uncertainty in the nature of scattering particles? So far these properties have only served to eliminate the idea of scattering by particles which are very small or very large compared with wave-length. The study of diffuse nebulae is not advanced enough to give us any detailed information. But we should notice that as a first approximation, the somewhat rough data at our disposal agree well with the predicted behaviour of grains of some thousands of angströms or more in diameter, as far as the bluish colour of scattered light and the very appreciable radial polarization are concerned. The scattering indicatrix of these particles (Fig. 34, page 209), shows clearly the concentration of scattered light towards the front, suggested by the study of general scattering of light in the Milky Way.

This study led Henyey and Greenstein (page 205) to think that the albedo of the scattering particles ought to be rather large $\left(\geqslant \frac{1}{3}\right)$ and to see in this an argument in favour of non-conducting grains. But Güttler [408] has shown that small metallic spheres could also lead to an albedo between 0·3 and 0·8. Thus we do not actually have enough observational data for solving the question.

Since we lack experimental criteria, we might remark that in the Sun, the stars and the planetary nebulae, light elements such as hydrogen, helium, oxygen and nitrogen, are much more abundant than metals, and particularly heavy metals (Table VI, page 81). The theory of metallic grains requires a quite different relative abundance in the solid and in the gaseous phases of interstellar space, since in accordance with the study of Galactic emission and absorption lines, the composition of interstellar gas appears similar to that of planetary nebulae and stars, at least as far as the light elements are concerned. But it is possible that by means of some unknown mechanism, only the heavy atoms of space might aggregate to form solid grains, whereas the light atoms would remain in the free state. What then would be the total composition of interstellar matter, solid and gaseous phases taken together?

The mean total mass of the solid grains, supposed metallic, contained in 1 c.c. has just been evaluated to be $7{\cdot}3.10^{-27}$ gm. (page 223), so that 1 c.c. would contain approximately $7{\cdot}8.10^{-5}$ atoms of atomic mass of about 56. Further, according to the work of Strömgren, the mean density of the hydrogen atoms can be of the order of 1·5 per c.c. (see page 125). The ratio of the total numbers of heavy atoms to the number of hydrogen atoms would thus be about $5{\cdot}2.10^{5}$. Now, the work of Unsöld permits the evaluation of the same ratio in the solar atmosphere. By considering as "heavy" all the atoms from titanium onwards (atomic weight 48), we find $5{\cdot}4.10^{-5}$. The agreement of these two numbers ought not to blind us to the uncertainty of the data. But this comparison shows that on the hypothesis considered, the *total* composition of the interstellar medium would remain similar to that of the Sun.

The argument based on the cosmic abundance of elements, invoked in favour of dielectric grains, is thus not entirely convincing by itself. The theoretical study of the formation of grains (Chapter XIV) should throw some light on this subject. Finally, the polarization of the light of the stars by interstellar clouds (Chapter XV) will provide us with information of an entirely new kind.

NOTES TO CHAPTER XII

[1]This conclusion is not incompatible with the possible existence of meteorites in interstellar space. But if they do indeed exist, they are not sufficiently numerous to produce an appreciable absorption.

[2]In the laboratory the scattering of light by free electrons has never been observed outside the X-ray range, but the theory cannot be doubted. It explains most of the continuous spectrum of the Solar corona very well.

[3]Experiments show that the light scattered by biatomic or polyatomic gases is not completely polarized for $\theta=\frac{\pi}{2}$. A residual depolarization ρ_0 of some hundredths or more cm. is observed for biatomic gases ($\rho_0 \sim 0.9$), but this may exceed $\frac{1}{10}$ in certain triatomic gases ($\rho_0 \sim 0.8$). This is due to the anisotropy of the molecules whose symmetry is no longer that of a sphere but of an ellipsoid of revolution. Cabannes has shown that in this case, the right hand

sides of (XII.3) and (XII.4) must be multiplied by the correction factors $\frac{6(1+\rho_0)}{6-7\rho_0}$ and $\frac{3(2+\rho_0)}{6-7\rho_0}$ respectively. In most cases the correction is hardly of the order of 10 per cent, and it may be neglected here.

[4]Some attempts were made to explain the colour excesses of the reddened stars by a mixture of particles, some scattering all radiation equally (large particles or free electrons), the others scattering proportionally to $\tilde{\nu}^4$. The absorption co-efficient would then be

$$k=p+q\tilde{\nu}^4.$$

It is always possible to get the correct colour excess for two given wave-lengths by a convenient choice of the constants p and q, i.e. by means of a particular proportion of the two kinds of particles. But it is almost impossible to reproduce an absorption curve resembling that of Stebbins and Whitford.

[5]If the existence of the very blue nebulae previously mentioned should be confirmed, in these particular cases scattering by the atoms of a gas might be conceivable.

[6]For small dielectric spheres of refractive index m and of volume V we obtain:

$$\frac{J(\theta)}{I}=\frac{9\pi^2\tilde{\nu}^4NV^2(1+\cos^2\gamma)(m^2-1)^2}{2(m^2+2)^2} \qquad \text{(XII.5)}$$

$$k=\frac{24\pi^3\tilde{\nu}^4NV^2(m^2-1)^2}{(m^2+2)^2} \qquad \text{(XII.6)}$$

It is easily seen that these are equivalent to (XII.3) and (XII.4). The solid particles in 1 c.c. effectively occupy the volume NV and constitute a medium of refractive index μ such that

$$\frac{(\mu^2-1)}{(\mu^2+2)}=\frac{NV(m^2-1)}{(m^2+2)}$$

Since μ differs little from unity, $\mu^2+2\sim 3$. By replacing in (XII.5) and (XII.6) $\frac{NV^2(m^2-1)^2}{(m^2+2)^2}$ by $\frac{(\mu^2-1)^2}{9N}$, we once more obtain (XII.3) and (XII.4).

[7]This is true both for non-conducting and for metallic particles.

[8]With these co-ordinates which attach considerable importance to the relatively uncertain point $\tilde{\nu}=0.48\mu^{-1}$, the solution D would become unacceptable.

Part Three

FROM ATOMS TO GRAINS AND FROM GRAINS TO STARS

CHAPTER XIII

CLOUDS OF GAS AND OF PARTICLES

1. Atoms and grains as constituents of the same clouds

Comparative dimensions of clouds of gas and clouds of grains

For some years it has been known that interstellar gas forms distinct clouds in space, which are characterized by different velocities (Chapter V, page 113). It is also known that the scattering medium constituting the "Trumpler layer" is essentially heterogeneous (Chapter VIII, page 154). Thus, the solid grains responsible for absorption also tend to concentrate in the form of clouds of particles. The existence of "windows", through which we can see distant stars whose light is hardly weakened, seems to suggest that the spatial density of the particles would be rather small outside the clouds. An important question immediately comes to mind: *do clouds of gas and clouds of grains exist independently of each other, or is there only a single kind of cloud containing both atoms or molecules and solid particles?* The fact that diffuse nebulae show the emission spectrum of a gas when they are illuminated by an O or a B0 star, and a spectrum arising from scattering by solid particles when they are illuminated by stars of more advanced spectral type, strongly suggest the simultaneous presence of gases and of grains in a given cloud.

In this respect it would be interesting to compare the dimensions of clouds of gas with those of clouds of particles. However, neither of them are as yet sufficiently well known.

In the gas clouds, Wilson and Merrill [382] first found a radius of the order of 700 pc., but the more recent researches of Whipple at Mount Wilson (page 119), based on spectra of high dispersion, have shown that a straight line 1 kpc. long, situated approximately in the Galactic plane, cuts through six to eight gas clouds. Their mean radius can, therefore, not exceed some tens of parsecs.

On the other hand, the statistical work of Ambarzumian and his collaborators [7, 166] has led to an evaluation of the average opacity, the number and the dimensions of the clouds of particles. To find the average opacity, Ambarzumian used the counts of extragalactic nebulae of

Shapley and of Hubble. At the same Galactic latitude, the number of these nebulae per square degree, brighter than a given magnitude m, shows large fluctuations. At first sight, we could be tempted to explain these fluctuations by a tendency of the Galaxies to group themselves in clusters. This tendency does in fact exist, but it is not sufficient to explain the observed fluctuation, because their importance becomes more pronounced the lower are the latitudes considered. Therefore, these fluctuations must mainly be attributed to absorption by those clouds which are abundant in the neighbourhood of the Galactic plane. Their numerical study leads to the evaluation of the mean absorption a_0 exercised by a single cloud. Ambarzumian found $a_0 \backsim 0{\cdot}27$ photographic magnitudes (1939) [7]. The dispersion of colour excesses of the external Galaxies at different latitudes, led Kukarkin to a comparable value. Similarly, Markaryan [184] derived a value for a_0 from the study of the fluctuation of star counts at low latitudes, and he obtained $a_0 = 0{\cdot}18\ m_{pg}$. From all this work it may be concluded that the average photographic absorption of a cloud of particles is of the order of 0·2 magn. (or at most 0·3 magn.).

By comparing a_0 with the average photographic absorption coefficient a, we obtain the average number j of clouds crossed by a straight line of 1 kpc. in the Galactic plane. If absorption is negligible outside the clouds, we have simply:

$$a = ja_0.$$

With $a = 1{\cdot}1 - 1{\cdot}3$ magn./kpc. and $a_0 = 0{\cdot}2$ magn., we obtain

$$j \simeq 5{\cdot}5 - 6{\cdot}5.$$

On the average, a line of 1 kpc. would thus traverse six clouds of particles. This number is almost identical with that which Whipple found later for the case of gas clouds.

The Armenian astronomers further managed to evaluate the average number N of clouds of grains per unit volume. In their work on the fortuitous association of stars and nebulae (page 195), Ambarzumian and Gordeladse [8] found that the fraction of space which is sufficiently illuminated by stars for clouds to become luminous, is only of the order of 1/2000. The total number of clouds of particles would, therefore, be approximately 2,000 times that of the diffuse nebulae, which leads to

$$N \backsim 10^{-4} \text{ clouds/pc}^3.$$

Roughly, there would be 1 cloud per 10,000 pc³.

Now it is easy to calculate the average radius R of the clouds, considered as being spherical. Let R be the radius of a cylinder of length L, and whose axis crosses j clouds contained in the volume $\pi R^2 L$ of the cylinder. The number N of clouds per unit volume is then

$$N = \frac{j}{\pi R^2 L}$$

an equation by means of which R can be calculated. With $j=6$, $L=1{,}000$ pc. and $N=100^{-4}$, we obtain $R=4.4$ pc. The average radius of the cloud would therefore be of the order of four to five parsecs. According to Spitzer, the hypotheses tend to over-estimate N, so that the radius obtained must be considered as a lower limit. Parenago [233] found the average radius to be $R=5$ pc.

We shall now present the arguments of Oort and Van de Hulst [225] who give a similar order of magnitude. Of the 160 stars of types O to B2, brighter than 7·5 magn. and situated higher than $-15°$ of declination, there are twenty-three (14 per cent) which, according to Hubble, are associated with diffuse nebulae. If we suppose that no relation exists between the spatial distribution of these nebulae and that of the stars, we can conclude that in the neighbourhood of the Galactic plane, diffuse nebulae occupy 14 per cent of space. Let us then assume that the volume of j spheres of radius R represents 14 per cent of the volume of the cylinder of the same radius and of length L:

$$\frac{j^4 \pi R^3}{3} = 0{\cdot}14\, \pi R^2 L.$$

From this we obtain $$R = \frac{0{\cdot}105\ L.}{j}$$

With $j=6$, $L=1{,}000$ pc., we find $R=17{\cdot}5$ pc. (Oort and Van de Hulst took $j=5$, which gives $R=10$ pc.).

If the nebulae in question were all gaseous line-spectrum nebulae, the closeness of Ambarzumian's values to those of Oort would lead to the conclusion that clouds of grain and clouds of gas have radii of the same order of magnitude (5 to 15 pc.). But, in fact, Oort and Van de Hulst used both line-spectrum nebulae (related to stars of the types O and B0) and diffuse nebulae, formed of solid grains (related to B1 and B2 stars). In any case, so far there is no evidence against the hypothesis of the identity of clouds of gas with clouds of grain. Their mean radius can, at most, be of the order of some ten parsecs, and thus is small compared with the distances which separate them.

Nevertheless, the study of dark nebulae (Chapter VI) has often shown that there are considerable thicknesses, all of which are certainly not illusory. According to Greenstein [106], the most prominent clouds of the Milky Way would have a diameter of a hundred parsecs with an absorption of approximately one photographic magnitude. A straight line in the Galactic plane, however, would meet only one of these clouds in 6,000 pc., so that their contribution to the general absorption would be very small (0·17 magn./kpc.). A recent work of Schatzman (1950) [263] has confirmed the existence of both small and large clouds. The author studied the fluctuations in the reddening of the B stars, according to the measurements of Stebbins, Huffer and Whitford. By supposing the clouds to be spherical, he showed that we can account for the observation by assuming the existence of two types of clouds: small ones, of which there are six in a line of 1,000 pc., would produce an average colour excess of 0·04 magn. (whence a_0=0·24 visual magnitudes, and 0·32 photographic magn.) and the large ones of which there are 0·6 per 1,000 pc., would produce average reddening of 0·33 magn. (a_0=2·0 vis. magn., and 2·7 photogr. magn.).[1] These large clouds of grains could be likened to the large clouds of gas of Wilson and Merrill.

Correlation between the colour excess and the intensity of interstellar lines

If the Galactic clouds contained both atoms and solid particles, the intensity of the interstellar absorption line would probably increase with the colour excess of the stars. At first sight, it would even appear that the two quantities should be strictly related, if the mixed clouds were of fixed composition.

It has been clearly shown that there exists a *marked correlation* between the intensity of the K line and the colour excess of the stars, but on the whole, their relation is rather loose. This is shown by the example of Fig. 40, in which abscissae are the intensities *I* of the K line estimated by Struve for thirty-four of the O6 to B2 types, and the ordinates are the photo-electric colour indices *C* measured by Bottlinger and Becker. (This group was sufficiently homogeneous so that the colour indices could be used directly, without calculating the colour excesses.)

On the average, *C* increases with *I*, but the dispersion of the points is considerable, and corresponds to the correlation co-efficient $Cr = +0{\cdot}65 \pm 0{\cdot}10$ (Dufay and Liau Ssu-Pin, 1933 [75]). During the same

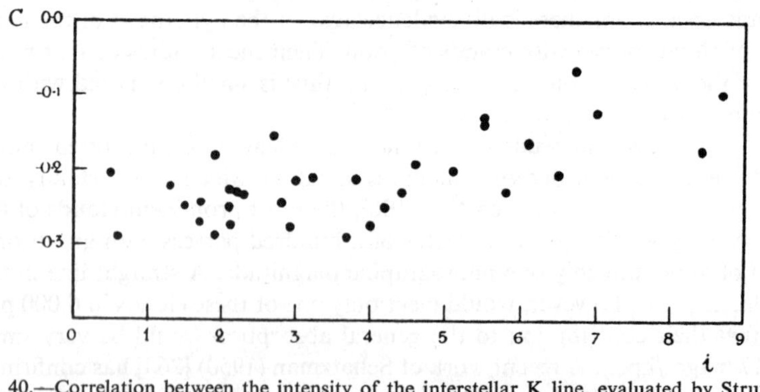

Fig. 40.—Correlation between the intensity of the interstellar K line, evaluated by Struve, and the photo-electric colour index measured by Bottlinger and Becker.

period, but with other data, Miss Westgate obtained a similar result ($Cr=+0{\cdot}55$), as also did Colacevich [400].[2] The dispersion becomes very great when, by correct photometric procedures, the equivalent breadth of the K lines (E. G. Williams [378], R. F. Sanford [247]), or that of the D lines is measured accurately. In this last case, P. W. Merrill and O. C. Wilson [201] obtained the correlation co-efficient $Cr+0{\cdot}61$.

Thus the general opinion seems to be that there is no *direct* correlation between colour excess and interstellar line intensities; the correlation found could simply have been the result of the fact that the two quantities, on the average, increase with distance. In fact, according to the statistical work of Evans [94], the intensity of the K and D lines seems to be much more closely related to the distance r of the stars (obtained from Galactic rotation) than to the colour excess. Between E and r, the correlation found is still weaker. To explain these results it would be necessary to suppose that the gas is distributed in space rather more uniformly than solid particles.

Spitzer [290] tried, nevertheless, to interpret observations on the hypothesis of mixed clouds of gas and of grains, by taking into account the different velocities of the clouds. Let us suppose that in all the clouds atoms and grains exist together in a fixed proportion. The passage of light across a cloud gives a colour excess E_0 and the intensity I_0 of the spectral line. The crossing of j identical clouds necessarily produces the colour excess jE_0 proportional to the number of clouds, but the intensity of the line does not necessarily increase in the same manner. If, after emerging from the first cloud, the D line is already almost *saturated* (see page 123), as Spitzer thinks it is, its intensity, after crossing the second cloud, depends essentially on the relative radial velocities of the two clouds. If these velocities do not differ much, so that the two lines produced emerge almost unchanged, the intensity I_0 does not increase by more than a very small quantity ϵ. When the radial velocities differ so much that the two lines, displaced by the Doppler effect, do not recover their original positions, the total intensity (or the equivalent magnitude) would be practically doubled. Thus, after crossing j clouds, the intensity of the D line can take any value between $I_0 + \epsilon$ and jI_0. Thus we can understand why the strongly reddened stars show both very strong and relatively weak interstellar lines. It is this which is mainly responsible for the dispersion of the points mentioned above.

On the basis of this interpretation, Spitzer developed rather complicated calculations which take into account the thermal agitation of atoms inside the clouds, the particular velocities of the clouds and those very important velocities which Galactic rotation imparts to them. Theoretical curves represent, for each distance, the variation of colour excess with the intensity of the D line, and agree well with observation. But, to account for the measurement of the ratio of the intensities of the D_1 and D_2 lines, Spitzer is led to assume that there are two kinds of sodium clouds: "opaque" clouds, associated with grains, would be involved in the case of the reddened stars, and "transparent" clouds, absorbing less of the D line and also containing fewer grains, would only be involved in the case of non-reddened stars. The first would contain three times more sodium atoms than the second ($1{\cdot}5.10^{13}$ and 5.10^{12} atoms respectively, in a cylinder of 1 square cm. section and 1 kpc. length), so that the mean density of sodium atoms in space would be of the order of 7.10^{-9} atoms per c.c. It is impossible to evaluate the density in the "transparent" clouds, whose dimensions are almost unknown. If we assume that "opaque" clouds have an average radius of 8 pc., and that a line of 1,000 pc. crosses six or seven of them, we obtain in the clouds a density of 4.10^{-8} to 5.10^{-8} atoms of sodium per c.c.

Recently Binnendijk [396] resumed the general study of the relation between interstellar gas and solid particles using very extensive data. The colour excesses are always those of Stebbins and Whitford, but following a suggestion of Oort, they are increased by 0·04 magn. On the other hand, all the known measurements of the equivalent magnitudes of the H, K and D lines are used, these having first been reduced to a uniform scale. Finally, the distances of the stars are taken from the unpublished absolute spectroscopic magnitudes of W. W. Morgan.

By means of the ratios of intensities of doublets, the author has constructed the *curves of growth* for Ca^+ ions and Na atoms. When *saturation* does not take place, they agree very well with the theoretical curves of Strömgren. An examination of radial velocities shows further that the effect of the turbulence of the gas is of the same order of magnitude as the Galactic rotation effect, which Spitzer seems to have over-estimated.

With these data, it is possible to substitute for the *total* study of the relations between the intensity of the lines, the colour excess and the distance, a *regional* study which is much more fruitful. For a certain number of selected areas in the Milky Way, the curves representing, as a function of the distance, the variation of colour excess and of the intensities of the K and D lines are constructed separately. The latter increase uniformly with distance, while the colour excess often remains constant over a considerable interval. Thus, the fact that the gas is distributed in a more uniform manner than solid particles is confirmed. Furthermore, curves of equal intensity of the K and D lines, traced separately in the Galactic plane, are very similar.

To Spitzer's model in which both gases and grains were entirely concentrated in discrete mixed clouds, it is thus necessary to add a *con-*

tinuous distribution of the gas, and to note that observations do not allow the distinction between truly continuous distribution, and the existence of a great number of transparent clouds of small dimensions.

If both the curve of growth and the relation between the intensity of a line and distance, for a particular Galactic area are known, we can evaluate, in each of these regions, the total number of absorbing atoms up to 100, 200, 300 pc. of distance, etc., in a column of 1 square cm. section and thus construct a schematic chart of the distribution of atoms in the Galactic plane. A similar chart, for the distribution of solid grains, shows that in most cases, the clouds of particles roughly coincide with the gaseous concentrations. The Table below gives the densities found for Ca^+ ions and Na atoms, in the gaseous continuum, in the concentrated regions and also the mean value for all space:

	Ca^+	Na
Continuum	$0{\cdot}8.10^{-9}$ cm^{-3}	2.10^{-9} cm^{-3}
Concentrations	6.10^{-9}	32.10^{-9}
Mean Value	3.10^{-9}	11.10^{-9}

For Ca^+, Jentzsch and Unsöld [411] had previously found a mean value of 1.10^{-9}. For sodium, the mean does not differ much from that of Spitzer (7.10^{-9}) and the density obtained in the concentrations differs little from that in the "opaque clouds" (50.10^{-6}). Finally, by regarding Spitzer's "transparent" clouds as spread throughout all space, the same value is obtained for the density of the continuum. Thus, the continuous distribution corresponds to the "transparent clouds" of Spitzer, from which it cannot be distinguished, while the "concentrations" correspond to "opaque clouds".

Binnendijk then took up the study of the relation between the intensity of the lines and the colour excess, by considering separately stars whose distances are between 0 and 300 pc., 300 and 600 pc., etc., up to between 1,500 and 1,800 pc., in all Galactic zones together. In each interval of distance he finds an almost linear relationship. On the whole, the intensities of the lines are both a function of the distance and of the colour excess. On a first approximation, the equivalent breadths b_K and b_D of the K and D lines can be expressed, in angströms, by the equations:

$$b_K=0{\cdot}18\,r+0{\cdot}7\,E_1 \qquad \text{(XIII.1)}$$

$$b_D=0{\cdot}27\,r+1{\cdot}0\,E_1 \qquad \text{(XIII.2)}$$

where r is the distance in kpc. and E_1 the colour excess on the Stebbins scale.

Let us finally plot, for each separate Galactic area, a graph of the equivalent breadth of the K (or D) line against colour excess. The points fall along a straight line when all the stars belong to the same open cluster and are, in consequence, at the same distance from the Sun. But this is not the case when the stars are at various distances; most often two distinct straight lines are found—one for near stars, the other for more distant stars which, for the same colour excess, show stronger interstellar lines than the former. This behaviour results from the presence of a gaseous continuum. In the ideal case, only two groups of stars, situated at two different distances, are involved, and we should, in fact, find two parallel straight lines, since the intensity of the lines of most distant stars would all be increased by the same quantity, because of the continuous distribution of the gas between the two groups of stars. As these conditions are not exactly fulfilled, the two straight lines are only approximately parallel. We could further try to correct the equivalent breadths for the effect produced by the gaseous continuum. According to equation (XIII.1), we could, for instance, take the quantity $b_K - 0{\cdot}18\ r$ as ordinate, which should generally give only one straight line. It represents the relation between the intensity of the K line and the colour excess in the mixed clouds of gas and particles.

The connexion between the reddening of the stars and the intensities of the interstellar lines is thus well explained, and we can easily account for the absence of a straight-line relationship when one has stars at different distances and longitudes of the stars. The co-existence of grains and of atoms within the same clouds is established, but we must assume, amongst other things, that outside these mixed clouds, there exists either a continuous distribution of atoms, or numerous small clouds of gas, practically without solid particles. Further, nothing proves that the proportion of grains, molecules and atoms would be exactly the same in all the mixed clouds, and it is possible that it varies within certain limits.[3] As we shall see forthwith, it is possible that around hot clouds radiation pressure modifies the composition of the clouds.

All the preceding remarks only apply to atomic lines (Ca^+, Na). But a much closer correlation exists between the reddening and the intensity of diffuse interstellar bands of unknown origin, and this obviates precautions in the choice of the stars.

In 1938, Merrill and Wilson [201] found that there is between the colour excesses of eighty-five stars and the intensity of the band 6284 A, a correlation co-efficient $Cr = +0{\cdot}81$ (instead of $+0{\cdot}61$ in the case of the D line). The very large band 4430 A behaves in the same manner. Greenstein and Struve [117] drew attention to a star (HD 147889), which is only a hundred parsecs distant, and which shows a considerable colour excess ($E = +0{\cdot}50$ magn.), a very strong 4430 A band, and only a trace of the K line. In a group of supergiants near to γ Cygni, W. W. Morgan [209] showed two very reddened stars ($E \sim 0{\cdot}55$), and the K line is weaker than in the non-reddened neighbouring stars, but where 4430 A band is intense.

The correlation has been confirmed by the examination of the spectra of twenty-three stars of various colour excesses (Miss Sherman, 1939 [282]).

More recently Duke [78] measured the absorption at the centre of the 4430 A band in the spectra of 405 stars, whose absolute spectroscopic magnitudes were determined on the same negatives. On the whole, he found a very close correlation between that absorption and the colour excess E_1. There exist, however, very reddened stars where the band is weak and, conversely, stars of weak colour excess where the band is intense. Duke thinks that the regions of space in which the band originates are not always exactly those which produce the reddening. There exists also a very marked correlation between the intensity of the 4430 A and the equivalent magnitude of the band 6284 A, measured by Merrill and his collaborators, a correlation which is less strong with the magnitude of the D lines. It is true that in the latter case, the progressive saturation of the sodium lines complicates the interpretation of the results.

Finally, the regional study of Binnendijk [396] shows that the points obtained by plotting equivalent breadth against colour excess for the 4430 A band always fall along a single straight line whose slope is the same for all the Galactic areas considered. Absorption at 4430 A is, therefore, not produced in the gaseous continuum, but only in the mixed clouds containing atoms and grains.

The idea mentioned previously (page 122), that diffuse interstellar bands could either be due to absorption by the grains themselves, or to the molecules somehow associated with them, or, having gone into their composition, is thus plainly confirmed by the most recent work.

2. Temperatures in interstellar space

Today it is very generally held that interstellar matter is largely constituted of clouds containing both atoms, molecules and solid particles. We can imagine that the grains are formed by progressive condensation of gases, or conversely, that gases are produced from the vaporization of grains. It is, in any case, quite probable that a certain state of equilibrium is established between the various constituents of the cloud, since, on the whole, the relative proportions of atoms and of solid particles would appear to be fairly constant.

In order to form some picture of the phenomena leading to this equilibrium, we must first study the physical conditions which prevail in interstellar space. The respective temperatures of the grains, and of the gases, ought to play an essential role and both depend mainly on the radiation due to all the stars inside the Milky Way.

Radiation density of all the stars

At each instant an element of volume of the Milky Way is crossed by radiation coming from all the stars. We suppose further, that no star is too near to have a preponderant effect. In the neighbourhood of the Earth, for instance, we must deduct Solar radiation in order to fulfil this condition. The count of stars of each magnitude and of each spectral type[4] permits the evaluation of the *energy density*[5] inside the element of volume. Eddington found, in 1926,

$$n = 7{\cdot}7.10^{-13} \text{ erg./c.c.}$$

The calculations of Dunham (1939) [81], based on more recent data on the influence of absorption on a number of stars of each category, have confirmed the accuracy of this order of magnitude. In fact, they have given

$$u = 12{\cdot}5^{-13} \text{ erg./c.c.}$$

It is often useful to know further how the energy coming from all the stars is distributed as a function of wave-length. By comparing the stars to black bodies: the total radiation is made up as follows:

> Five per cent of stars at 18,000°K, 10 per cent at 12,000°K, 20 per cent at 9,000°K, 40 per cent at 6,000°K and 25 per cent at 3,000°K

(Eddington's evaluation). The distribution curve of the aggregate, in the visible and the near ultra-violet is not very different from that of a black body at 10,000°K. We can thus, as a first approximation, compare the complex radiations which traverses space to that of a black body at 10,000° K, *suitably reduced.* The energy density in the Milky Way is, in fact, much weaker than that of radiation in equilibrium inside an enclosure at 10,000°K. Let us define the *dilution factor* w by the equation

$$u = wu_0.$$

The energy density u_0 in the enclosure at the temperature T is, according to Stefan's law (page 34):

$$u_0 = \frac{4\sigma T_0^4}{c}$$

With $\sigma = 5{\cdot}7.10^{-5}$ C.G.S., $u = 7{\cdot}7.10^{-13}$ erg./c.c., $T_0 = 10{,}000$°K, we immediately find

$$w \simeq 10^{-14}\ ^{(6)}.$$

Temperature of solid particles

Let us consider a spherical particle, isolated in space and exposed on all sides to radiation from all the stars. Its equilibrium temperature—shown, for instance, by a thermo-element placed inside the particle—is reached when the energy radiated by the particle is equal to that which it absorbs.

If interstellar radiation were that of a black body (non-diluted) at a temperature T, the particle, irrespective of its optical properties, would assume the temperature T. But this is not the case since the radiation received is not that of a black body and *the temperature of the particle depends essentially on its absorbing properties*. This was shown in 1916 by Ch. Fabry in a remarkable paper [96], which has become of great importance today.

Let us first suppose that the particle of radius R, is perfectly *black*, in other words, that it completely absorbs all radiation. The energy absorbed is the product of the apparent surface πR^2 of the particle by the total energy of illumination cu. The power radiated is the product of the actual surface $4\pi R^2$ by the energy radiance of the particle, given by Stefan's law $\mathbf{R}=\sigma T^4$ (page 34). Equating these two quantities we obtain

$$T=\left(\frac{uc}{4\sigma}\right)^{\frac{1}{4}}$$

for any radius R.

In numerical calculation it is convenient to replace u by

$$wu_0=4\sigma w\frac{T_0^4}{c}.$$

We thus obtain

$$T=T_0 w^{\frac{1}{4}}$$

With $T_0=10{,}000°$K, $w=10^{-14}$ we find

$$T=10^{\frac{1}{2}}=3{\cdot}2°\text{K or } -269{\cdot}8°\text{C}.(^7)$$

Our "black thermometer" placed in space, very far from the star, would show a temperature of 3° to 4° above absolute zero.

It would change nothing in the preceding result if the particle were a *grey* body, absorbing for all wave-lengths the same fraction $\mathbf{A}$ of the incident energy, since both sides of the equilibrium equation are multiplied by $\mathbf{A}$.

Let us now take another extreme case, and suppose that the particle possesses a single (and relatively narrow) absorption band. Outside the band the absorptivity and, in consequence, the emissivity are zero, and therefore, the energy exchanges would be produced in the band only, i.e. for approximately monochromatic radiation, of wave-length λ_1. Let us represent by $f(\lambda_1, T_0)$ the spectral radiance of the black body at a temperature T_0, for wave-length λ_1. The radiation density of the stars for this wave-length is

$$u=wu_0=4wf(\lambda_1, T_0)/c.$$

By denoting by $\mathbf{A}$ the average absorptivity of the particle in the band, the energy absorbed (per sec.) is:

$$\pi R^2\mathbf{A}_1 cu=4\pi R^2 w\mathbf{A}_1 f(\lambda_1, T_0).$$

Further, the spectral radiance of the particle, at the temperature T, is $\mathbf{A}_1 f(\lambda_1, T)$ and the power emitted is

$$4\pi R^2\mathbf{A}_1 f(\lambda_1, T).$$

Thus, the equality of absorbed and emitted energy leads to

$$wf(\lambda_1, T_0)=f(\lambda_1, T), \qquad \text{(XIII.3)}$$

which permits the calculation of T, if the function $f(\lambda, T)$ is made explicit.

For simplicity, let us take the approximate formula of Wien (0·16) (page 36) $f(\lambda, T)=C_1\lambda^{-5}e^{-C_2/\lambda T}$. The relation (XIII.1) becomes

$$we^{-C_2/\lambda_1 T_0}=e^{-C_2/\lambda_1 T}$$

and, by taking Napierian logarithms of both sides,

$$\frac{1}{T}=\frac{1}{T_0}-\frac{\lambda_1}{C_2}Lw \qquad \text{(XIII.4)}$$

The temperature T depends on the position of the absorption band, not on the absorptivity. This very approximate formula suffices to show that the temperature varies within large limits, falling by some thousands of degrees to only a few degrees above absolute zero, when the absorption band considered changes from the near ultra-violet to the distant infra-red.

Van de Hulst [342] has calculated the temperatures by distinguishing metallic spheres from dielectric spheres. In the first case, at low temperatures ($T<30$°K), the super-conductivity of the metal is involved, which lowers their resistivity enormously. Owing to this, spheres which are large compared with wave-length radiate proportionally to T^7 and spheres which are small compared with their wave-length, proportionally to T^{11}. Finally, Van de Hulst concluded that the temperature of the large spheres ($R>10\mu$) would be of the order of 10 to 30°K, and that of the small spheres ($R\sim 0, 1\mu$) of the order of 60 to 120°K.

In the case of dielectrics, Van de Hulst considers two extreme possibilities. He first supposed (as was done above) a single narrow absorption band, of wave-length λ_1 (Case a). Then he supposed that the particle absorbs all the wave-lengths above a certain limit λ_1, equally (Case b). Here are some temperatures calculated on these two hypotheses:

λ_1	0·25	0·50	1·0	4·0	6·0	32	μ
Case a	1500	840	450	121	32	16·6	°K
Case b	3·6	3·3	2·5	1·0	0·35	0·22	°K

We see that the temperature can take all possible values between 0 and 1,500°K! But the extremes are most improbable. The absorption bands are probably in the region of 10–50μ (for ice of 5μ to 1,000μ perhaps). The second hypothesis then gives a temperature of 10 to 50°K and discussion of all the facts led Van de Hulst to suppose 10 to 20°K as the most plausible value.

Temperature of the gas

(a) *Electron Temperature*—Let us suppose for a moment that interstellar gas is made up only of atoms of hydrogen. These absorb all the photons of frequencies above the limit $\tilde{\nu}_0$ of the Lyman Series (page 62). The absorption of a photon tears from the atom an electron with a velocity which varies as the frequency of the photon. The initial distribution of velocities is thus determined by the energy distribution in the exciting radiation. If the final distribution of the velocities obeyed Maxwell's law (and this hypothesis can be considered well verified), we can characterize it by a single parameter T_e, which is the electron temperature (page 76).

It is essential to note that the initial distribution of velocities depends only on the relative intensities of the various wave-lengths, not on their absolute intensities. If all are weakened in the same ratio, there would be less photons absorbed per unit time, less electrons torn away, but the distribution of velocities would not be changed. In other words, the factor of dilution is not, in fact, involved. Thus in interstellar space the velocities of photo-electrons are distributed as in the atmosphere of a star which is crossed by black-body radiation of about 10,000°K. It would thus appear that the electron temperature in space is very high, and that we could, without committing any important error, suppose it to be in the neighbourhood of 10,000°K (Eddington, 1926). It would naturally be higher in the neighbourhood of very hot stars (O, B) and lower far from these.

(b) *The Equipartition of Energy*—In colliding with atoms, electrons impart a certain velocity to them. Because of the rarefaction of interstellar space, these encounters are rare, but it is not absurd to think that on the scale of astronomical time, a final state of equilibrium between electrons and atoms could be reached.

This equilibrium would be characterized by the *equipartition of the energy of translation.* Denoting by $\overline{V_H^2}$ the mean square velocity of the hydrogen atoms of mass m_H, and that of the electrons, of mass m_e, by $\overline{V_e^2}$, we would thus have

$$\frac{m_e\overline{V_e^2}}{2}=\frac{m_H\overline{V_H^2}}{2}=\frac{3kT}{2} \tag{XIII.5}$$

In fact, this is always the case when many kinds of particles of different mass are found together. The mean square velocities are thus inversely proportional to the masses.

At not very small pressures, the collisions, which are extremely frequent, rapidly establish not only equipartition of the energy of translation between molecules of different masses, but also the equipartition of energy between the various *degrees of freedom* of the molecules (3 degrees of freedom for translation, 2 for rotation, 1 for vibration, in the case of biatomic molecules considered as two spheres placed on the same axis). Thus the temperatures derived either from the energy of translation, the energy of vibration, or from the energy of rotation of the molecules, are equal. The kinetic temperature of the gaseous mixture is unique and well defined. Is this also the case for the very low pressures which are obtained in interstellar space?

Let us suppose that due to a collision (or else to the absorption of a photon), a molecule is raised to a certain level of rotation or vibration, or to an excited electron level. The lifetime of these levels—even if they are metastable—is always very small compared with the time interval of some weeks by which, on the average, two collisions are separated. Thus, between two successive collisions, the molecule spontaneously drops to its normal level, emitting permitted or forbidden radiation.

This explains the well-established fact (page 124) that ions, atoms and molecules of space are practically all found in their ground state, as is the case at low temperatures. The temperatures of rotation or vibration of the molecules, for instance, are very near absolute zero, but their temperature of translation might be different.

It is not absurd to think that a given gas, very rarefied, can appear as very cold, if we consider its energy of vibration or rotation (or else its electron energy), and hotter, if we evaluate its energy of translation.

In interstellar space, only the energy of translation follows the mechanism of equipartition. Thus, on the astronomical scale of time, equilibrium between atoms, ions and electrons is rapidly established. The equipartition probably even extends to solid grains at the end of 10,000 to 100,000 years.

(c) *Temperatures of the H II regions*—Until very recently, a high temperature of translation was almost always assigned to interstellar gas (8,000 to 10,000°K, for instance), in contrast to a very low temperature for solid grains (10 to 20°K, according to Van de Hulst). The work of Spitzer (1947, 1948) [288], and that of Spitzer and Savedoff (1950) [294] show that this rather oversimplified conception ought to be completely revised. We must take into account cosmic radiation, the presence of atoms other than hydrogen, and finally, solid grains.

In the regions where hydrogen is ionized (regions H II of Strömgren, page 125), the forbidden doublet O II 3727–3729 A generally is of an intensity which is comparable to that of Hα. As its excitation by the electrons requires the energy 3·31 eV (Table II, page 55), the electron temperature is certainly some thousands of degrees (at least 7,500°K). In fact, the calculation of Spitzer and Savedoff leads to a temperature of this order, taking probable values for the density of protons and other ions. Thus, with 10^{-2} protons per c.c., we find, far from all stars, 6,000°K or 11,000°K if we assume 2.10^{-6} or 2.10^{-5} ions per c.c. With 1 proton per c.c. and 2.10^{-3} ions, we obtain 5,000°K in the neighbourhood of a B star, and 6,900°K in the neighbourhood of an O star. When the density of ions is ten times smaller (2.10^{-4} c.c.), these temperatures become 9,500 and 12,000°K respectively. Finally, the temperature of translation of gases in the H II region is probably between 7,500 and 13,000°K. Thus, the value of 10,000° assumed previously remains valid.

In these regions, equilibrium of temperatures is established rapidly. Outside the clouds (with 10^{-2} protons/c.c.), the time for this establishment is of the order of 2.10^5 to 2.10^6 years. Since it is inversely proportional to the density of the protons, equilibrium is attained even more quickly inside the clouds.

(d) *Temperature of the H I regions*—In the regions where hydrogen is not ionized, the heating is due largely to cosmic radiation, since the density is very small, whereas in higher densities, the capture of electrons by ions (carbon, magnesium, sulphur, iron) followed by emission, plays an important role. But there are many causes of cooling. They are: excitation of the lowest levels of the C^+ and Si^+ ions, excitation of the first levels of rotation of molecules of hydrogen (H_2) by collisions of H atoms, and finally the inelastic collisions against the solid grains. The equilibrium temperature is almost always lower than 1,000°K (even if they are only 10^{-2} H atoms per c.c.). In the clouds, where the density of hydrogen can reach 10 atoms per c.c., the temperature is probably between 30° and 100°K. We may assume an average value of 60°K.

The time to establish equilibrium is again the shorter, the greater the density. From the order of 1,000,000 years or more in the clouds, it becomes extremely long for low densities. If an H I region outside the clouds was initially at a high temperature—if, for instance it had previously been an H II region subsequently abandoned by the exciting star—it is doubtful whether it could have cooled appreciably in 1,000,000,000 years. *Thus there ought to exist in the Milky Way large local differences of temperature capable of giving rise to enormous convention movements.*

3. Role of radiation pressure

Ratio between pressure and gravitation

We know that when electromagnetic radiation falls normally on a plane black surface of dimensions large compared with wave-length, it exerts on this surface a pressure which is numerically equal to the energy density in the incident beam. If, instead of being completely absorbing, the receptive surface is a perfect reflector, the reflected beam exerts a pressure which is equal to that of the incident beam. The radiation pressure is doubled, as is the energy density in front of the surface.

The pressure of light is always very small. Let us evaluate, for instance, that exerted by Solar radiation on a surface of 1 square cm. at the distance of the Earth. Above the atmosphere, the Sun produces an energy of illumination of 0·138 watts per square cm., or $1{\cdot}38.10^6$ erg. per sec. per square cm.[8] We obtain the density u by dividing this illumination by the speed of light $c=3{\cdot}10^{10}$ cm./sec. (footnote 5 this chapter), which gives $u=4{\cdot}6.10^{-5}$ erg./c.c. Thus the radiation pressure is $4{\cdot}6.10^{-5}$ bars. A surface of 1 square cm. is submitted to a force of $4{\cdot}6.10^{-5}$ dynes, a surface of 1 hectare to approximately 4·6 grams weight. Thus the radiation pressure predicted by Maxwell in 1874 could only be demonstrated by means of extremely delicate experiments (Lebedew, Nichols, Poynting).

In spite of its smallness, it is this radiation pressure which largely determines the movement of the solid grains of space, because of their very small mass. It naturally assumes special importance in the neighbourhood of giant and supergiant stars, which are much brighter than the Sun.

But the simple theory of Maxwell can no longer be applied when the dimensions of the illuminated surfaces become of the order of magnitude of the wave-lengths. The force F which light exerts on a small sphere of radius R is then of the form

$$F = \pi R^2 u \Psi(x),$$

where x as before denotes the parameter $\frac{2\pi R}{\lambda}$ (page 211) and u the energy density. $\Psi(x)$ is a new efficiency factor which can be expressed in principle, by means of Mie's theory (page 211). If R is large compared with λ (x large), $\Psi(x)=1$ and we return to Maxwell's law. Further $\Psi(x)$ tends towards zero with R. In 1909, Debye calculated the function $\Psi(x)$. The results were used to study the influence of radiation pressure on solid grains of space (Gerasimovic [101]; Schoenberg and Jung [265]; Greenstein [105]; Schalén [257]).

Not far from a star a particle experiences a force of repulsion F due to radiation pressure and a force of attraction due to gravitational attraction G. Both vary inversely with the square of the distance of the star, such that the ratio $\frac{F}{G}$ is independent of that distance. The particle is repelled by the star if $\frac{F}{G}$ is larger than unity; it is attracted if this ratio is smaller.

The ratio $\frac{F}{G}$ does not vary with radius in the same way for absorbing as for perfectly reflecting particles. In the first case it increases more and more slowly as R decreases, in the second case it passes through a maximum when the radius is approximately 10^{-5} cm. (1000 A).

Almost all stars repel particles of specific gravity 1. Only the dwarfs of the "advanced" types (G, K) attract particles of specific gravity equal to that of iron (7·8) (Schalén). Dielectric particles of refractive index between 1·1 and 1·5 are also repelled by stars when their radius is between certain limits (Greenstein) (approximately 10^{-3} and 10^{-6} cm., for instance, in the case of the B3 stars).

An isolated grain in space is submitted to the force $F-G$, inversely proportional to the square of the distance. The study of its motion shows that, if it starts from a distant region with a particular initial velocity, it cannot approach closer to the star than a certain easily calculated limit.[9] By supposing that the initial velocity is of the same order of magnitude as the particular velocity of the star, Schalén found that the radius of the sphere forbidden to the grains is of the order of many parsecs around the B stars. It is already very small around the A stars (0·01 pc. at most).

As the radiation pressure is weaker on atoms than on solid grains, the very hot stars (O and B) could thus expel the particles far away, and around these stars only clouds of gas would remain. These clouds excited by ultra-violet radiation from the stars, emit the characteristic line-spectra of the nebulae (Greenstein). On the other hand, the radiation pressure of the B stars could play an important role in the distribution of clouds (Schalén). The study of certain dark nebulae seems to have confirmed this conclusion. Thus, a very small number of B stars are projected inside the apparent surface of the dark cloud of Auriga-Taurus, while these stars are much more numerous around this cloud. Conversely, the distribution of stars of the classes F, G and K seems the same inside and outside the cloud. Similar remarks also apply to the clouds near θ and ρ Ophiuchi. *Everything apparently takes place as if the radiation of the B stars produced the boundaries of the clouds* (Schalén).

Nevertheless, in accordance with the more recent work of Spitzer [286], the preceding calculations, exact only if we consider a cloud of particles, ought to be modified to account for atoms. The latter are more numerous than the grains, which constantly collide with them and change their motion. Around giants of the advanced type, the interaction of atoms and particles is sufficiently small not to affect the results, but this is no longer the case around the B stars. Hydrogen is ionized there (H II region) and the equipartition of the energy of translation between protons and grains is established so quickly that the composite cloud behaves like a homogeneous gas, repelled *in its entirety* by the radiation of the star. But the effect produced is *considerably smaller* than in the case of a cloud made up exclusively of solid grains.

Spitzer also studied the action of the radiation pressure of all the stars of the Milky Way on grains at a great distance, and he found that it cannot expel solid particles beyond the Galaxy.

Application to stars of the type T Tauri and to emission-line nebulae associated with dwarf stars

Dielectric grains of all diameters are attracted by dwarfs of the advanced type, which have only a weak radiation pressure. They can either reach the atmosphere of the star or else approach it sufficiently to be vaporized.

It is in this manner that Greenstein [111] tried to interpret the spectral peculiarities of stars of the type T Tauri. Joy [159] had classified twelve irregular variable stars of this type (average amplitude three magnitudes), apparently inside absorbing clouds, and often associated with diffuse nebulae. Thus R Coronae Australis and R Monocerotis become the nuclei of two variable nebulae that look like comets (NGC 6729 and 2261); T Tauri is associated with the variable nebula of Hind (NGC 1555). To this group belong also the star BD—6° 1253, situated in the diffuse nebulae NGC 1999 (135, 210]. Almost half of these stars are very closely grouped in the dark cloud of Taurus.

Their spectra show, superimposed on absorption lines of a dwarf of the type dF5 and dG5, a great number of weakly excited emission lines (H, Fe II, Ti II, Fe I, etc.). The energy necessary for their production could be liberated by the fall of the grains in the atmosphere of the star or even by their evaporation. The nebulae associated with R Monocerotis and R Coronae Australis show sharp and deep absorption lines, similar to those of the A and F supergiants; without doubt, they originate in an envelope spread around the star (Greenstein).

T Tauri also possesses a nebular envelope where emission lines of O II and S II appear. In the neighbourhood of NGC 1999, Herbig [136] recently discovered three small nebular objects and he managed to photograph the spectra of two of them. We find there, on a very weak continuous background (probably due to the star) bright lines of gaseous nebulae; Balmer lines of O I (6300–6364 A), O II (3727 A), and weakly those of O III and of Ne III. The doublets of S II 4068–4076 and 6717–6731 A are particularly intense. To explain, as is customary, the excitation of these nebular spectra by ultra-violet radiation from a hot star, it would be necessary to suppose the somewhat improbable existence of a blue companion near T Tauri, and of stars of weak luminous intensity, but at high temperature, in the two nebular objects of Herbig. It is more probable, as Herbig himself thinks, that the excitation is produced through an entirely different and still unknown mechanism: the action of interstellar matter on dwarfs of advanced types.

NOTES TO CHAPTER XIII

[1]It should be noted that all these together would give a very strong mean absorption ($6\times0.32+0.6\times2.7=3.54$ magn./kpc.).

The statistical study of the fluctuations of the apparent density of extragalactic nebulae, resumed by Constance Warwick [366] seems to confirm the existence both of small and very large clouds. The distribution of galaxies observed in high latitudes (between $\pm40°$ and $\pm60°$) clearly demonstrates the existence of clusters of galaxies, but gives no evidence in favour of irregular absorption. Between latitudes $\pm15°$ and $\pm30°$ the distribution favours small clouds of diameter not larger than 15 pc., but the large darkened regions can only be explained by large clouds. Finally, bearing in mind the tendency of galaxies to be grouped in clusters, C. Warwick finds a much smaller value than Ambarzumian for the opacity of small clouds (about 0.08 magn.).

[2]It is true, however, that other authors found no correlation at that time.

[3]The fact that the straight lines obtained by plotting for various Galactic regions b_K—0.18 r against E_1, do not all have the same slope may be due to a difference in the proportion of atoms to grains, but also to density fluctuations inside the gaseous continuum.

[4]Spectral types would not have to be taken into consideration if the bolometric magnitudes (giving the energy) of all stars could be determined by means of a thermal detector (such as a bolometer) equally sensitive to all wave-lengths. The knowledge of spectral type allows us to pass from visual or photographic magnitudes to bolometric magnitudes (page 24).

[5]The energy density u is connected with energy of illumination in a simple manner. By definition the latter measures the energy flux crossing unit surface per second. The flux may be considered as being localized at any one instant in a cylinder of section 1 square cm., and of length equal to the velocity of light (c), i.e. in volume c. Thus each c.c. of the cylinder contains the energy $\mu=\frac{E}{c}$.

[6]A better approximation is obtained by calculating, for each wave-length, the temperature $T\lambda$ of the black body at which the density of radiation at equilibrium would be equal to the radiation density of the Milky Way, for the wave-length under consideration. This equivalent temperature decreases rapidly with increase in wave-length. For $\lambda \geqslant 0.2\mu$ it is, according to Swings [329], given by

$$T\lambda=389^{-0\cdot95} \text{ (or } \log T\lambda=2\cdot59-0\cdot95 \log \lambda\text{)},$$

which gives respectively for $\lambda=0\cdot3$, $1\cdot0$ and 10μ,

$$T\lambda=1{,}200°,\ 389° \text{ and } 44°\text{K}.$$

[7]This temperature corresponds to the energy density $u=7\cdot7.10^{-13}$ erg./c.c. of Eddington. With $u=12\cdot5.10^{-13}$ erg./c.c. (Dunham) we should have $T=3.6°$ K.

[8]This is called "solar constant". It is more often evaluated in gram-calories per square cm. per minute. 0.138 watt/cm.^{2}=0.138×60/4.18=1.98 cal. gm./cm.^{2}/minute [217].

[9]Let F_0 and G_0 be the pressure and the force of gravity at the surface of a star of radius R_0. The force acting at a distance r is

$$F-G=(F_0-G_0)\frac{R_0^2}{r^2}$$

The variation in potential energy of a particle of mass m is equal to the work corresponding to a very small displacement dr;

$$d\left(\frac{mV^2}{2}\right)=(F_0-G_0)\ R_0^2\frac{dr}{r^2}$$

By integrating from infinity, where the initial velocity is V_1, to the distance r_0 where it is zero, we immediately obtain

$$\frac{mV_1^2}{2}=(F_0-G_0)\frac{R_0^2}{r_0}$$

which, if the other quantities, and especially V_1 are known, enables us to calculate r_0.

CHAPTER XIV

FORMATION OF GRAINS. EQUILIBRIUM BETWEEN GAS AND SMOKE

Lindblad's hypothesis

We shall now try to find out by what mechanism there is established an apparent equilibrium in which an almost invariable proportion of gas and solid particles is maintained within the various Galactic clouds. We shall start from the basic hypothesis that grains result from the condensation of gas. Atoms and molecules crossing the cloud are captured by very solid grains to which they become attached.

This hypothesis, first put forward by Lindblad (1935) [176], has shown itself to be very fertile and has served as the point of departure for much theoretical works on the part of Jung, Spitzer and mainly, in the last few years, of the Dutch physicists Oort, Van de Hulst, Kramers and Ter Haar. It is hardly necessary to underline the highly hypothetical character of most of this work of which we are going to give a short résumé.

Formation of condensation nuclei

We must first try to understand the origin of the nuclei on which the atoms condense. Thus, the first problem which we must consider is the formation of biatomic molecules; it was treated by Ter Haar [334], and later by Kramers and Ter Haar [172].

Under conditions in which the laws of chemical kinetics prevail, molecules may be formed when three atoms meet. Interstellar gas is, however, much too rarefied for these triple collisions to play an appreciable role (their frequency is up to 10^{24} times smaller than that of double collisions). Molecules can only be formed by *radiative association*; that is to say, association accompanied by the emission of a radiation, in accordance with

$$A+B \rightarrow AB+h\nu.$$

As the atoms A and B approach each other, the system effectively forms a dipole whose moment varies with time. The probability of the formation of a molecule at the moment of double collision is extremely small, when the capture is not accompanied by electron transitions. This is the case for the molecules NH, OH, NO, NaH, etc. It becomes much greater if electron transitions take place at the same time, as for the molecules CH^+, CH, CN^+, CN, N_2^+, CO, BH. Thus, the predominance in interstellar gas of the molecules CH, CH^+ and CN, showing the most distinct absorption lines (page 122), is explained qualitatively.

Quantitatively, we can endeavour to evaluate the number of CH and CH^+ molecules per c.c. when equilibrium is established for C and H atoms, C^+ ions and electrons. At the time when Kramers and Ter Haar made their calculations (1946), however, only very uncertain values of the velocity co-efficients of the different reactions were available; it was still supposed that the temperature of the gas was near 10,000°K and further the concentration of the CH and CH^+ molecules in space was greatly overestimated. The whole problem was re-examined by Bates and Spitzer (1951) [24] using much better values.

Dunham [82] measured the equivalent breadth of the lines CH 4300·3 A and CN 3874·6 A in the spectra of some stars (six and three respectively), and the equivalent breadth of the line CH 4232·6 is estimated to differ little from that of the line 4300·3. Starting from this, Bates and Spitzer evaluated the average density of the molecules between the stars and the observer, by taking the most probable values of the "oscillator strengths" of the transitions considered. They obtained

$$n(\mathrm{CH}) = 3.10^{-8} \text{ molecules per c.c.,}$$
$$n(\mathrm{CH}^+) = 1{\cdot}5.10^{-8} \text{ molecules per c.c.,}$$
$$n(\mathrm{CN}) \sim 1{\cdot}9.10^{-9} \text{ molecules per c.c.}$$

On the other hand, the numerous visual estimates of the intensity of the molecular lines made by Adams [2] can be calibrated by means of some measurements of equivalent breadths made by Dunham. Bates and Spitzer found that there exists a marked correlation between the colour excess E_1 of Stebbins and the equivalent breadths of the lines CH 4300 and CH^+ 4232 A, *in the case of stars of types O*5 *to B*1 (when the effect of distance is not appreciable). It leads to the equation

$$n(\mathrm{CH}) = 2{\cdot}9.10^{14}\, E_1$$

where n(CH) represents the total number of CH molecules contained in a column of 1 square cm. section, between the star and the observer. If we take as the average

$$E_1 = 0.13 \text{ magn./kpc. (page 179)}$$

we get a new value for n(CH):

$$n(\mathrm{CH}) = \frac{2{\cdot}9.10^{14} \times 0.13}{3{\cdot}08.10^{21}} = 1.2{\cdot}10^{-8} \text{ cm.}^{-3}$$

$$(1 \text{ kpc.} = 3{\cdot}08.10^{21} \text{ cm.}).$$

Similarly, for CH^+, we have

$$n(CH^+)=6.10^{-9} \text{ cm.}^{-3}$$

(Bates and Spitzer take E_1=0·17 magn./kpc., which gives

$$n(CH)=1{\cdot}6.10^{-8} \text{ cm.}^{-3},\ n(CH^+)=8{\cdot}10^{-9} \text{ cm.}^{-3}).$$

These values are almost of the same order of magnitude as those previously derived from the simple measurements of Dunham.

Bates and Spitzer tried thereupon to calculate theoretically the density of the CH and the $CH+$ molecules, at equilibrium, by taking into account the different possible reactions between molecules, C and H atoms, C^+ ions and electrons. The radiative associations

$$C+H \rightarrow CH+h\nu,$$
$$C^+ + H \rightarrow CH^+ + h\nu$$

remain the essential processes, but account is also taken of the photo-ionization of CH and of the converse reaction (radiative recombination of CH^+ and an electron), the dissociative recombination of CH^+, which gives an atom of neutral carbon and an atom of neutral hydrogen, finally the photo-dissociation of CH^+ which gives, together with an H atom, either a C atom, or a C^+ ion. As certain of the co-efficients involved in these reactions are not well known, the authors successively give them a number of plausible values, taking 100°K as the temperature of the gas. The results of these calculations are very discouraging: in all the cases the calculated densities for CH and CH^+ are much lower than the observed densities (roughly 10,000 times smaller), if we assume, with Strömgren, that the number of atoms of hydrogen inside clouds is of the order of 10 to 20 per c.c. To reconcile theory with observation, one would have to suppose that at the centre of the clouds, the density may be as high as a thousand atoms per c.c., which seems to be altogether impossible. Perhaps solid grains play a part in the production of CH and CH^+ molecules, but then it would be necessary to discard the point of view that the latter are the original condensation nuclei of the grains.

Further, Bates and Spitzer showed that, in the case of the B6 to B9 stars, there is no longer any correlation between the intensity of the CH^+ 4232 A line and the colour excess. These stars, relatively near, are not reddened, but nevertheless, many of them show, in their spectrum, the line 4232 A. They all belong to the cluster of the Pleiades.[1] But, on examining the stars of the types B2 to B5, we find that the line 4232 A can appear in non-reddened stars in all regions of the sky.

These curious results seem to corroborate an idea of Merrill's [198], according to which the line 4232 A could, in certain cases, originate, not in the atmosphere of the stars, but in its immediate neighbourhood. We are no longer dealing with an interstellar line, but, as Merrill has said, with a *circumstellar* line. This opinion is also confirmed by the examination of radial velocities. Bates and Spitzer obtained a formula for the difference between the radial velocity of the cloud and that of the star in which the line 4232 A appears.[2] For the reddened stars (principally of the O5–B1 types), the difference is negative in most cases (twenty out of twenty-eight), and results from Galactic rotation (since the star is twice as far removed as the cloud on the average) and also from the unequal distribution of the stars in Galactic longitude. Conversely, for non-reddened stars near enough for the influence of Galactic rotation to be negligible, the difference is positive in fifteen out of sixteen cases, as if the cloud were approaching the star.[3]

Bates and Spitzer tried to explain the production of CH^+ molecules in the neighbourhood of B stars by sublimation, followed by dissociation, of CH^+ molecules contained in the grains. The results of their calculations favour this interpretation, but it still seems difficult to account in a similar manner for the presence of the CH 4300 line established in the spectra of at least two non-reddened stars.

The problem of equilibrium between the CH and CH^+ molecules has definitely not been solved. We are obliged to bypass it and to conclude, with Bates and Spitzer, that the theory of the formation of grains, at least as far as the first stages are concerned, is still essentially qualitative. Although the densities of the CH and CH^+ molecules formed by atomic and molecular reactions are very small, they are perhaps large enough to initiate the condensation.

If this point is granted, it would then be necessary to examine the capture of a second atom of hydrogen or of carbon by the CH molecule. Because of the abundance of hydrogen, the first is more probable, but the calculation of the probability of the formation of the CH_2 molecule is very difficult (Kramers). It is possible that some more hydrogen atoms attach themselves to the CH_2 molecule ($CH_2 \rightarrow CH_3 \rightarrow CH_4$?) but before long it is mainly atoms of much larger mass (C, N, O, . . .) which are captured, because the large molecule formed is not sufficiently cold to prevent the evaporation of hydrogen. Thus, little by little, due to a mechanism which is difficult to evaluate in detail, minute grains which seem to serve as centres of condensation are built up.

Growth of grains

If atoms, ions and electrons in movement meet a previously formed grain which is at a low temperature, a number of them attach themselves to the surface; others can escape by thermal evaporation or photo-dissociation of the molecules formed. If further, the particle carries an electric charge, the electrostatic forces which it exercises on the ions and on the electrons must also be taken into account.

All these points have been discussed by Van de Hulst. Experiments made at different temperatures on surfaces of different kinds show that the proportion of reflected atoms diminishes rapidly as the temperature becomes lower. The temperature of the particle ($\sim$10 to 20°K) seems too high for the condensation of hydrogen and helium atoms, but sufficiently low to fix all the other atoms. For the latter, thermal evaporation is negligible.

There only remains the question of electrostatic forces. A grain supposed initially neutral, can acquire an electric charge by two different mechanisms. It becomes positively charged by the photo-electric effect when interstellar radiation knocks electrons from it (Jung). Further, it becomes charged positively if it captures more ions than electrons, negatively in the opposite case. But the photo-electric effect is probably very small if the surface of the particle is not metallic, so that the second phenomenon is usually of greater importance than the first. We shall soon see that the frequency of the capture is proportional to the average velocity of the atoms, ions and electrons compared with that of the particle. In accordance with the relation expressing the equipartition of the energy of translation (page 241), this average velocity varies inversely with the square root of mass. Electrons of very small mass (1/1849 that of an atom of hydrogen) have an average velocity forty-three times larger than that of protons, and 172 times larger than that of O^+ ions. They are thus much more frequently captured than ions, and the grain becomes negatively charged. It then repels electrons and attracts ions, so that its excess charge rapidly decreases. According to Spitzer it is always very small.

Thus we can very probably obtain a correct order of magnitude for the velocity of growth of a particle by supposing that it retains all the atoms it encounters, with the exception of hydrogen and helium. Thus, a very simple calculation permits the evaluation of the duration t_1, necessary for the formation of a spherical grain of radius R_1. We find:

$$t_1 = \frac{4sR_1\mathbf{N}}{n_1\sqrt{3\mathbf{R}AT}} \qquad \text{(XIV.1)}$$

In this expression, s is the specific gravity of the grain, and n_1 the number of atoms per c.c. We suppose them all identical with one another, of the same atomic weight A, and having the same velocity, equal to the average square velocity at the translation temperature T. $\mathbf{N}$ and $\mathbf{R}$ respectively denote Avogadro's number and the gas constant for a perfect gases.[4]

With $\mathbf{N}=6{\cdot}02.10^{23}$ and $\mathbf{R}=8{\cdot}313.10^{7}$ C.G.S., we have $t_1=1{\cdot}52.10^{20}$

$$\frac{sR_1}{n_1\sqrt{AT}} \text{ seconds} = 4{\cdot}82.10^{12} \frac{sR_1}{n_1\sqrt{AT}} \text{ years.}$$

The times of formation calculated on various hypotheses are given in Table XII.

TABLE XII

Time of Formation of Grains

s	R_1 (cm.)	A	n_1 (cm.$^{-3}$)	T (°K)	t_1 (years)
7.8	3.10^{-6}	55·8	5.10^{-5}	10,000	3.10^{9}
				64	$3{\cdot}8.10^{10}$
1	1.10^{-5}	14	4.10^{-3}	10,000	$3{\cdot}2.10^{7}$
	3.10^{-5}				$9{\cdot}6.10^{7}$
	1.10^{-5}			64	$4{\cdot}0.10^{8}$
	3.10^{-5}				$1{\cdot}2.10^{9}$

The first two results are for iron grains ($s=7{\cdot}8$, $A=55{\cdot}8$). In this case, in conformity with Schalén's results (page 215) values of $R_1=3.10^{-6}$ cm. (300 A) and 5.10^{-5} atoms of iron per c.c. have been taken (corresponding to specific mass $n_1m=4{\cdot}6.10^{-27}$ gm./c.c. for the "iron gas"). The duration of 3,000,000,000 years, calculated on the temperature $T=10{,}000$°K is just the age attributed to the Milky Way today.* Galactic clouds probably originated at the same time as our entire stellar system and the grains probably continued to increase until the reserve of metallic atoms of space had been completely used up. At first this remarkable coincidence could be taken as a confirmation of the theory. But the temperature assumed is much too high for the H I regions. With the value $T=64$°K which, according to recent works of Spitzer and Savedoff, is much more probable we find a time of formation of above 30,000,000,000 years, which is hardly permissible.

The other results are for dielectric grains similar to ice, of specific mass near unity and of radii $R_1=10^{-5}$ cm. (1000 A) and $R_1=3.10^{-5}$ cm. (3000 A). After hydrogen and helium which do not attach themselves to grains, the most abundant elements in space are probably oxygen (O=16), carbon (C=12) and nitrogen (N=14) (page 224). Thus, an average atomic mass A=14 has been taken. The number of these atoms per c.c. in the clouds, can be of the order of 2.10^{-3} to 6.10^{-3}; one may take $n_1=4.10^{-3}$.

* It is now believed to be much longer, following Baade's restandardization of the period-luminosity curve of the Cepheid variables. A.J.P.

If the temperature is near 10,000°, as Oort and Van de Hulst supposed it to be, some tens of millions of years would suffice to form grains of radii between 1000 and 3000 A, and this time is very small compared with the age of the Milky Way. Now, in accordance with observation, the proportion of atoms and of grains and also of their diameters vary very little on the whole from cloud to cloud. This presents a very serious difficulty, since the rapid growth of the grains probably exhausts the reserve of interstellar gas very quickly. Oort and Van de Hulst emphasize the fact that the simultaneous existence of atoms and of particles should be an altogether transitory phenomenon and that it could only be due to a most improbable accident that we should be living just at the epoch when the ephemeral coexistence of gas and grains can be observed in all the regions of the Milky Way.

The difficulty is considerably reduced if we adopt the low temperature calculated by Spitzer and Savedoff, since the time of formation of grains then becomes of the order of 500,000,000 to 1,000,000,000 years.

Destruction of grains: Evaporation by collisions

In accordance with Oort and Van de Hulst's calculations (with T=10,000°K), all the clouds actually observed could be considered as being of recent formation. It would then be necessary to suppose that there is some cause of destruction of the solid grains which opposes their progressive growth and would permit the maintenance of a statistical equilibrium between the gas and the cloud.

Oort and Van de Hulst [225] have seen this cause of destruction in the encounter of two clouds of different velocities. The mutual collision of two particles can, in fact, release a quantity of heat sufficient to vaporize them both entirely, if the relative velocity is higher than a certain limiting value V_m, which, with reasonable hypotheses, Oort and Van de Hulst evaluate to be 2·90 km./sec.[5]

This limit is clearly lower than the average relative velocity of the two clouds, which Oort and Van de Hulst estimated to be of the order of 22·5 km./sec., a value which is perfectly compatible with the more recent work of Whipple (page 119). If two clouds meet, a particle of one of them penetrates more or less deeply into the second cloud, where it is progressively braked by the collisions with hydrogen and helium atoms. The vaporization is produced if it hits a particle of the second cloud with a velocity higher than the limiting velocity V_m.

If each cubic centimeter of a cloud contained $7{\cdot}6.10^{-12}$ particles of radius 10^{-5} cms. (a density corresponding, in the case of ice, to a photographic absorption of 1 magn./kpc.), one finds that the probability for a particle which penetrates into the cloud to be vaporized is 0·087 (one in twelve cases approximately). On the other hand, the data on the velocity of clouds, on their number per unit volume and on their dimensions show that the probability for one particle to penetrate into a cloud is of the order of 0·11 for 1,000,000 years. The probability of vaporization of a particle in 1,000,000 years would be $0{\cdot}11 \times 0{\cdot}087 = 0{\cdot}010$. The mean life of a grain would thus be of the order of 100,000,000 years and, in the course of its existence, a grain would encounter ten distinct clouds on the average.

Vaporization by collisions would explain the fact that the radius of the grains can never be greater than some thousands of angströms. The probability of vaporization increases very quickly with the radius and the

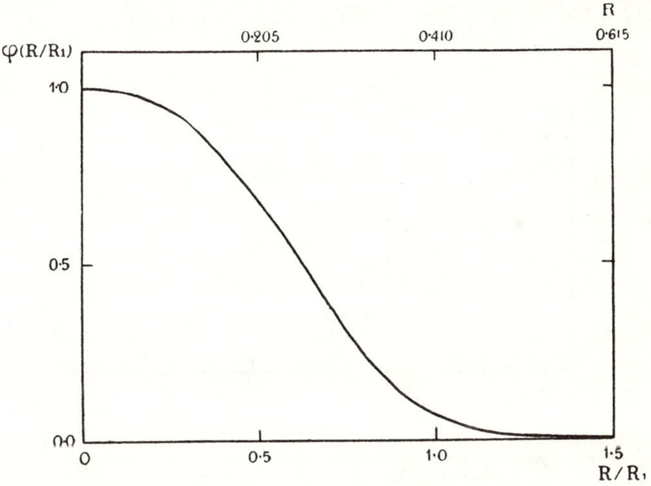

Fig. 41.—The distribution function of the radii $\phi\left(\frac{R}{R_1}\right)$ of Oort and Van de Hulst. On top are the values of R for $R_1=0.410\mu$.

largest particles can be considered as being completely absent. We thus understand why, in all the clouds, the gas can subsist indefinitely in the presence of grains.

The short times of formation of the grains calculated by Oort and Van de Hulst (with $T=10{,}000°$) definitely require their rapid destruction. This hypothesis is not as imperative for the case of longer times of formation obtained for the low temperatures by Spitzer. Vaporization by collision remains, nevertheless, quite probable and the encounter of two clouds can also explain other phenomena as we shall see later on.

Distribution function of radii

If the particles formed within the gas are sufficiently numerous and have reached a sufficient size, future formation and growth can be compensated by vaporization by collision. We can then imagine that a statistical equilibrium would be established. On this hypothesis, which is perhaps realized in the Milky Way, Oort and Van de Hulst have tried to determine the distribution function $\phi(R)$ of the radii of the particles. Their very complicated calculations have led to the curve shown in Fig. 41.

Let us recall that in it $\phi\left(\frac{R}{R_1}\right)$ is plotted against $\frac{R}{R_1}$. R_1 is a constant which we can vary from one solution to the next, but which is of the order of magnitude of the wave-lengths. It is this function which we previously used (page 220) in discussing the absorption curve, with R_1=4100 A, and which has led to a very satisfactory agreement with observation. It is

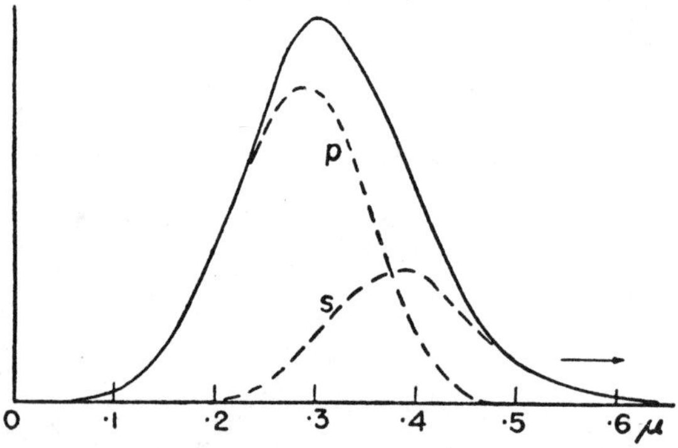

Fig. 42.—Contribution of particles of different radii to the absorption for $\lambda=0.44\mu$, according to Van de Hulst. The dotted curves p and s are for primary and secondary particles, the full curve is for all the particles together.

characterized by a rapid decrease in the number of particles for R between $0{\cdot}1\lambda$ and λ.

In Fig. 42 the curve p, in dotted lines, represents, according to Van de Hulst, the relative contribution of the particles of different radii to the total absorption for the wave-length 4400 A. Because of the large value of the corresponding efficiency factor, grains of radius $0{\cdot}3\mu$ play the largest part in spite of their restricted number.

Fusion of grains and formation of secondary particles

If two particles collide with too small a relative velocity to give rise to total vaporization, it may come about that they fuse to form one particle. Oort and Van de Hulst suppose that this takes place when the kinetic energy liberated by the collision suffices to carry the total mass to the temperature of fusion and to melt half of it. With a reasonable value of the fusion temperature (see note 5), they find that the necessary energy is

70 calories per gram ($0.29 \cdot 10^{10}$ erg./gm.). The minimum relative velocity corresponding to this would therefore, be

$$V = \sqrt{8 \times 0{\cdot}29.10^5} \text{ cm./sec.} = 1{\cdot}52 \text{ km./sec.},$$

for the case of two particles of the same mass.

Calculations similar to those made for vaporization, then, permit the evaluation of the frequency of the formation of these *secondary particles* and to study their influence on the distribution function of the radii.

The number of secondary particles should represent only 6 per cent of the total number of grains, but would contain 26 per cent of their mass. The curves and dotted lines p and s of Fig. 42 respectively show the contribution of primary particles (p) and secondary particles (s) to the total absorption for $\lambda = 4400$ A. The full curve corresponds to the contribution of all the particles, primary and secondary, together. All the secondary particles would be responsible for 30 per cent of the total absorption.

Luminous phenomena produced by the encounter of two clouds

The encounter of clouds responsible for the vaporization of grains ought to produce another directly observable, and even almost spectacular, phenomenon. Between two clouds which come into contact, there ought to be produced a compression to a temperature high enough to ionize the hydrogen atoms. The encounter is generally produced with a velocity larger than sound and ought to engender shock waves.

According to Oort's calculations [223], if the relative velocity of the two clouds, supposed to be of the same density and at the same temperature, reaches 40 km./sec., the density and the absolute temperature of the region between them ought to be multiplied by three. If the two clouds have different densities, we can assume the existence of a compressed zone in the external layer of each one of them. In these zones, the temperatures are inversely proportional and the densities directly proportional, to the original densities of the two clouds. It is, therefore, the thinnest cortical layer of the cloud which is carried to the highest temperature and easily becomes the most luminous. It is most probable that luminosity is strongly increased at the surface separating the two compressed layers where a sudden increase in density is produced.

Oort proposes to explain in this way the very bright fringe which frequently borders dark nebulae in contact with diffuse nebulae of moderate brightness. The latter would correspond to the compressed region of the thinnest cloud; the very bright fringe would represent the interface seen in section. A good example of these configurations is also provided by the nebula near ζ Orionis (Plate XX), another by the nebula North America (Plate X). According to Oort the energy liberated by the encounter of two clouds is of the order of magnitude required to produce all the radiation of the nebula. Perhaps a similar mechanism provokes the luminescence of diffuse nebulae in which no exciting stars have as yet been discovered.

Conclusion

Without appealing to too arbitrary hypotheses, the theory which we have just outlined accounts, in a logical manner, for all the known facts. It gives a plausible picture of the phenomena which lead to equilibrium between gas and smoke.[6] Is this picture really representative? It is only future observations which will decide this question.

At the end of Chapter XII we have indicated that the total composition of the interstellar medium, gaseous and solid phases taken together, would be almost identical with that of the planetary nebulae and the stellar atmospheres, if the heavy atoms are joined together to form solid metallic particles. But up to now, the theoretical study of the formation of grains has not revealed a mechanism capable of producing such a selection amongst the atoms of space. A priori, the formation of grains of a dielectric nature remains more probable.

Short of any such modifications, we must at least make some alterations in the scheme. First of all, it will be necessary to put the ideas of Oort and Van de Hulst in harmony with the recent calculations of Spitzer on the kinetic temperature of the gases of space. We have further supposed that the velocities of the clouds were distributed at random. It is possible that in reality, they result either from enormous convection movements, induced by strong differences in local temperatures, or by immense turbulent motions which are continually formed in the interstellar medium due to the differential rotation of the Milky Way. It is true that Oort and Van de Hulst [225], who have examined the latter possibility, have found no traces of turbulence on *a large scale* either in bright or in dark nebulae. Nevertheless, the possibility of turbulent movements is indicated by the high value of the Reynold's number for interstellar space (Zwicky, 1941 [392]). The theory has been developed in the last few years, in connexion with problems of aerodynamics, by Von Karman [164], von Weizsäcker [368] and Chandrasekhar [57]. S. von Hoerner [410] has devised a statistical method for the study of the turbulence in the interior of diffuse gaseous nebulae, and this has been applied to the Nebula of Orion. The study of the distribution of brightness of the Hα line has led Aller [5] to show the existence of strong turbulence in the extended region of Cygnus. Hardly begun today, the study of the turbulence of interstellar matter will, no doubt, give rise to important developments in the near future.

If the great majority of astrophysicists seem favourably inclined to the current theory of the formation of the grains, it is nevertheless necessary to mention the views of Cernuschi [55], who, after having criticized the ideas of Oort and Van de Hulst, has suggested an entirely different mode of evolution which is in some way quite opposite. According to him, instead of originating in the condensation of the gas, the solid grains are the result of progressive evaporation of larger grains, formed probably in the early history of the Milky Way, an epoch when the average density could have been much higher than its present value. Thus, they remain only as vestiges of the *very dense state* which, in the theory of Canon Lemaitre, is envisaged as the first stage of the expansion of the giant atom in which all the matter of the Universe is supposed to have been originally concentrated.

NOTES TO CHAPTER XIV

[1]For this interpretation to be wrong, all these stars would have to be intrinsically much bluer than they are assumed to be, but reddened by an absorbing cloud.

[2]They take as the radial velocity of the cloud that determined by means of the H and K lines, since this is more accurate and, according to Adams, identical with the determined velocity of the 4232 A line.

[3]This is sufficient to prove that it is not made up of atoms and molecules ejected by the star.

[4]The atoms of velocity v and mass m which hit the particle of radius R in the time dt are at any one instant contained in a cylinder of section πR^2 and of length $v\,dt$. Their total mass represents the increase in mass dM of the particle in the time dt:

$$dM=\pi R^2 n_1 mv dt. \qquad \text{(XIV.2)}$$

On the other hand, the mass of the grain of radius R is

$$M=\frac{4\pi R^3 s}{3}$$

and the increase in radius dR corresponding to the increase in mass dM is obtained by differentiating this expression $dM=4\pi R^2 s dR$. (XIV.3)

By eliminating dM between (XIV.2) and (XIV.3) the velocity of increase of the radius is obtained

$$\frac{dR}{dt}=\frac{n_1 mv}{4s}$$

Thus the rate of growth of the grain is uniform so long as it does not appreciably impoverish the surrounding gas (by diminishing n_1). Since we can neglect the original radius which was assumed to be very small compared with R_1, the time t_1 necessary for the formation of a grain of radius R_1 is thus

$$t_1-\frac{4sR_1}{n_1 mv} \qquad \text{(XIV.4)}$$

The mean velocity v differs little from the root mean square velocity $\bar{V}$ $\left(v=\sqrt{\frac{8}{3\pi}}\bar{V}=0{\cdot}921\,\bar{V}\right)$.

By substituting for v the mean square velocity at the temperature T

$$V=\sqrt{\frac{3kT}{m}} \qquad \text{(XIV.5)}$$

Equation (XIV.4) becomes

$$t_1=\frac{4sR_1}{n_1\sqrt{3kmT}}$$

It is convenient to replace k and m by $\frac{\mathrm{R}}{\mathrm{N}}$ and $\frac{A}{\mathrm{N}}$ respectively, which gives equation (XIV.1).

[5]With the data below:

Specific heat of the solid phase	0·2 cal./gm.
Specific heat of the liquid phase	0·3 cal./gm.
Latent heat of fusion	30 cal./gm.
Latent heat of vaporization	130 cal./gm.

we find that the heat necessary for the vaporization of particles is about 250 calories per gram ($250\times 4{\cdot}18.10^7=1{\cdot}05.10^{10}$ erg./gm.).

When two particles of equal mass m and of equal and opposite velocities $\frac{V}{2}$ meet, their total kinetic energy is $\frac{mV^2}{4}$. Let us assume that this is all used in vaporizing the two particles i.e.

$$mV^2=1{\cdot}05.10^{10}\times 8\text{m}.$$

from this we obtain

$$V=\sqrt{8\times 1{\cdot}05}.10^5 \text{ cm./sec.}=2.90 \text{ km./sec.}$$

[6]The term "smoke" (*fumée*) today replaces the term "dust" (*poussière*), since smoke results from the condensation of a gas, and dust from the dispersal of larger particles.

CHAPTER XV

POLARIZATION OF THE LIGHT OF THE STARS BY INTERSTELLAR MATTER

1. Observations

Discovery of polarization in certain stars

It is known that the light coming from stars, such as the Sun, is generally completely unpolarized (page 199). Now, in 1949, two American astrophysicists, W. A. Hiltner [138] and J. S. Hall [122] announced that the light of *certain* distant stars was partially polarized and that the phenomenon was apparently due to the passage of the light through Galactic clouds. This entirely unexpected discovery is of the greatest interest today.

The first work of Hiltner and Hall had the aim of showing an entirely different polarization effect, that predicted by Chandrasekhar [56] under strictly defined conditions. In the course of his theoretical work on the equilibrium of stellar atmospheres, Chandrasekhar predicted that, if scattering by free electrons plays an important role, radiation from the edge of the star ought to contain 11 per cent of polarized light (1946). This ought to be demonstrable for stars of the first spectral types (W, O, B), *provided that the radiation emitted by the edge of the disc could be isolated.* This can only be done in a single case: let us suppose that the star is part of a double system in orbital movement, and that it is totally (or almost totally) eclipsed once per period, by the dark (or much less luminous) second component. Under these conditions, just before the beginning and just a little after the end of the total eclipse, that light which is emitted from the edge of the star can be received by itself. It is in a similar manner that we observe the radiation of the edge of the Solar atmosphere (and especially the flash spectrum) during the total eclipses of the Sun, since the largest part of the disc is hidden by the moon in the short times which precede and which immediately follow upon the total eclipse. Thus, to demonstrate the effect predicted by Chandrasekhar, it would be necessary to investigate an eclipsing binary, of which the most luminous component belongs to the classes W, O, or even B, and which is totally or almost totally eclipsed by the second and colder component. It is further necessary to study the system just before or just after totality. The possibilities of investigating the theory are thus singularly limited.

Some of the photographic observations of Miss Jansen (1946) [156] on the system of U Sagittae, where a component of the B9 type is eclipsed by a giant G2 indicated the probable presence of weak polarization. A

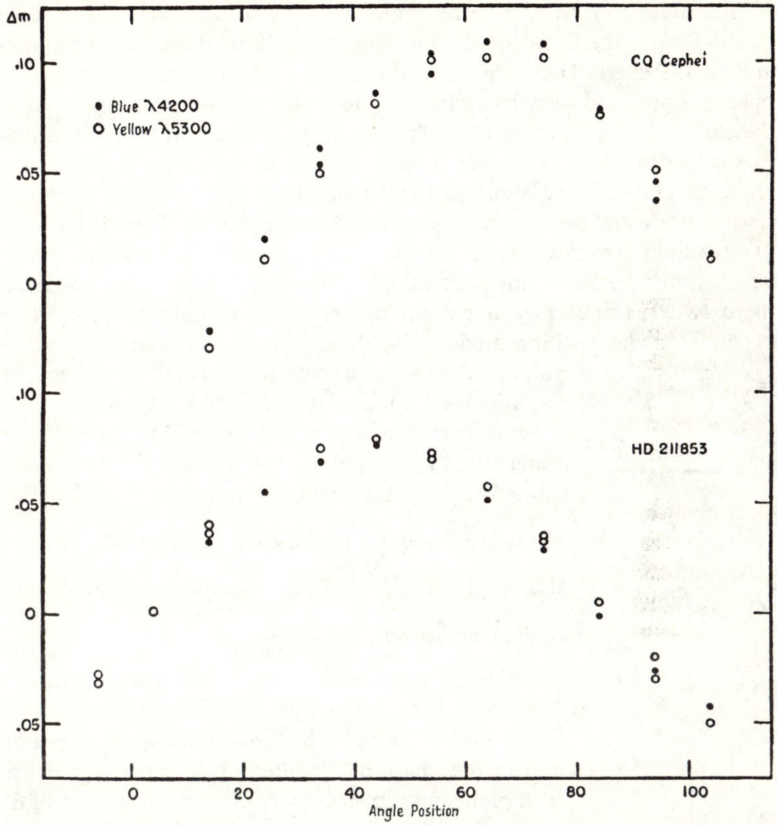

Fig. 43.—Polarization of the stars CQ Cephei and H D 211853 for two spectral regions centred on 0.420 and 0.530μ, after Hiltner. The ordinate is the difference Δm in magnitude for two rectangular orientations of the analyser, and the abscissa is the position angle θ.

similar result was obtained by Hiltner (1947) [137] for the case of RY Persei.

The star CQ Cephei whose most luminous component is of the WN6 type also shows almost 6 per cent of polarized light, but Hiltner [138] noticed that polarization continues outside of eclipses. *This can, therefore, not be the Chandrasekhar effect.* This conclusion was confirmed by the discovery of a weaker, but yet perfectly demonstrable polarization, in the case of stars which do not belong to eclipsing systems of the W type, or to more advanced types such as A0, or even to the Cepheid Z Lacertae (F8–K2).

Methods of observation

A. *Method of Hiltner.* Hiltner and Hall substituted the more precise photo-electric methods for the photographic observations on U Sagittae and R Y Persei). The simplest method, that of Hiltner [139], consists of placing a polaroid (which can be given two orientations differing by 90° in the same plane), in front of a photo-multiplier connected to a sensitive galvanometer. The apparatus is placed into the Cassegrain focus of a telescope and can be completely turned about the optical axis of the latter. Its positions are marked on a graduated circle. For each position angle two readings are taken, by giving the polaroid the two desired orientations which differ by 90° from each other. If the light is partially polarized, these two readings give a certain difference of magnitude Δm, which depends on the position angle θ. By doing this for a certain number of values of θ, we can plot the curve representing the variations of Δm as a function of θ (Fig. 43).

Δm is evidently a maximum when the vibration transmitted by the polaroid is parallel to the privileged vibration I of the light of the star. In that case $\Delta m = -2.5 \log \left(\frac{i}{I}\right)$. We thus obtain both the position angle of the privileged direction and the proportion of polarized light $p = \frac{(I-i).(^1)}{(I+i)}$

B. *Hall's Method.* The more complicated method of J. S. Hall [124] employs an analyser which turns with a constant velocity.[2] The measurement of amplitude and of the phase of the alternating current produced by a photo-multiplier permits the determination of the proportion of light polarized and the position angle of the privileged vibration.

The optical arrangement is shown in Fig. 44. A small diaphragm O (diameter approximately 0.5 mm.), in the focal plane of the telescope, isolates the light of the star. The beam, rendered parallel by the

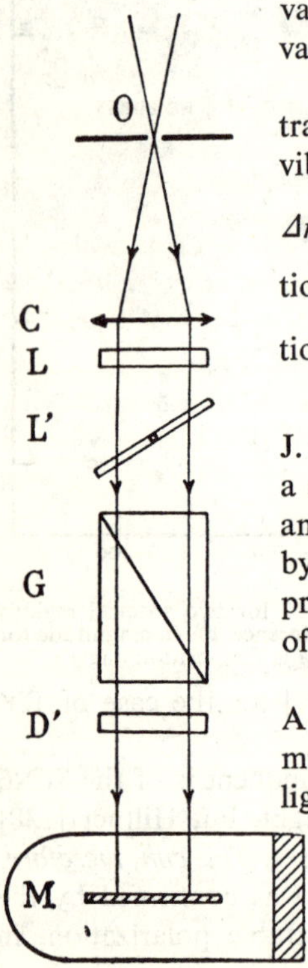

Fig. 44.—Diagram of the optics of Hall's apparatus. O diaphragm, C collimating lens, L glass plate or the Lyot Depolarizer, L′ thin glass plate that may be turned, G Glan-Thomson prism, D′ Lyot Depolarizer, M cathode of photo-multiplier.

lens C, traverses the plate L, the inclined plate L′, then the analyser G (Glan-Thomson prism) and finally the "depolarizer" D′, before falling on the cathode M of the photo-multiplier.

The partially polarized light coming from a star can be considered as two incoherent vibrations at right angles, of intensities I and i ($I>i$). Let ϕ be the angle which the privileged vibration of intensity I makes at any instant with the direction Og of the vibration transmitted by the analyser (Fig. 45).

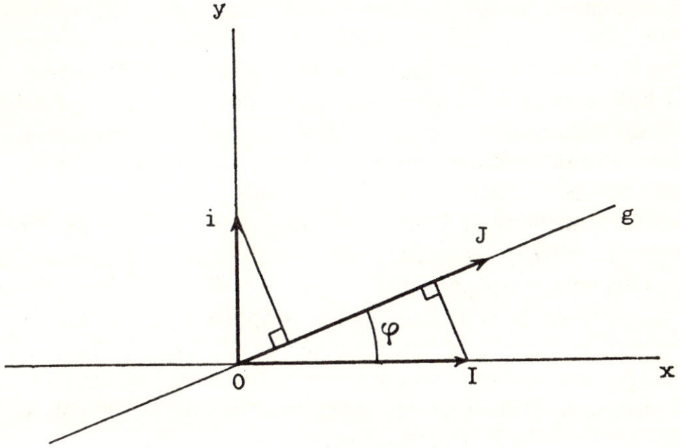

Fig. 45.

Apart from some losses, the intensity J of light transmitted by this is

$$J=I\cos^2\phi+i\sin^2\phi=\tfrac{1}{2}\,[(I+i)+(I-i)\cos 2\phi]=\tfrac{1}{2}\,[I+i)\,(1+p\cos 2\phi)]$$

It contains a term proportional to $\cos 2\phi$. Thus when the analyser is turned with the constant speed of 15 revolutions per second by a small synchronous motor, the multiplier gives an alternating current of double the frequency (30 cycles per second), which is sent into a synchronous amplifier with a narrow pass band. There the current is mixed with a current of the same frequency, formed by a succession of square pulses, so that the phase compared with that of the photo-electric current is varied progressively and in such a way that it describes a complete cycle in two minutes. The resulting current, emerging from the amplifier after detection, operates a miliammeter pen recorder.

In principle, each measurement consists of the three following operations, of which each lasts two minutes (Fig. 44):

(a) The light is received from the star with the plate L′ placed normally to the beam. The plate L is of glass of low absorption and its function will be appreciated in a moment. The pen of the milliammeter describes one complete oscillation every two minutes. We note its amplitude and the phase corresponding to the maximum;

(b) We repeat the observation substituting for the glass L a Lyot "depolarizer" formed of two quartz plates fixed parallel to the optical axis, but whose own axes are inclined at 45° to each other and whose thicknesses are in the ratio of 2 to 1 (for instance 1 mm. and 2 mm.). When measurements are made over a very broad spectral interval (some hundreds of angströms), the light emerging consists of two vibrations at right angles of equal amplitude and behaves exactly like natural light, even if the incident light is completely polarized. All trace of polarization is thus eliminated, and if there were no side effects, the milliammeter record would be a straight horizontal line. The only aim of the operation is to correct the recording (a) for side effects. (The glass L, put in place of the depolarizer in (a), is there simply to reproduce the losses of light by reflection and absorption.)

(c) We turn the plate L′ (a simple microscope cover slip of thickness 0.25 mm.) to a particular angle, about an axis situated in its plane, so as to polarize the light under perfectly known conditions. The classical formulae of Fresnel permit the calculation of the proportion of polarized light once the angle of incidence and the refractive index of the plate for the effective wave-length used ($\backsim$4500 A) are known. Thus, for inclinations of 20.25° and 30.25° which Hall generally used, the percentages of polarized light are $P=1{\cdot}50$ and $P=3{\cdot}54$ per cent respectively.

As the response of the apparatus is linear, a comparison of amplitudes registered in (a) and (c) immediately gives the proportion of polarized light in the radiation of the star. Further, in the operation (c), the privileged vibration is in the plane of incidence. The comparison of the phases of the maxima in (a) and (c) thus gives the position angle θ of the vibration I in the ray from the star.[3]

The second depolarizer D′, similar to the first one, but placed *permanently* in front of the multiplier (Fig. 44) prevents the photo-electric current varying if the direction of the vibrations transmitted by the analyser is turning simultaneously. This effect may be produced when the cathode is obliquely illuminated.

Hall mounted his apparatus on the aplanatic Ritchey-Chrétien telescope of the Naval Observatory of Washington (diameter 100 cm., focal length 680 cm.). The multiplier, with a caesium-antimony cathode, is sensitive to ultra-violet radiation starting from the red. Hall determined the effective wave-length corresponding to the B stars ($\backsim$4500 A) by passing their spectrum on to a cathode by means of a wire grid used by him in the previous work [121] (page 173).

Both the methods of Hiltner and of Hall permit the observation of three to five stars per hour. Their accuracy seems to be nearly equal. Hall [124] estimated that the probable error of P increases from 0·1 per cent, when the polarization is weak, up to 0·3 per cent when it is stronger ($P>2$ per cent). Hiltner [141] evaluated it at 0·1 per cent approximately, for all P. Uncertainty in the angle of position naturally depends on the proportion of polarized light: θ is undetermined when the polarization is zero. According to Hall and Mikesell it is measurable to within a few degrees when polarization is marked ($\pm 3{\cdot}5°$ for $P>1$ per cent, $\pm 5{\cdot}5°$ for 0·5 per cent $<P<0.9$ per cent), Hiltner suggested a probable error a little less than 1° for all his measurements.

The two methods lead to very concordant results. They do not show any systematic differences in the proportion of polarized light (228 stars are common to the two series of observation), but they do show a difference in the position angles of a little more than 3°, whose origin is unexplained.

2. General results of the measurements

Today, Hiltner's observations and those of Hall and Mikesell have important statistical consequences. First we must consider as well established, the fact that the polarization of light is not an intrinsic property of certain stars. Polarization can, in fact, be found in stars of very different spectral types in the same region of the Milky Way. Neither stars showing emission lines nor "c" stars which show very fine absorption lines, nor even certain supergiants of absolute magnitude—seven show a polarization which is abnormal compared with stars of the same colour index (Hall and Mikesell [124]). However, the location of the stars in the Galactic plane plays an essential role and the relationship which has been found to exist between polarization, on the one hand, and the colour excess or the intensity of the interstellar lines, on the other, forces us to conclude that light is polarized by its passage through interstellar matter.

Value of the proportion of polarized light and orientation of the privileged vibration

Hall and Mikesell (1950) [124] published results of more than 1,300 measurements made on 551 stars. Most are brighter than the ninth magnitude and only some attain the tenth. They are, in general, chosen from the list of 1,332 stars of the first spectral types of Stebbins, Huffer and Whitford [297], from those whose colour excess E_1 exceeds $+0{\cdot}14$ magn. Some stars, however, whose colour excess are weak or nil, or which present a particular interest, have been added to these.

Hiltner (1951) [141], on his part, published measurements made on 841 stars, with a telescope of 2·05 m. aperture at the McDonald Observatory. They were mainly taken from the list of very luminous B stars of Nassau and Morgan; the weakest were also of the tenth magnitude. Since there are 228 stars common to the two series of measurements, we thus have information on 1,164 different stars. We shall soon see that there exists a certain correlation between polarization and colour excess, so that the stars thus selected have, for the most part, strong chances of showing an appreciable polarization. The proportion of stars polarized would certainly have been smaller if they had been chosen at random. Nevertheless, for more than half of the stars studied by Hall and Mikesell, the proportion of polarized light does not surpass 1 per cent and it is inappreciable in approximately 15 per cent of the cases. Less than 1/5 of the stars studied give a polarization larger than 3 per cent; and there are only seventeen stars for which P attains 5 per cent, and four stars with $P \geqslant 6$ per cent. The strongest polarization observed ($P = 7{\cdot}2$ per cent) is for a supergiant of visual magnitude $m_V = 6{\cdot}9$ (HD 183143), whose spectrum (cB9 eα) shows the Hα emission line. It is situated in Sagitta quite near the Galactic plane ($b = -0{\cdot}5°$). The brightest star of appreciable polarization ($P = 0{\cdot}5$ per cent) is α Cygni ($m_V = 1{\cdot}33$, spectrum cA2eα). Hiltner [461] very recently published the results of measurements on 405 new stars. In Cassiopeia he shows the existence of a star of the thirteenth magnitude where the proportion of polarized light exceeds 8 per cent.

Fig. 46, taken from Hall and Mikesell's paper, represents, in Galactic co-ordinates, the distribution of 551 stars studied. Each one of them is represented by a line whose length measures the proportion of polarized light, and whose direction is parallel to the vibration i, and thus perpendicular to the privileged vibration I.[4] The points correspond to stars showing no polarization, in general rather distant from the Galactic plane.

In the lower latitudes we immediately notice that, around longitude $l = 105°$, situated in the region of Perseus, the lines are almost parallel to each other, and roughly perpendicular to the Galactic plane (the privileged vibration I would thus be parallel to the plane); their lengths denote, in general, a relatively strong polarization. If we move from this region along either direction of the Galactic plane, the lines tend to incline progressively. But we encounter two regions, towards $l = 40°$, in Cygnus, and $l = 335°$, in Sagittarius, where the orientations seem to be distributed almost at random, as in the case of some polarized stars in higher latitudes. On the average, the proportion of polarized light is also weaker than in Perseus. The general appearance of the diagram would thus suggest that *polarization is produced by discrete absorbing clouds*, mainly near the Galactic plane, rather than by a homogeneous medium.

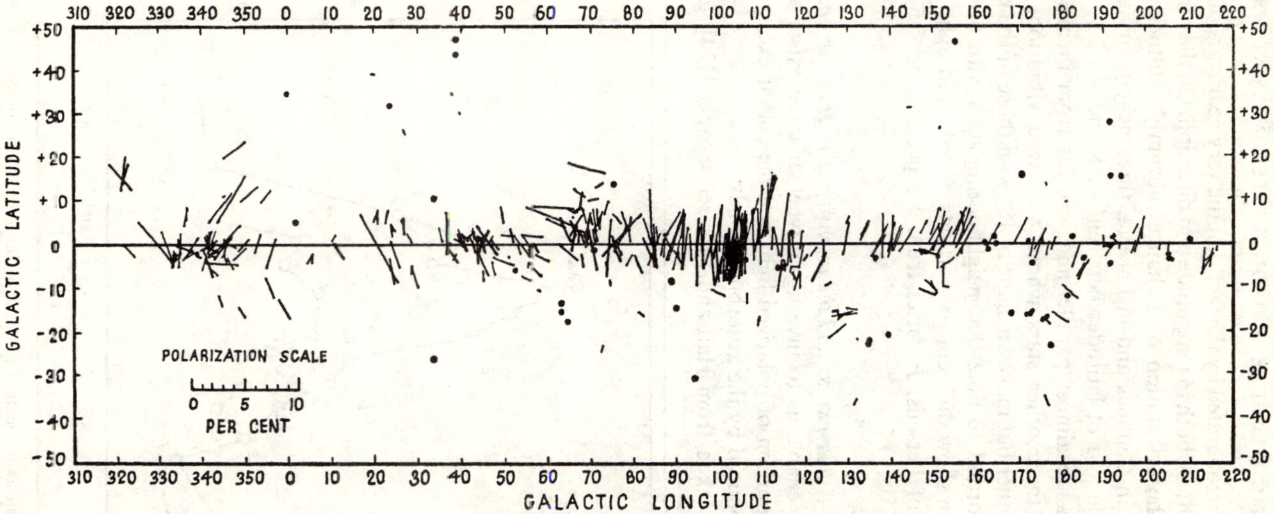

Fig. 46.—Chart of the observations of Hall and Mikesell in Galactic co-ordinates. Each line corresponds to a star. Its direction is that of the least intense vibration *i*, and its length is proportional to the proportion of polarized light. Points correspond to stars without polarization.

This idea is corroborated by the fact that the very near stars often show the same polarization, both in magnitude and direction. On the other hand, we notice that the orientation of I changes abruptly: Hiltner [139] for instance showed in Cepheus and in Lacerta, three neighbouring stars of 70·3° of longitude and of latitudes between −0·8 and +1·2°, the directions I of which are almost parallel and, not far from them (l=73·2°, $-1{\cdot}7° < -0{\cdot}9°$), three other stars whose privileged vibrations are still parallel to each other but make an angle of 18° with those of the preceding group. It is important to note that neighbouring stars with orientations parallel to I often show different proportions of polarized light. Thus, in the double star of Perseus, P varies from 1·8 per cent to 5·5 per cent (Hiltner).

Relation between colour excess and the intensity of interstellar lines

These results enable us to investigate whether there exists a relation between the proportion of polarized light and the colour excess which is known to be produced by the absorbing clouds.

Fig. 47, also taken from Hall and Mikesell's paper [124], gives as a

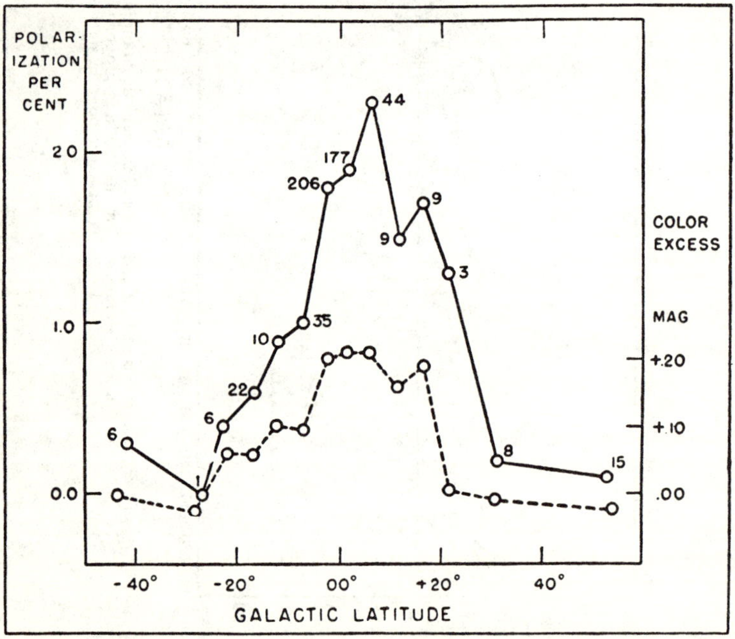

Fig. 47.—Distribution as a function of Galactic latitudes of the proportions of polarized light (full line) and of colour excesses (broken line). The figures beside the points of the first curve indicate the numbers of stars taken for each average (according to Hall and Mikesell).

function of Galactic latitude the distribution of percentages of polarized light P and of colour excesses E_1 of Stebbins, Huffer and Whitford.

The two curves are remarkably parallel and this shows that a certain correlation between these two quantities does exist. But if, for various zones of longitude, we plot P against E_1, we see that the dispersion of the points is large. Further, the average slope of the line clearly varies with the zone of longitude considered. It is larger between 110° and 150° longitude than, for instance, between 90° and 110°, or between 325° and 360°. Finally, if we take all the stars situated less than 5° from either side of the Galactic plane, irrespective of longitude, we find that the proportion of polarized light increases on the whole with colour excess when the latter is lower than +0·35 magn., but thereafter it tends to diminish. For a group of fifteen stars of the first type, centred on co-ordinates $l=107{\cdot}8°$, $b=+2{\cdot}3°$, Hiltner [141] found a good correlation between P and the colour excess E_1 which, for one of these stars, exceeds 0.8 magn.

In the double cluster h and χ Persei, the colour excess is rather variable, although the stars are practically all at the same distance. In this particular case, Hiltner found no clear correlation between E_1 and P (thirteen stars).

These stars of the cluster in Perseus present a notable polarization and an intense interstellar K line. The line is also strong in the spectrum of the star HD 14653, situated almost at the same distance as the cluster, but which is not reddened and the light of which shows no trace of polarization (Hiltner). The measurements of Hall and Mikesell show, nevertheless, that there still exists a certain correlation between the proportion of polarized light and the equivalent breadth of the interstellar K or D lines, but this is even weaker than for the colour excess. Conversely, we find a correlation (at least as strong as with the colour excess) between P and the intensity of the diffuse non-identified interstellar lines 4430 A (Greenstein and Aller [112]), Hall and Mikesell [124] and 6284 A (Hall and Mikesell). We already know that the intensity of the diffuse lines is directly proportional to the reddening (page 236).

Hall and Mikesell put forward an explanation of these statistical results. When the light of a star crosses numerous successive clouds, the colour excess produced by each one of them is added, but this is not necessarily also the case with the proportions of polarized light. The direction of the privileged vibration in the transmitted beam can, in effect, vary from one cloud to the next. Thus, in the region of Perseus, the strong average polarization and the parallelism of the vibration can be the results of the crossing of a single large cloud, or even of the concordant effects of numerous clouds. On the other hand, in Sagittarius and in Cygnus, successive clouds seem to produce discordant polarizations. Hiltner draws attention to the behaviour of two neighbouring stars HDE 237090 and

237091, only 0·7′ distant from each other. The angle of position of the privileged vibration differs by 8° and the reddest star is the least polarized. It is thus probable that the light of the more distant star has been depolarized by the cloud situated half way between the two stars.

In support of this point of view, Hall and Mikesell recall that the stars with multiple interstellar lines are mainly found in Orion, Sagittarius and Cygnus. Their light has thus crossed numerous clouds of different velocities, and we find that it is in general, much less polarized. Conversely, the interstellar lines are more often simple in the relatively near stars of Perseus and in Scorpio where the average polarization is stronger and where the privileged vibrations are almost uniformly orientated. There can be no doubt that the study of interstellar lines shows us the velocity of the atoms only, but Spitzer has shown that the small solid particles follow the movement of the gas.

We also know that the various regions of the Milky Way emit, with very different intensities, radio frequency waves whose wave-lengths are of the order of one meter. Grote Reber [241] found a minimum in Perseus, whereas the strongest emission comes to us from Sagittarius and Cygnus. Reber suggested that, in each direction, the intensity roughly measures the total quantity of matter contained between the observer and the confines of the Galaxy. He thinks that, towards Cygnus, we see the extremity of an arm of the spiral (Ryle and Smith (1948) have shown that the radio emission from Cygnus is not polarized). On the other hand, it is well known that Sagittarius corresponds to the direction of the centre. The regions of strong radio-frequency emission—those where interstellar matter would be seen as thickest—are just those where polarization is weak and the direction of vibrations unco-ordinated. Thus, according to Hall and Mikesell, the regions of intense radio-emission, of multiple interstellar lines and of weak and incoherent polarization would be essentially the same.

Polarization of various wave-lengths

In order to construct a theory for this phenomenon, it is obviously important to know how polarization varies with the wave-length of the transmitted light. Hiltner [139] and Hall [122] both stated, in 1949, that the orientation of the privileged vibration and the proportion of polarized light were appreciably the same for all wave-lengths.

Hall found in ξ Persei almost identical proportions of polarization (1·8 per cent and 2·3 per cent) in the ultra-violet (λ=3700 A) and in the red (λ=6200 A). Figure 43 represents the results of Hiltner on the star CQ Cephei and HD 211853. The measurements were made through colour filters corresponding to the effective wave-lengths 4200 and 5300 A and show no systematic difference whatever. More recently Hiltner [140] extended his observations up to the near infra-red; the star BD+58° 607 gives exactly the same proportion of polarized light in the three spectral regions centred on 4200, 5500 and 8000 A. In the case of BD+62° 643, the polarization is perhaps a little weak at 8000 A.

3. Attempts at a theoretical interpretation

General

In spite of immediate efforts of theoreticians to interpret the effects as they were discovered, we do not yet know for certain the mechanism responsible for the polarization of the light of the stars. We shall, therefore, limit ourselves to an examination of the general ideas which have served as the bases of the first tentative explanations.

In all the proposed hypotheses, the polarization of light received from stars results from scattering by the particles of space. It is thus necessary that these particles scatter preferentially vibrations parallel to a particular direction. Let us assume, for instance—which is far from being the case in general—that the privileged vibration is parallel to the Galactic plane. To explain this effect it is necessary to suppose that the vibrations perpendicular to the Galactic plane are more strongly scattered than are the parallel vibrations. Thus the latter would dominate in the transmitted light.

Now, it is inconceivable that spherical particles should have such a property. It is true that the light which they scatter individually is often polarized; but, since the overall phenomenon necessarily has a symmetry of revolution, the light transmitted remains unpolarized. Thus the scattering particles are not spheres—which is not at all surprising. The theory of scattering has been developed for the case of spheres, for reasons of simplicity only, but it is perfectly possible that the grains are, in fact, more or less elongated.

But this is still not enough. If particles having, for instance, the symmetry of an ellipsoid of revolution or of a rod have their large axes orientated at random, the transmitted light would still not be polarized. It is further necessary that a sufficiently large number of particles have their axes parallel; they must be orientated by forces whose origins should be investigated.

Let us first of all try to see what approximate ratio of the absorptions in magnitude A_2 and A_1, corresponding respectively to privileged vibration of intensity I and to the perpendicular vibration of intensity i, could account for the observed polarization. We can write

$$I=I_0\, e^{-0.92\, A_2}\; i=I_0\, e^{-0.92\, A_1} \text{ with } A_1>A_2$$

and the depolarization ρ of light received is

$$\rho=\frac{i}{I}=e^{-0.92(A_1-A_2)}\simeq 1-0.92\,(A_1-A_2),$$

since ρ differs little from unity.

By introducing the proportion of polarized light $p=\frac{(1-\rho)}{(1+\rho)}$ this relation becomes

$$A_1-A_2=1{\cdot}086\,(1-\rho)=1{\cdot}086\times 2\,\frac{p}{1+p} \qquad \text{(XV.1)}$$

Observations suggest further that on the average, p is proportional to the colour excess E_1, which in turn is proportional to the total photographic absorption A:

$$\rho=\gamma E_1,\; E_1=\frac{A}{\chi}=\frac{(A_1+A_2)}{2\chi}.$$

From this we get

$$A_1+A=2Mp, \text{ with } M=\frac{\chi}{\gamma} \qquad \text{(XV.2)}$$

By dividing both sides of equations (XV.1) and (XV.2) we obtain (remembering that the constant M is large compared with unity)

$$\frac{A_1}{A_2}=1+2\times 1{\cdot}086\frac{1-p}{M}$$

The results of Hiltner and of Hall lead to a value $\gamma=0{\cdot}08$; we know further that $\chi=8{\cdot}16$ (page 179), whence $M=102$ and

$$\frac{A_1}{A_2}=1+0{\cdot}0213\,(1-p)\simeq 1{\cdot}021.$$

In order to interpret the observations it would thus suffice for the absorptions to differ by 2 per cent. But, since the proportion of polarized light seems to vary little with the wave-length, the difference A_1-A_2 should, according to equation (XV.1), do likewise. A priori, it would seem very difficult to satisfy this condition theoretically.[5]

Scattering of light by needles of ice

The scattering of light by ellipsoids which are *small compared with wave-lengths* was studied in 1912 by Gans [99]. The theory does not allow a polarization independent of wave-length in the transmitted light. Van de Hulst [345, 346] dealt with a very much more complicated case of long, circular, dielectric cylinders, aligned parallel to each other, whose diameter is of the order of the wave-lengths. Very complicated calculations permitted him to draw separately the curves representing, as a function of the parameter $y=2x\,(m-1)$, the efficiency factors $\mathbf{E}_1$ and $\mathbf{E}_2$ of the vibrations parallel and perpendicular to the axis of the cylinders respectively. m is again the refractive index of the particles and $x=\frac{2\pi R}{\lambda}$. We obtain the law of reddening of space with entirely identical cylinders if y is almost equal to the wave number (i.e. with $m=1{\cdot}50$, $2R\simeq 0{\cdot}32\mu$, and with $m=1{\cdot}25$, $2R=0{\cdot}64\mu$). Apart from some small local accidents which are probably strongly mitigated if we deal with a mixture of cylinders of different radii, the difference $\mathbf{E}_1-\mathbf{E}_2$ is almost independent of y in the entire interval which interests us. Thus the proportion of polarized light ought to be practically the same for all wave-lengths, which is what is found, as we have already stated.

We can finally read from the curves, in the region corresponding to violet light ($\lambda=0{\cdot}41\mu$) the value of the ratio $\frac{\mathbf{E}_1}{\mathbf{E}_2}$. For $m=1{\cdot}25$ $\frac{\mathbf{E}_1}{\mathbf{E}_2}=1{\cdot}07$. *Thus small crystals (needles) of ice can account for the polarization observed*, but it would be necessary that they are all very long and conveniently orientated, for there is little margin for an imperfect alignment, or for a reduced length, observation having given $\frac{A_1}{A_2}=1{\cdot}02$.[6] Ellipsoids of axial ratio 2 would give a proportion of polarized light twice as small as long cylinders. With $m=1{\cdot}50$ (a much more improbable value), we find $\frac{\mathbf{E}_1}{\mathbf{E}_2}=1{\cdot}22$ and the margin is clearly larger. Ellipsoids having the refractive index of glass, all perfectly orientated, would still give $\frac{\mathbf{E}_1}{\mathbf{E}_2}=1{\cdot}06$, if their axial ratio were only 2·5.

Smith, Elske and Van de Hulst [285] have examined the cases of slightly absorbing elongated cylinders of refractive index $m=1{\cdot}50-0{\cdot}10\,i$, in the hope that this would allow a larger margin for an imperfect alignment or a reduced length. This hope was not fulfilled: the margin is still smaller than for purely dielectric cylinders. Similar calculations of Elske and Grootendorst, for metallic particles of index $m=1{\cdot}414-1{\cdot}414\,i$ have not given more encouraging results.

F. D. Kahn [412] has recently studied the possible formation, in an electric field, of crystals of impure ice, very elongated and electrically polarized. Such crystals may exist in thunder clouds according to Rossmann, who managed to produce them at a low temperature (1950). Kahn estimated that, in the H I regions where the temperature is probably approximately 60°K, conditions are favourable for their formation. The initial condensation nucleus would be formed of heavy atoms and the author showed that, in the presence of stellar radiation, equilibrium between the nucleus and the free electrons creates, by progressive fixation of hydrogen and oxygen atoms, an electric field sufficient for the formation of an impure ice crystal. Magnetic impurities contained in these elongated needles could offer an explanation of the orientation in a magnetic field, by the mechanism suggested by Davis and Greenstein, which we shall examine later on. Furthermore, Kahn later proposed that, by virtue of their electric polarization, these crystals could constitute a source of Galactic radio waves.

Thus we could, without doubt, arrive at an interpretation of interstellar polarization without having to invoke particles other than those whose existence is revealed by the reddening of the stars. Nevertheless, Hall and Mikesell are more inclined to the point of view that the two phenomena are caused by two sorts of distinct particles which are found mixed in most of the clouds, but not necessarily in all. How otherwise could we interpret the fact that certain stars, without any colour excess, nevertheless, show a very clear polarization?[7]

Orientation of particles in a magnetic field

Let us now return to the main problem of the orientation of the scattering particles. What can be the nature of the forces capable of exercising so directive an action? It is natural to think that it is a magnetic field. In 1943, Alfvén had proposed the presence of such fields in interstellar space, and more recently, Fermi [98] used this idea to explain the origin of cosmic rays. It is by travelling considerable distances in interstellar fields that the primitive particles of cosmic radiation acquire their enormous energy. We can envisage a relaxation time which is so long that, if a magnetic field in the Milky Way existed at the time of its formation, it would still continue today. Further, Fermi has shown that the turbulent movements in the interstellar conducting medium could give rise to magnetic fields. The magnetic induction B ought, in fact, to grow until the density of magnetic energy $\left(\frac{B^2}{8\pi}\right)$ becomes equal to the density of kinetic energy $\left(\frac{\rho v^2}{2}\right)$. With an interstellar density of 1 atom of hydrogen per c.c., largely exceeded in the clouds, and an average velocity of 22·5 km./sec. (value adopted by Oort and Van de Hulst), the equipartition of energy leads to an induction of the order of 10^{-5} gauss.[8]

All the modern theories involve the action of the magnetic field on solid particles.[9] It evidently depends on the nature of these. Two quite different theories have recently been developed: Spitzer and Tukey [295] suppose that they are *ferromagnetic* grains, Davis and Greenstein [65, 66] think they are *paramagnetic* grains.

Formation and orientation of ferromagnetic grains

The theory was published in 1951, before the publication of Güttler's paper [408] which provided evidence for the metallic nature of the solid grains. The authors tried to interpret the phenomena by means of the dielectric grains of Oort and Van de Hulst, formed principally of ice and some other compounds of hydrogen. According to the data on the cosmic abundance of elements, we can assume that these granules contain nearly 12 per cent of iron by weight (approximately 1 atom in 100), mainly in the state of iron oxide Fe_2O_3.[10] But the iron atoms and the molecules of iron oxide dispersed in the solid grains cannot explain the directive action of the field. It is thus necessary to imagine a mechanism capable of assembling the atoms of iron and of constituting a new ferromagnetic nucleus of sufficient dimensions. For this Spitzer and Tukey make use of mutual collisions of grains.

According to Oort and Van de Hulst (page 254), these collisions can have as their effect the complete vaporization of the grains, if the relative velocity is sufficiently large, but the Dutch astrophysicists have not concerned themselves with these heavy atoms. With weaker relative velocities, the collisions can lead, by fusion, to the formation of "secondary particles" (page 257). Spitzer and Tukey took up once more the study of collisions and found that with relative velocities between 2·0 and 5·2 km./sec., they can produce total fusion, with at least the partial evaporation of water and other volatile constituents.[11] Thus, the Fe_2O_3 molecules contained in the two particles could, after fusion, become a monocrystal which, after cooling, would without doubt be surrounded by a sheath of volatile molecules such as H_2O. An analagous phenomenon is to be predicted if the grains contained atoms of free iron. The authors also examined the chemical reactions capable of taking place at the time of the encounter of the grains. Finally, they considered as very probable the formation of a ferromagnetic crystal of iron, of magnetic oxide Fe_3O_4 or of Fe_2O_3 (Fe_2O_3 is also ferromagnetic if is it formed by precipitation from an aqueous solution). Once equilibrium by mutual collision is established, the proportion of ferromagnetic grains could be as much as 25 per cent.

The ferromagnetic grains thus formed tend to orientate themselves in the field, in the manner of a compass needle. Nevertheless a field of the order of 2.10^{-3} gauss would still be necessary to account for the observed polarization. Larger grains which are capable of being orientated in a weaker field, could be formed, either by the attachment of ferromagnetic particles to each other thus forming elongated rods, or by further collisions of ferromagnetic grains with the ordinary grains. But, to be effective, the first process would require that the clouds should be more dense than is commonly supposed. These large ferromagnetic particles would produce a neutral absorption, which Spitzer and Tukey estimate to be of the order of 0·1 magn./kpc. They think that it would be difficult to detect by the usual observations of absorption.[12]

Orientation of paramagnetic grains

In the preceding theory great importance is attached to mutual collisions of the grains, which are relatively rare. The collisions of grains with atoms and ions, mainly hydrogen, are much more frequent and have as their effect the rapid rotation of the grains about axes orientated at random. Basing their argument on the probable equipartition between the rotational energy of the grains and the translational energy of the atoms, Davis and Greenstein [66] calculated the average rotation velocity of the granules to be about 10^5 revs. per second in the H I regions and about 10^6 revs. per sec. in the H II regions, where the kinetic temperature is higher. It becomes difficult to understand how, under these conditions, the magnetic field manages to maintain the alignment of the elongated grains.

Davis and Greenstein [65] at first thought that the rotation of the particles could be braked by some hysteresis phenomenon in the interval between the collisions with the atoms, such that their rotational energy would always remain smaller than the kinetic energy of the atoms. But this hypothesis still requires that the grains contain ferromagnetic crystals. The authors abandoned this view afterwards and constructed a new theory, which this time involved *paramagnetic* grains, which is what the grains of Oort and Van de Hulst would be if they contained approximately 12 per cent of iron by weight, chiefly in the state of oxides. This theory depends on the phenomenon known by the name of *paramagnetic relaxation* [66].

Let us consider a top rapidly spinning about an axis inclined to the vertical. Gravity gives it a circular precessional movement. The small friction on the pivot, however, progressively diminishes the amplitude of the precession and tends to right the axis of the top. Since an elongated paramagnetic grain is similar to an ellipsoid of revolution, it rotates about an axis which is orientated at random in the field, and *paramagnetic absorption* plays a role which is similar to the friction on the pivot of the top. The rotating particle acquires a magnetic moment whose magnitude and direction depend on the imaginary part of its susceptibility, which is a complex quantity. From this there results a very small couple, which tends to make the grain rotate more and more about its short axis and to bring the latter into the direction of the field.

Thus, the co-efficient of absorption is larger for waves whose electric vector is parallel to the long axis of ellipsoids which is orientated perpendicular to the magnetic field. An approximate statistical calculation based on the time of relaxation ($\sim 10^{13}$ seconds) of the orientation of the grains, gives the distribution attained by the directions of the axes, when the bombardment of the grains by interstellar gas counterbalances the tendency to alignment. Elongated ellipsoids, small compared with their wave-lengths, and whose axial ratios are between 1·2 and 1·7, are sufficiently aligned in a field of 10^{-5} to 10^{-4} gauss to produce the observed ratio between the proportion of polarized light and colour excess. The calculations were made, in fact, as were those of Spitzer and Tukey, by supposing the ellipsoidal grains to be small compared with the wave-length (average radius 10^{-5} to 3.10^{-5} cm.), so that they cannot quite account for the variations of the proportion of polarized light along the spectrum.

Discussion

It will be noted that the two theories lead to orientations of the magnetic field differing by 90°. For Galactic longitudes between 80° and

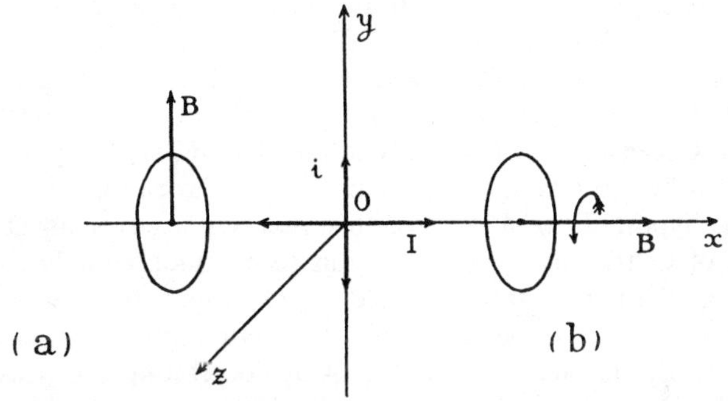

Fig. 48.—Orientation in the magnetic field of ferro-magnetic and paramagnetic particles. To account for the observations, the ellipsoidal particles must have their long axes perpendicular to the Galactic plane xOy. In the case of ferromagnetic particles (a), the magnetic field B is normal to xOy; in the case of paramagnetic particles (b), it is situated in the Galactic plane xOy (but it is not necessarily directed along the Ox axis as the diagram shows).

120° (and perhaps 170°), the privileged vibration *I*, always perpendicular to the long axis of the particle, is almost parallel to the Galactic plane. According to Spitzer and Tukey, the long axes of the ferromagnetic ellipsoids orientate themselves in the direction of the field: thus, the latter is normal to the Galactic plane (Fig. 48a). On the other hand, according to Davis and Greenstein, the long axes of the paramagnetic particles are orientated perpendicular to the field, which is thus situated in the Galactic plane (Fig. 48b).

The latter idea seems to be in better accordance with that of the production of the field by turbulent movements. Because of the rotation of the Milky Way, great turbulence is most probable parallel to the Galactic plane. It is possible that in the region between longitudes 80° and 120°, the lines of magnetic force follow the direction of the arm of the spiral, containing clouds of gas and of grains;[13] these lines of force would be roughly perpendicular to the line of vision (Davis and Greenstein). On the other hand, in the direction of Cygnus, the line of vision could be almost parallel to an arm of the spiral and to the lines of force. The component of the magnetic field normal to the line of vision would thus be very small and its direction would vary with distance. Thus, both the feeble polarization of the stars of Cygnus and the disordered orientation of the privileged vibration are explained.

The theory of Davis and Greenstein has the further advantage of not being based on the ingenious but somewhat flimsy, hypothesis of the formation of ferromagnetic grains, and it contents itself with a magnetic field 10 to 100 times smaller. It would perhaps be possible to choose between the two mechanisms if we knew whether the polarization is produced only in the H I regions of space, or in both the H I and H II regions. The calculations of Spitzer and Tukey were in fact only made for the low kinetic temperatures of the H I regions and their ideas would be highly confirmed if we could show that in the H II regions colour excess is produced without polarization. On the other hand, the calculation of Davis and Greenstein apply equally to the H I and H II regions, both of which could be responsible for producing the polarization of light.[14]

On the hypothesis of metallic grains, necessarily ferromagnetic, which the work of Güttler no longer allows us to overlook completely, the interpretation of the polarization would, a priori, seem to be easy. But it is then necessary to revise the theory of the formation of the particles entirely and to find the mechanism capable of agglomerating the heavy atoms only, and of producing elongated needles. It is still too early for us to decide whether this conception offers a reasonable explanation.[15]

NOTES TO CHAPTER XV

[1]An elliptical polarization, which is the resultant of two vibrations at right angles to each other and having a phase difference, would give a similar result. In this case a quarter-wave plate, conveniently orientated, and placed in front of the polaroid, would enable us to observe total linear polarization. Hall found that this was not the case (for HD 183143). We are therefore dealing with two *incoherent* vibrations at right angles, i.e. with partial linear polarization (see page 198).

[2]Similar apparatus had already been used by Öhman for astronomical observations, and by others for laboratory measurements (Hardy, Williams and Hiltner).

[3]If the sky is illuminated by the Moon, its light is partially polarized and to account for this, it may be necessary to repeat operation (a) on the sky near the star. Further, in certain cases, the current of the cell may be recorded in darkness.

[4]Fresnel's "luminous vector" corresponds to the *electron vector* of the wave. The lines of Fig. 46 are directed according to the *magnetic vector*, perpendicular to the electric vector of the privileged vibration I. The lines are thus parallel to *i*.

[5]If A_1 and A_2 varied in the same way as a function of the wave-length, we should have, representing two constants by C_1 and C_2,

$$A_1 = C_1 g(\lambda), \quad A_2 = C_2 g(\lambda).$$

The ratio $\frac{A_1}{A_2} = \frac{C_1}{C_2}$ would be quite independent of λ, but not the difference $A_1 - A_2$.

The depolarization $\nu=1-0.92(C_1-C_2)g(\lambda)$ would vary appreciably with the wave-length. If conversely we supposed polarization to be strictly independent of λ, we could write

$$A_2=g'(\lambda),\ A_1=g'(\lambda)+2\eta$$

where η is constant. From this it follows that the mean absorption

$$A=\frac{(A_1+A_2)}{2}=g'(\lambda)+\eta$$

would also contain a constant term η representing neutral absorption. The latter, however, would be very small. For weak polarizations the value of η in magnitudes would practically equal that of p in hundredths ($\eta=0.01, 0.02$ magn. for $p=0.01, 0.02$, etc.). There would therefore be no contradiction with the almost total absence of neutral absorption (page 178).

[6]Van de Hulst obtained $\frac{A_1}{A_2}=1.06$ from the observations, so that hardly any margin remained, but he took a very exceptionally limited case with $\gamma=0.23$ instead of $\gamma=0.08$ mentioned above.

[7]To the known examples they added HD 14633 which is the *bluest* in the list of Stebbins, Huffer and Whitford, and for which Markowitz [185] found a polarization of 0.5 per cent by photography.

[8]The hydrodynamic theories of Batchelor [23] and of Schlüter and Biermann [264] lead in fact to an equipartition of energy but to a value of B somewhat lower than 10^{-5} gauss.

From point to point in space the fields due to turbulent movements seem to vary rapidly. The interpretation of polarization, however, requires fields uniform over 100 to 200 pc. approximately.

[9]Only Hiltner [140] thought of quite a different solution. He asked if polarization could not be due to the scattering of light by the free electrons of space. The vibrations of the latter under the action of the electric field of the incident wave could be orientated by the interaction of the magnetic field and electron spin. But this mechanism would require too high an electron density ($\simeq 17$ electrons per c.c.). Further, according to Fermi, there ought to be no interaction of field and spins.

[10]We should recall that hydrogen would only be fixed in the grains as a molecular compound; most of the hydrogen and all the helium would escape condensation.

[11]The requisite energy for vaporizing the grains completely must have been under-estimated by Oort and Van de Hulst.

[12]In any case it is difficult to imagine the existence of a neutral absorption of this order in *all* the Galactic clouds (page 179).

[13]We shall see that in spiral galaxies similar to the Milky Way, the clouds are in fact aligned along the arms (page 297).

[14]In any case, if the mechanism proposed by Kahn is the cause of the formation of elongated crystals, the latter could only develop in the H I regions.

[15]Cayrel and Schatzman [462] recently suggested that polarization of the light of stars could be attributed to small hexagonal graphite crystals. In the laboratory, colloidal graphite particles suspended in water and heated gelatine, will align themselves and become fixed on cooling, if they are slightly rotated in an intense magnetic field. Such a preparation strongly polarizes the light it transmits; the extinction is three times stronger for the vibration parallel to the plane of the hexagons than for vibration perpendicular to this plane. According to Cayrel and Schatzman only three to four graphite crystals in every hundred interstellar particles would suffice for the polarization observed in the stars.

This theory, on which judgement must still be suspended, has the further advantage of only needing a weaker magnetic field ($\sim 10^{-6}$ gauss) for aligning the particles.

CHAPTER XVI

GLOBULES AND THE BIRTH OF SUPERGIANTS

Globules

If we look at a photograph of the nebula M8 in Sagittarius (Plate XXI), our attention is immediately drawn to the presence of a very large number of tiny black patches projected on the bright background of the nebula. Some of them have an irregular form: they are often thin filaments and look as if they were blown about by a wind. Others again are round or oval; Bok and Edith Reilly [42] have called the latter "globules".

The most characteristic globules of M8 have an apparent diameter of the order of 10″ to 20″ and we can form an idea of their true dimensions, since they cannot be further away from us than the nebula on which they are projected. Now this distance is very well known. According to Hubble, the spectrum of the nebula is excited by two stars of the type Oe5 and B0 whose apparent and absolute magnitudes we can evaluate. Further, the nebula seems associated with the Galactic cluster NGC 6530 whose distance has been determined by Wallenquist. This leads to an *apparent* distance of the nebula of approximately 1,600 pc. The true distance is probably appreciably smaller. From the colour excesses of four stars in this region measured by Stebbins, Huffer and Whitford, Bok and Edith Reilly estimate the total photographic absorption to be 0.6 magn. The true distance would thus be 1,260 pc. To apparent diameters 10″ and 20″ there would thus correspond the linear *maximum* diameters 12,600 and 25,200 astronomical units (or 0·061 and 0·126 pc.). Thus the smallest globules known have a maximum diameter of 0·06 pc.

Bok and Reilly made a systematic search for globules in other diffuse nebulae and in photographic records of the Milky Way. They found almost fifty of them unevenly distributed along the Galactic equator. The emission-line nebula near the southern star η Carenae is almost as rich in globules as M8. The Trifid Nebula on the other hand near to M8 and clearly divided by dark filaments (Plate XXII), shows only three or four of them. The globules seem particularly abundant in Sagittarius, Ophiuchus and Scutum where, it is true, the search for them is facilitated by the abundance of bright nebulae and of large stellar clouds against which they can be more easily distinguished. The small dark nebulae of Barnard Nos. 68, 69, 70 and 255 (in the region of θ Ophiuchi) are very similar to the globules of M8, and if we suppose that their distance is equal to that of the large absorbing cloud of Ophiuchus, we find that their diameters are of the order of 30,000 A U (0·15 pc.). A dozen similar objects appear in the region of Scutum where the diameters of the nebulae Barnard 117 and 118 probably never exceed 20,000 A U (0·1 pc.). On the other hand, we find nothing in the direction opposite to the centre of the Galaxy. Not a single globule is visible against the bright background of the Great Nebula in Orion (Plate I), nor against the California Nebula in Perseus. Barnard 34, in Auriga, can be considered as a "very large globule" of diameter 100,000 A U, that is, approximately 1/2 parsec.

It is difficult to estimate the opacity of very small globules which are visible on the bright background of the large nebulae. The comparison of their brightness with that of the neighbouring regions gives, for globules of diameters 10,000 to 35,000 A U, total photographic absorptions of at least two to five magnitudes. The opacity of some "large globules" across which stars could be photographed, can be evaluated more accurately by means of stellar counts. Thus Stoddard [308] obtained for the nebulae Barnard 34, 201, 226 and 229, of which the diameters are of the order of 100,000 A U (0·5 pc.), a photographic absorption of the order of 1 magnitude.

Inside these dense clouds, the absorption co-efficient reaches a higher value. For a sphere absorbing 1·5 magn., along a diameter of 0·5 pc., $a_{pg}=3$ magn./pc. and for a small globule of diameter 0·06 pc. absorbing 5 magnitudes, $a_{pg}=83{\cdot}3$ magn./kpc. These values are approximately 2,400 and 66,000 times larger than the average assumed for the neighbourhood of the Galactic plane ($1 \leqslant {\cdot}25$ magn./kpc.). The density of the solid grains ρ_s is increased in the same ratio, if we suppose them to be the same grains in every case. With dielectric particles of the previously discussed maximum efficiency (page 222) $\left(R_1=4{\cdot}07.10^{-5}\text{ cm.},\ m=\frac{4}{3}\right)$, we would have, in gm./c.c.,

$$\rho_s=4{\cdot}17.10^{-27}\, a_{pg},$$

if we express a_{pg} in magn./kpc. Thus we find in the interior of large globules of diameter 0·5 pc.,

$$\rho_s \backsim 1{\cdot}3.10^{-23}\text{ gm./c.c.}$$

and, in the interior of small ones (diameter 0·06 pc.),

$$\rho_r \sim 3{\cdot}5.10^{-23}\text{ gm./c.c.}$$

For purposes of comparison, let us give figures for the *Southern Coal Sack* which, with a diameter of approximately 8 pc., can be considered as an average type of cloud. The total photographic absorption is of the order of 1·5 magn. From this we get,

$$a_{pg}=0{\cdot}19 \text{ magn./pc.}, \quad \rho_S \backsim 7{\cdot}7.10^{-25} \text{ gm./c.c.}$$ [1]

Schoenberg and Tolmar [427, 428] photographed in the blue, the yellow and the red, twenty-six small absorbing clouds in the southern regions of the Milky Way. The diameters which they obtained range between 0·8 and 5 pc. approximately. Thus, these objects appear to be intermediate between globules and ordinary clouds. Densities ρ_S, evaluated on the hypotheses that the cloud consists of metallic particles, are between 10^{-24} and $^{-25}$ gm./c.c.[2] and the total masses between 0·1 and $3{\cdot}5.10^{33}$ gm. (average $1{\cdot}3.10^{33}$ gm.). The latter could thus attain, or even exceed, the mass of the Sun ($2{\cdot}98.10^{33}$ gm.).

Condensation of absorbing clouds: Formation of dense clouds

The globules appear to have a radius of about $\frac{1}{100}$ that of typical clouds (R=5 to 15 pc., page 231) and the density of matter ought to be at least some hundreds of times larger.[3] Thus it seems natural to suppose that they are derived from the condensation of ordinary clouds.

Which mechanism is responsible for the condensation? Whipple [372] has shown that radiation pressure ought to play an essential role. To simplify the problem, let us suppose, with Whipple, that the radiation from all the stars arrives equally from all directions. Then, an isolated particle in space is under equal pressure in all directions and radiation cannot affect its motion. If, however, the particle is in the neighbourhood of an absorbing cloud, stellar radiation crossing the cloud is weakened and the symmetry of pressures is destroyed. The particle is evidently pushed towards the cloud. A chance concentration of solid grains thus tends to become the centre of attraction for neighbouring particles.

If we now consider spherical clouds, all the grains in their periphery are pushed towards the centre and the entire cloud of particles tends to contract. It is interesting to evaluate the duration of this contraction. We shall give Spitzer's results [289] for a well-defined case (without entering into the details of the very complicated calculations). Let us suppose that just outside the cloud the number of particles per unit volume is such that the absorption becomes 1 magn./kpc. Inside, the density is initially 1·58 times the external density. We suppose further, that there is 1 atom of hydrogen per c.c. and that the mean square velocity of hydrogen is $1{\cdot}6.10^5$ cm./sec. (1·6 km./sec.), a value which corresponds with the temperature 100°K. Under these conditions we find that the radius of the cloud tends towards zero in $3{\cdot}9.10^7$ years—a time small compared with the age of the Milky Way. (On taking 10,000°K as the temperature of translation of hydrogen, as Whipple did, the time becomes ten times as long.)

This time seems just short enough for concentration to take place before the cloud becomes broken by the shearing stress of Galactic rotation. If in the neighbourhood of the Sun we imagine the existence of an initially cubic mass, differential rotation would impress on the two opposite faces a displacement equal to the side of the cube, in 3.10^7 years. Although the duration of the concentration does not depend explicitly on the original dimensions of the cloud, the calculation is not valid if the initial radius is between certain limits: it must be higher than 1 pc. and may not exceed 100 pc.

Further condensation: Possible birth of a star

Thus the condensation of clouds, produced by radiation pressure, seems to be able to explain the formation of small and very dense clouds similar to globules. What occurs afterwards?

During the first stages of concentration, the radiation pressure pushes the grains through the gas, and only acts very weakly on the atoms. Thus the composition of the cloud changes progressively:[4] the proportion of solid particles to atoms increases, and there comes a time when inside the dense cloud, the total mass of the granules approaches that of the gas.[5] Thus the entire cloud starts to contract under the action of radiation, granules and atoms now being pushed towards each other. At the same time the opacity of the cloud increases and, since light practically no longer penetrates into the interior, the radiation pressure ceases to act.

Nevertheless, the grains are now sufficiently close to each other for their mutual attraction to become important, and it is quite possible that the contraction now continues under the action of gravitational forces. The final result of this evolution could very well be the formation of a star. For this it would be necessary that the surface temperature gradually increase up to some thousands of degrees, and this implies the ionization of hydrogen atoms inside the protostar. Now, the opacity of this star is too great for radiation from the stars of the Galaxy to penetrate it. Spitzer thinks that particles of large energy which constitute the cosmic rays will still traverse the protostar freely provided that its internal density does not exceed 10^{-17} gm./c.c. Ionization of interstellar gas by cosmic rays is generally held to be negligible compared with that which is produced by the ultra-violet radiation of the stars. But its role can become important if all other causes of ionization are absent. Evidently heating would have the effect of revaporizing the grains inside the protostar.

The probable existence of systematic movements within the interstellar medium presents a further serious obstacle to the ultimate condensation of Galactic clouds. The existence of a rotational movement of the clouds would be particularly fatal. Since the angular momentum of a mass in rotation—product of the velocity and the radius—remains constant, the velocity is in inverse proportion to the radius, and it is strongly believed that the protostar would not be broken up by the centrifugal force. It is possible that the rotation would be braked by turbulent movements in the interstellar gas, movements on which we have no information whatever. Spitzer thinks that braking by Foucault currents produced by a Galactic field is also involved.

Finally it must be noted that the mass of the globules known is very small compared with that of the stars, so that the theory of the formation of stars from Galactic clouds actually has no solid basis in observation. Thus it is mainly a speculative idea which, however, is of great interest since it is capable of giving a plausible explanation for the formation of supergiant stars [289].

The age of normal stars and that of supergiants

It is not idle to recall the ideas on stellar evolution of thirty years ago. The determination of parallaxes and of absolute magnitudes had led to the tracing of the famous Hertzsprung-Russell diagram. In the so-called "advanced" spectral classes, from F to M, there exists two categories of stars, the giants and the dwarfs, of almost equal temperatures,[6] but differing in their luminous intensities, their masses, their diameters and their densities. The giants, very much more luminous and also much more massive, owe their spectral characteristics to the low density of their atmospheres. One could thus consider two series of distinct stars going from the classes B and A to the class M, or rather one continuous series from the class M to the class B, and finally back again to the class M.

Further, Einstein had just discovered the transformation of matter into energy. Thus, in principle, the source of radiation of the stars had been found: i.e. the progressive destruction of stellar matter as a source of greater radiation. The stars in evolving, therefore, lose their mass bit by bit. The most massive, which are also the most luminous in accordance with the "mass-luminosity" relationship of Eddington, ought to be considered as the youngest; the lightest as the oldest. The sense of the evolution along the Russell diagram (Fig. 1) was thus fixed: the stars ought to be born as red giants and then pass through the branch of the giants up to the classes A or B, becoming progressively hotter due to their contraction, before descending by gradual cooling along the dwarf branch.

The progress of nuclear physics and that of astronomy have forced us to abandon this theory of evolution completely. Bethe elucidated the elementary mechanism according to which stellar matter becomes transformed into energy. It is the formation of a helium atom from four hydrogen atoms. The mass of 4 gram-atoms of hydrogen ($4 \times 1{\cdot}0078$ gm.)

exceeds by $4 \times 0{\cdot}0078$ gm. that of the gram-atom of helium formed (4·0000 gm.). This fraction of the original mass is transformed into energy. This is the principle of the "hydrogen bomb". In the stars, the nuclear reaction

$$4H \rightarrow He + h\nu$$

is produced under the catalytic action of carbon and of nitrogen which take part in the intermediate reactions, but are left intact at the end of the cycle.

From this it follows that the loss of mass of a star, due to radiation, is really very small. Even if the star were almost entirely composed of hydrogen, the transformation of all its hydrogen into helium would consume less than $\frac{8}{1000}$ of its total mass. It is thus impossible to hold this mechanism responsible for the formation of red dwarfs, of mass lower than that of the Sun, from giants whose mass can exceed twenty times the Solar mass. If this evolution had indeed taken place, an entirely different mechanism, much more active in the destruction of mass must have completely escaped the notice of nuclear physicists—and this is very improbable (Spitzer [289]).

On the astronomical side, the age of the Earth was determined by the study of the disintegration products of uranium which are destroyed spontaneously in the course of time, finally to be transformed into lead. It was found to be a little more than 3,000,000,000 years. The existence of certain somewhat unstable, and therefore transitory, clusters of stars has led us to attribute to them an almost equivalent age. Finally, the expansion of the universe also requires a "short scale" of time. Thus, we think today that the age of the universe probably does not exceed 3,000,000,000 (or perhaps 4,000,000,000) years. "If the universe had not been created at that time, it must at least have undergone a profound reorganization due, perhaps, to some cosmic explosion" (Spitzer). Now, for our Sun to be considered as a typical dwarf star, it would require some hundreds of thousands of million years to lose an appreciable fraction of its mass by radiation. We are thus forced to conclude that most of the stars at present in the sky have not evolved appreciably for the last 3,000,000,000 years. They must have been born much as we observe them, with the same differences in temperatures, masses and densities.

The only exception are the supergiants. Less numerous, but visible from much further away, these stars, because of their extraordinary luminosity, are placed well above the branch of the giants on the Russell diagram. Their absolute magnitude can be up to -6 or -7, in other words, their luminous intensities are some tens of thousands of times that of the Sun. Such radiation rapidly consumes that fraction of the mass which can be transformed into energy according to the Bethe cycle, and all the hydrogen would probably be transformed into helium within a hundred million years at most. These stars are thus of recent formation and it is very possible that they continue to be formed under our very eyes.

If then the condensation of a dark cloud finally tends to create a star, this will often be a supergiant. This, then, would be the mechanism capable of giving birth to these ephemeral stars. We shall see further on (Chapter XVIII) how the study of the distribution of absorbing clouds in the external galaxies has rendered this hypothesis extremely probable.

NOTES TO CHAPTER XVI

[1]These examples are taken from Bok [41]. But Bok obtained densities almost four times greater, for instead of considering grains of maximum efficiency, he took spheres of radius 10^{-5} cm. (1000 A) whose efficiency is less. Obviously the ratios are not changed. The densities given here can be considered as minimum values.

[2]This order of magnitude would not be changed at all if we considered the clouds as made up of dielectric grains (see page 223).

[3]For his twenty-six clouds, Schoenberg states the mass to be proportional to the square of the radius, the density inversely proportional to the radius. If then small clouds are due to the condensation of larger ones, this phenomenon ought to be accompanied by loss of mass.

[4]This resulted particularly from Schatzman's calculations [262].

[5]This is probably the case for the typical globules examined (page 281).

[6]The giants are a little colder than the dwarfs, but the difference of the colour temperatures in only 1,000° to 1,500°, according to the spectral type.

CHAPTER XVII

MATTER CONCENTRATED IN THE STARS AND DIFFUSE MATTER

Average densities of gas and of smoke

We have seen (page 223) that to account for the general absorption of light in the neighbourhood of the Galactic plane, it is sufficient for the average density of the solid grains to be approximately 10^{-26} gm./c.c. ($1{\cdot}5.10^{-5}$ times the Solar mass/pc^3). This, let it be stressed, is only an average order of magnitude. Outside the clouds, solid particles are certainly much rarer, but they are very much more numerous inside them. Let us recall that in the Southern Coal Sacks the specific mass is almost eighty times greater ($7{\cdot}7.10^{-25}$ gm./c.c.) and, for globules of diameters 0·5 and 0·06 pc., the values were respectively 1,300 and 3,600 times stronger (page 282).

The total mass of the grains in the clouds is relatively small. With the above densities, we find that it is three times the Solar mass in the Southern Coal Sacks and, in the two globules, 0·01 and 0·0006 times the Solar mass. The mass of a large cloud of diameter 40 pc., absorbing two photographic magnitudes ($c_s=2{\cdot}08.10^{-25}$ gm./c.c.) would only be about 100 times that of the Sun.[1]

If we assume smoke to have an average density $\rho_s=10^{-26}$ gm./c.c., then, if it filled a very flat cylinder having roughly the radius of the Milky Way (13,000 pc.) but of a height of only 200 pc., its total mass would be 15,000,000 to 16,000,000 times that of the Sun. Compared with the total mass of the stars of the Milky Way, this is a relatively small quantity. The mass of the grains contained in a volume equal to that of the Earth would hardly exceed 10 gm.

Let us compare this density with that of the gases of space. Strömgren (page 208) estimates that each cubic centimeter of a typical cloud contains approximately ten atoms of hydrogen, i.e. mass $1{\cdot}67.10^{-23}$ gm. Other much rarer atoms all have greater mass, and if we take them into account, the preceding density ought to be increased appreciably. If the composition of the gases of space were identical with that of the Sun and the planetary nebulae, this increase ought to be more than 50 per cent, because of the great abundance of helium atoms (see Table VI, page 81). In the absence of all information on the proportion of helium contained in the Galactic clouds, we have here taken 20 per cent which may not be sufficient. Thus we suppose that the specific mass inside a typical cloud is approximately 2.10^{-23} gm./c.c. The masses of spherical clouds of 5, 7 and 10 pc. radius would be of the order of 153, 420 and 1215 Solar masses respectively, and thus much greater than that of the most massive stars.

To evaluate the average density of gases in space we should still have to know the fraction f of the Galactic volume occupied by the clouds, and the specific mass of space between the clouds. We suppose, with Oort and Van de Hulst (page 231) that the fraction f is 0.14 and that the specific mass is 100 times smaller outside the cloud (Strömgren, page 125). The average density of the gas would thus be

$$\rho_g = (2 \times 0{\cdot}14 + 0{\cdot}02 \times 0{\cdot}86)\ 10^{-23} = 3.10^{-24} \text{ gm./c.c.},$$

or $4{\cdot}4.10^{-2}$ the Solar mass per pc^3; this is 300 times greater than that of smoke. The mass of the gas contained in a volume equal to that of the Earth would be 3·24 kg. (2·86 kg. for hydrogen alone). In our cylinder of radius 13,000 pc. and of height 200 pc., the mass of the gas would attain 46,000,000,000 Solar masses. This is an impressive mass and it is necessary to examine if it represents a large fraction of the total mass of the Milky Way.

Total density of matter in the neighbourhood of the Sun

Stellar dynamics permits us to evaluate the *total* density of matter condensed in the stars and dispersed in space in the neighbourhood of the Sun. The problem first treated by Kapteyn (1922) [163], was taken up again in an entirely different manner with more numerous and more precise data by J. H. Oort, in a paper which has become a classic (1932) [220].

We assume that in the neighbourhood of the Sun the number of stars per unit volume and the distribution of their velocities perpendicular to the Galactic plane are in a state of equilibrium. This plausible hypothesis seems well corroborated by the fact that at various distances from the Galactic plane, we find almost the same number of stars approaching as receding.

Each star is attracted by all the others. Near the Sun, the thickness of the Milky Way is only some hundreds of parsecs, hardly 1/20 of the distance of the Sun from the centre of the system. Parallel to the Galactic plane, the movement of the stars is mainly determined by the attraction of the distant masses, but perpendicular to this plane, their movement is essentially governed by the distribution of the near masses.

Under the action of gravitational forces, the stars execute, on either

PLATE XXI.—The Nebula M 8 (NGC 6523) in Sagittarius on which there are projected many small dark clouds, and particularly tiny globules. Photographed by the light of the H α line. 80 cm. telescope (J. Dufay).

PLATE XXII.—The Trifid Nebula, M 20 (NGC 6514) in Sagittarius. 80 cm. telescope (M. de Kérolyr).

side of the Galactic plane, oscillations of a period of the order of 200,000,000 years. Since the formation of the Milky Way, they have had time to go through fifteen complete oscillations which have no doubt smoothed out the initial irregularities in their distribution, and so led to the setting up of an approximate dynamic equilibrium.

The total density of matter in the neighbourhood of the Sun is simply related to the gravitational field. For a medium of uniform density, Poisson's theorem states that the divergence of the field is equal to -4π times the product of the gravitational constant G, and the density ρ. If $H(z)$ is the field perpendicular to the Galactic plane, at a height z above it, we have

$$-\frac{\partial}{\partial z}[H(z)]=4\pi G\rho \qquad \text{(XVII.1)}$$

The formula can be corrected to account for the fact that the density is not strictly uniform.

Thus, the essential problem is to find the variation of the field $H(z)$ as a function of z, so that the gradient $\frac{\partial}{\partial 3}[H(z)]$ can be evaluated. Oort managed to do so by using the distribution of stellar velocities perpendicular to the Galactic plane, and also the distribution of the stars per unit volume as a function of z.

Let Z be the component along the z axis of the particular velocity of a star, Z_0 its value when the star passes the Galactic plane ($z=0$). By the theorem of the conservation of energy, the variation of kinetic energy of a unit mass between the heights 0 and z is equal to the work done by the field.

$$\frac{1}{2}(Z_0^2-Z^2)=\int_0^z H(z)dz,$$

whence

$$Z_0=\sqrt{Z^2-2\int_0^z H(z)dz}. \qquad \text{(XVII.2)}$$

Let us denote by $\Phi(z, Z)\, dzdZ$ the number of stars contained in an element of volume of thickness dz, at the height z and having velocities between Z and $Z+dZ$. The state of equilibrium requires that $\Phi(z, Z)$ should be a function of Z only. If then we calculate Z by means of equation (XVII.2), we ought to have

$$\Phi(z,Z)=\Phi(0, Z_0)=\Phi\left(\sqrt{Z^2-2\int_0^z H(z)dz}\right).$$

In particular, when the distribution of velocities for $z=0$ is represented by a Gaussian function

$$\Phi(0, Z_0)=\Delta(0)\frac{l}{\sqrt{\pi}}e^{-l^2Z_0^2}, \tag{XVII.3}$$

where $\Delta(0)$ and l are constants; the distribution law for Φ (z, Z) is

$$\Phi(z, Z)=\Delta(0)\frac{l}{\sqrt{\pi}}e^{-l^2Z^2+2l^2\int_o^z H(z)dz}. \tag{XVII.4}$$

Now, the total number of stars $\Delta(z)$ contained in unit volume, at a height z, is evidently the sum

$$\Delta(z)=\int_{-\infty}^{+\infty}\Phi(z, Z)dZ.$$

By substituting, in this integral, the expression for $\Phi(z, Z)$ taken from (XVII.4), we obtain

$$\Delta(z)=\Delta(0)e^{2l^2\int_o^z H(z)dz}. \tag{XVII.5}$$

This formula solves the problem in principle, since from it we can obtain $H(z)$ from $\Delta(z)$ and the modulus l of the distribution of velocities.[2]

Actually, the distribution of velocity Z is not quite a Gaussian function, but we can always represent it by the sum of two or three Gaussian functions of different moduli, which complicates the calculation a little. In the case of an arbitrary function of the distribution of velocities, Oort shows that we can derive $H(z)$ from $\Delta(z)$ and from the mean square $\overline{Z^2}$ of the velocities, using the equation:

$$H(z)=\frac{1}{\Delta(z)}\frac{\delta}{\delta z}[\Delta(z)\overline{Z^2}]$$

The number of stars per unit volume $\Delta(z)$ at different heights above or below the Galactic plane is taken from a work of Van Rhijn (1925). The distribution of the component Z of the velocity is studied separately for each spectral type and certain intervals of absolute magnitude. Oort used both radial velocities of the stars (except B) of known absolute magnitudes and of Galactic latitudes between $+40°$ and $+90°$ and between $-40°$ and $-90°$, and also the velocities of dwarfs of classes K to M. He found that the distribution of velocities of the stars of each category (i.e. for a given spectral type and absolute magnitude) was generally well represented by the sum of two Gaussian functions. Using these data, Oort traced (up to $z=600$ pc.) the curve representing the variations of $H(z)$ as a function of z. From 0 to 200 pc., $H(z)$ increases as z, and above this, more slowly. The linear part, between 0 and 200 pc., leads to the gradient

$$\frac{\delta}{\delta z}[H(z)]=5{\cdot}62.10^{-30}\ \mathrm{sec}^{-2}.$$

We can thus calculate an approximate value of the total density ρ of matter in the neighbourhood of the Sun by the application of Poisson's theorem. By putting in equation (XII.1) $G=-6{\cdot}67.10^{-8}$ cgs., we find

$$\rho=-\frac{\delta}{\delta z}\frac{[H(z)]}{4\pi G}=6{\cdot}70.10^{-24}\text{ gm./c.c.}$$

Actually, equation (XVII.1) assumes that the Milky Way is infinite and that the density of matter in it is uniform, which is not the case. If it were distributed uniformly in an ellipsoid of revolution, the right-hand side must be multiplied by a factor x, a little less than unity and depending on the axial ratio.[3]

But the central part of the Galactic system certainly has a larger density, and a small portion of the gradient of $H(z)$ must be due to the attraction of the central mass. To evaluate the limits of uncertainty which result therefrom, Oort examined two extreme cases.

First case: Let us assume that the Galaxy consists of a *local system*, centred on the Sun, and a strong attracting mass at the centre of the Milky Way. The curves of equal density of the local system are taken to be ellipsoids of revolution of axial ratio 5. For the "large system" we assume the following:

(a) It is spherical and its mass is $2{\cdot}04.10^{11}$ Suns;

(b) It is an ellipsoid of semi-axes 9,000 and 900 pc., of mass $0{\cdot}93.10^{11}$ Suns (these masses have been chosen to balance the centrifugal force due to Galactic rotation).

Second case: There is no local system. In the outer regions of the Galaxy, surfaces of equal density are ellipsoids of revolution symmetrical with respect to the centre. Their semi-axes are chosen so as to reproduce, near the Sun, the distribution of the stars observed along the z axis, and to extend, in the direction away from the centre, the same distance from the Sun as do the ellipsoids of the local system envisaged in the first case. We again make two hypotheses for the inner part of the Galaxy:

(a) A spherical mass of $1{\cdot}71.10^{11}$ Suns;

(b) Ellipsoids of semi-axes 9,000 and 900 pc.; mass equals $0{\cdot}81.10^{11}$ Suns.

Here then are the densities obtained on the four hypotheses:

	a	b
First case	$7{\cdot}40.10^{-24}$	$6{\cdot}37.10^{-24}$ gm./c.c.
Second case	$6{\cdot}10.10^{-24}$	$5{\cdot}41.10^{-24}$ gm./c.c.

Oort concluded that the most probable value of the total density near the Sun is

$$\rho=6{\cdot}3.10^{-24}\text{ gm./c.c.}$$

or

$$\rho=0{\cdot}092\text{ Solar masses per parsec}^3.$$

(1 Sun per parsec^3=$6{\cdot}85.10^{-23}$ gm./c.c.).

This value, a little lower than that obtained by Kapteyn ($\rho=0{\cdot}099$ Suns/pc.$^3=6{\cdot}78.10^{-24}$ gm./c.c.) can be considered as fairly accurate.

The "Oort Limit"

The density which we have just calculated includes both matter contained in the stars and that dispersed throughout space. If we could evaluate exactly the average density of matter condensed in the stars, the difference of the two densities should give the average density of interstellar matter in the neighbourhood of the Sun.

Starting from the data we have on the spatial distribution of the stars in the vicinity of the Sun, and on their masses, Oort calculated the average density of matter contained in the stars of absolute magnitude M_V lower than 13.5. By grouping these stars according to absolute magnitudes, he obtained the numbers below:

M_V	ρ_*	M_V	ρ_*
B stars	0·0016	$+5{\cdot}5<M_V<+7{\cdot}5$	0·0079
$M_V<+1{\cdot}5$	0·0012	$+7{\cdot}5<M_V<+9{\cdot}5$	0·0072
$+1{\cdot}5<M_V<+3{\cdot}5$	0·0022	$+9{\cdot}5<M_V<+11{\cdot}5$	0·0055
$+3{\cdot}5<M_V<+5{\cdot}5$	0·0074	$+11{\cdot}5<M_V<=13{\cdot}5$	0·0049 $\odot$/pc^3.

The total corresponds to 0·038 Solar masses per parsec^3, or $2{\cdot}60.10^{-24}$ gm./c.c. It is still impossible to estimate the mass corresponding to stars fainter than absolute magnitude 13·5. From $M_V>7{\cdot}5$, the partial densities for each interval of two absolute magnitude only decrease slowly; the total must be increased to account for the less luminous stars, the contribution of which is considerable. Two white dwarfs are already included in the last line of the Table: the Canis Minor of mass 0·39$\odot$ and Van Maanen's star whose mass is many times that of the Sun. Oort thinks that the interval of two magnitudes from 13·5 could easily give, for unit volume, a mass twice that of the preceding interval. We should thus get approximately 0·048$\odot$/pc^3, or $3{\cdot}3.10^{-24}$ gm./c.c.

Thus for interstellar matter there is left the density $0{\cdot}092-0{\cdot}048=0{\cdot}044\odot$/pc^3 or $6{\cdot}3-3{\cdot}3=3{\cdot}0.10^{-24}$ gm./c.c.

As the matter contained in the stars is very probably underestimated, we have thus only obtained an *upper limit* to the average density of interstellar matter in the neighbourhood of the Sun. This is the "Oort Limit".[4] *The density of interstellar matter cannot exceed* 3.10^{-24} gm./c.c.

Discussion

This is precisely the same as the value obtained from Strömgren's work. Previous evaluations were either comparable (Struve, 1939) or a little higher (Dunham, 1939). Thus the Oort limit would appear to be reached in the vicinity of the Sun, and the average density of interstellar matter seems to be almost half of the total density (47 per cent). In other words, the mass of the Milky Way is almost equally divided between the clouds and the stars.

Let us first note that Oort's calculations of the total density ρ and the evaluation of the density ρ_* of the stars apply exclusively to the neighbourhood of the Sun; it would therefore be dangerous to extend its validity to the entire Galaxy. Despite its restricted application, the preceding conclusion has been adopted perhaps a little too hastily by certain authors, and it is with good reason that Vorontzov-Velyaminov has opposed it [360].

In 1932, Oort took the precaution of indicating that the density of interstellar matter could in fact be *much lower* than the limit which he had just calculated. Vorontzov-Velyaminov observed that as the techniques of observing faint stars have increased, so more and more of them have been found: this is equally true of white dwarfs and of stars at very low temperature radiating mainly in the infra-red. The total mass of the stars contained in unit volume might thus be *very much under-estimated*, and it is still not possible to know to what extent. In a more recent work, A. Danjon and P. Couderc (1950) [64] showed that inside a sphere of a radius of a few parsecs around the Sun, the number of stars of each absolute magnitude increases *at least* up to $M_V=15$. The sphere of 3·915 pc. radius, centred on the Sun, contains at least 32.10^{33} gm. of matter in the form of stars, which leads to an average density

$$\rho_*=4{\cdot}3.10^{-24} \text{ gm./c.c.}$$

(instead of the value $3{\cdot}3.10^{-24}$ gm./c.c. adopted above).

An increase in the density of matter concentrated in the stars lowers the Oort limit. To the extent that the small volume studied by Danjon and Couderc is representative, the density of interstellar matter does not exceed $(6{\cdot}3-4{\cdot}3)\times10^{-24}=2{\cdot}01.0^{-24}$ gm./c.c. and its contribution to the total density ρ is just less than one-third. The average density previously adopted for interstellar gas ($\rho_g=3.10^{-24}$ gm./c.c.) would thus be 50 per cent too high, which is not impossible, for its evaluation was very uncertain.

The calculations have been based on the following data: the density ρ_N inside a cloud-type, the fraction f of space occupied by the clouds (0·14), and finally the density outside the clouds. The latter hardly matters: if space were completely empty between clouds, the average density would hardly be diminished. It would be

$$\rho_g=0{\cdot}14\times2\times10^{-23}=2{\cdot}8\times10^{-23} \text{ gm./c.c.}$$

The average density, however, is affected by any change in the other two quantities. Table XIII gives the average density obtained by varying ρ_N and f within reasonable limits. We have also indicated, in each case, the average corresponding stellar density ρ_* considered as unknown, calculated from $\rho_* = 6{\cdot}3.10^{-24}$ gm./c.c.$-\rho_g$. The figures in parentheses indicate the proportion of the total mass contained in interstellar matter.

TABLE XIII

Average Densities ρ_g and ρ_* (gm./c.c.)

ρ_N	f	0.14	0.10	0.07
2.10^{-23} gm./c.c.	ρ_g	$3{\cdot}0.10^{-24}$	$2{\cdot}2.10^{-24}$	$1{\cdot}6.10^{-24}$
	ρ_*	3·3	4·1	4·7
		(47%)	(35%)	(25%)
$1{\cdot}10^{-23}$	ρ_g	$1{\cdot}5.10^{-24}$	$1{\cdot}1.10^{-24}$	$0{\cdot}8.10^{-24}$
	ρ_*	4·8	5·2	5·5
		(24%)	(17%)	(13%)

Again we find that with the same value of the density inside a cloud type as before (2.10^{-23} gm./c.c.) we obtain stellar densities similar to those found by Danjon and Couderc, provided we take $f < 0{\cdot}10$, or, if we do not change the fraction f, by halving the density in the cloud ($\rho_N \sim 10^{-23}$ gm./c.c.).

We can provisionally conclude that, although the average density of interstellar matter is not yet known exactly, it is probably of the order of 1.10^{-24} to 2.10^{-24} gm./c.c. It is perhaps a fifth, or a quarter (or at any rate an important fraction) of the total density of matter in the neighbourhood of the Sun.[5]

NOTES TO CHAPTER XVII

[1]Let us recall that these are the minimum values, since we have attributed to the grains a diameter corresponding to the maximum efficiency factor. The actual values might be five to ten times greater.

[2]From (XVII.5), equation (XVII.4) may be written

$$\Phi(zZ) = \Delta(z) \frac{l}{\sqrt{\pi}} e^{-l^2 Z^2}.$$

The distribution of the velocities at the height z is thus also a Gaussian function having the same modulus as that for $z=0$.

[3]Denoting the semi-axes of the ellipsoid by a and b we have

$$x = \frac{a^2}{a^2-b^2} - \frac{a^2 b}{(a^2-b^2)^{3/2}} \tan^{-1} \sqrt{\frac{a}{b^2} - 1},\ (b<a),$$

for $\frac{a}{b} = 5, x = 0.75$, and for $\frac{a}{b} = 30$, $x = 0.95$.

[4]Sometimes the Oort limit refers to the *total* density ($6{\cdot}3.10^{-24}$ gm./c.c.), due both to stars and to diffuse matter.

[5]Vorontzov-Velyaminov has proposed that it is $\frac{1}{10}$th at most.

Part Four

DIFFUSE MATTER OUTSIDE THE MILKY WAY

CHAPTER XVIII

DIFFUSE MATTER IN EXTERNAL GALAXIES

Classification of external galaxies

Before examining the behaviour of diffuse matter in external galaxies, it is essential to review Hubble's classification briefly [148]. The galaxies are divided into three categories according to their appearance: elliptical, spiral and irregular.

Photographs of the first, which are circular or vaguely elliptical, do not show a spiral structure. In the most elongated, the ratio $\frac{a}{b}$ of the axes is never larger than 3. Thus their *ellipticity* $\frac{(a-b)}{a}$ is between 0 and 0·7 Hence the eight classes are designated by the symbols E0, E1, E2, . . . E7, according to whether their ellipticity is 0·0, 0·1, 0·2 . . . 0·7.

The true spirals, if seen in profile, are always more flattened than elliptical nebulae. Seen face on, their spiral structure shows a clear symmetry of revolution. Hubble divides them into three classes: Sa, Sb and Sc. The Sa spirals have a large central nucleus around which the much less bright arms are closely wound. In the Sc group, the nucleus is very reduced and the whorls (or arms of the spiral) are much more widely separated from one another. They often present very clear nodule-like condensations. The intermediate class Sb is less homogeneous and consists of objects both with a large central nucleus having very open whorls and of others with a much reduced nucleus, surrounded by serrated arms. Plate XXIII shows an Sc galaxy seen face on (NGC 5194), Plate XXIV, an Sa nebula (NGC 4594) and two Sb nebulae (NGC 891 and NGC 4565) seen almost in profile. The Great Nebula M 31 in Andromeda (Plates II and III), and very probably the Milky Way, both belong to the class Sb.

In all these spirals the arms start from two points opposite to each other at the periphery of the nucleus. But there also exist galaxies, less numerous, where a straight nebular bar extends diametrically right across the nucleus. It is at the extreme ends of this bar that the two arms of the spiral start abruptly. Hubble subdivided the *barred spirals* into three classes denoted by the symbols SBa, SBb and SBc, according to the same characteristics as in the *normal spirals* Sa, Sb and Sc.

Finally, the group of irregular galaxies consists of objects of very diverse shapes, and not having the symmetry of revolution common to spirals and elliptical nebulae. It is to this group that the Magellanic Clouds are assigned (even though the large cloud appears to belong to the barred spiral type [279]). On Plate XXIII, the small irregular galaxy NGC 5195, partially covered over by an arm of the Sc NGC 5194 spiral can be distinguished.

Since the time of Hubble's classification, a fourth class of normal spiral, designated by Sd, has been added. It represents an extreme type, characterized by the smallness of the nucleus and by very open arms, almost entirely reduced to those nodulous condensations which are already apparent in the class Sc. Further, a new large category, designated by SO, was created for galaxies similar to elliptical nebulae without a spiral structure, but which differs from them by the presence of a lenticular nucleus, surrounded by an amorphous mass which is less bright and more flattened. They are probably a series parallel to that of normal spirals, for the same degrees of flattening can be observed in them, while the central lens seems to shrink with greater flattening. The SO nebulae are particularly abundant in certain clusters of galaxies, as in those of Coma Berenices and of the Corona Borealis. Table XIV summarizes the above classification.

TABLE XIV

Classification of External Galaxies

Symmetry of Revolution.	No spiral structure.	*Elliptical galaxies* from E0 to E7 according to ellipticity.
		SO galaxies with a lenticular central nucleus and amorphous peripheral mass.
	Spiral structure.	*Normal spirals* with arms coming out of the nucleus. Sa, Sb, Sc, Sd, according to size of nucleus, and the openness of the spirals.
		Barred spirals, with arms starting from a diametrical bar SBa, SBb, SBc.
No Symmetry of Revolution.	*Irregular* galaxies.	

Absorbing clouds in the spirals

The equatorial plane of the spirals can naturally take any orientation with respect to the line of vision. When almost perpendicular to it, the nebula is seen "face on" and the spiral structure is projected without any deformation on to the plane tangential to the celestial sphere. This is, for instance, the case with the well-known nebula M 51 of the Canes Venatici (NGC 5194), of which photographs obtained with two different time exposures are reproduced on Plate XXIII. As the angle i between the equatorial plane and the plane tangential to the celestial sphere increases, the projection on the latter deforms the structure of the nebula, the apparent contour of which can no longer be inscribed in a circle, but only in a more or less flattened ellipse. Thus the axial ratio of the ellipse permits the absolute evaluation of the angle of inclination.[1] For the Great Nebula in Andromeda, i is approximately 19°. When the angle of inclination approaches 90°, it becomes impossible to distinguish the arms of the nebula, which, seen "in profile", take the form of a characteristic spindle. This is the case with the nebulae NGC 891, 4594 and 4565, photographs of which are reproduced in Plate XXIV on the same scale. Since, however, there is a perfect continuity in their appearance, it is almost certain that they are still spirals, but seen in cross-section.

All the views from different perspectives show that the spirals are almost circular in their equatorial plane and very flattened in the perpendicular direction. One could inscribe them in an ellipsoid of revolution whose minor axis would be of the order of 1/5 the equatorial axis.

Now, on photographs of the spirals seen in profile, attention is immediately drawn to the presence of a narrow dark band, which stands out against the bright regions and particularly against the nucleus of the nebula, and which indisputably indicates a very strong absorption.

In 1918, H. D. Curtiss [63] showed that this phenomenon, observed at that time in seventy-seven objects, could be considered as a general characteristic of the class of spirals. When the nebula is seen almost exactly in cross-section, we have the impression that it is surrounded by an absorbing ring, either continuous as in NGC 4594, or partially resolved into distinct clouds as in NGC 891 and 4565 (Plate XXIV). When the inclination is a little less, black bands are still frequently visible on one side of the long axis (NGC 3623 and 3169). If it becomes less still, half of the nebula, on one side of the great axis, is often less bright (NGC 2683). Finally, Curtiss observed that signs of absorption appear even in nebulae seen almost exactly face on. Inside NGC 5194, dark lines are seen next to the bright arms; on the photographs of Plate XXIII, we particularly distinguish those which follow the contours of the most external arms, one to the top, the other to the bottom.

Thus, absorbing matter is not confined uniquely to the periphery of the spirals. A thin absorbing layer, covering the entire equatorial plane, similar to the "Trumpler layer" could, in certain cases, give the illusion of a peripheral crown (NGC 4594). In the nearest objects like the Great Nebula in Andromeda, we distinguish a multitude of absorbing clouds, almost right up to the nucleus; they are particularly abundant in the south-west region (Plate III).

The elliptical nebulae behave in an entirely different manner. They show no signs of absorption, even if seen in profile as is NGC 3115. Theoretical considerations have led Spitzer [287] to conclude that, in almost spherical systems, the total mass of diffuse matter ought to be extremely small. In elliptical systems such as NGC 3115 it could be greater but closely localized in an extremely thin equatorial layer (only some parsecs in thickness). Now, if these formations do include diffuse matter (of which observation shows no trace), its distribution differs profoundly from that observed in the spirals.

Photometric and colorimetric measurements

In general, it is very difficult to evaluate the absorption due to clouds of the external galaxies. In our Milky Way this study is essentially based on star counts inside and outside the dark zones (Chapter VI). In principle one can also measure the opacity of the small clouds which are projected on the luminous background of a large diffuse nebula. The first method is inapplicable to spirals, of which only some can be resolved into stars, and then generally only in the outermost regions. The second can be adapted to the study of spirals by measuring the contrast between the dark regions and the bright adjacent regions. Thus, photometric measurements, which give the distribution of brightness over the surface of the nebulae, can have interesting results bearing on absorption. The interpretation of the results, however, is very delicate since the distribution of brightness does not only depend on the presence of absorbing clouds, but also, and more significantly, on the distribution of the luminous masses due to the accumulation of stars. Thus it involves the structure of the spiral in all its complexity.

In principle, the technique of the measurements is simple: we photograph the nebula at the focus of the telescope and, by means of the recorder of a microphotometer, curves are traced representing variations in darkening of the plate along lines across the nebula. Furthermore, if the curve of darkening of the plate has been found for standard illuminations, we can pass from the densities recorded by the microphotometer to the brightnesses themselves. This photometric scaling can be realized in various ways in the laboratory, and even for the sky, by means of extra-focal images of stars of known magnitudes.

Measurements of this kind have been made successively by Hubble [149], by Redman [242], then by Redman and Shirley [243], by G. de Vaucouleurs [354] and mainly by the astronomers of the Swedish School. Lindblad, with various collaborators, traced the photometric profiles of numerous nebulae in two or three colours, as also did Holmberg whose most recent work (1950) [145] was carried out on thirty-nine nebulae.

From all this work we shall here give only some of the results for the opacity of the absorbing clouds.

If a very narrow dark band crosses the bright nucleus of a spiral seen in profile, we can endeavour to evaluate the corresponding minimum absorption by extrapolating inside the dark band that curve which represents the brightness on either side of the nucleus. The graph is often uncertain, because of the rapid increase in brightness towards the centre of the nucleus. But if the measurements are made in two colours, the colour excess due to absorption is obtained with greater certainty. Thus, in NGC 4565, the measurements of Lindblad and Brahde [179] show that the weakening produced by the dark band is much more marked in the blue than in the red (the difference is of the order of 1 magnitude). *Thus, the particles which constitute the cloud absorb the shorter wave-lengths much more strongly just as do those of the clouds of our Milky Way.*[2]

For a direct evaluation of the opacity of the clouds, we generally lack a sufficiently uniform luminous background. Holmberg [144] found a case in which such a background was accidently present, almost independent of the nebula itself. We can see on Plate XXIII, that the outermost arm south of NGC 194 partially covers the irregular galaxy NGC 195, situated further away. It is permissible to suppose that the distribution of brightness in the irregular galaxy is almost symmetrical. Therefore the eastern part, covered by the arm of the spiral, before undergoing absorption ought to have had a brightness comparable with that of the western part. On this hypothesis, photometric profiles obtained along a certain number of straight lines in the east-west direction would, at each point, give the opacity of the absorbing screen. (Actually we must bear in mind that some light is emitted by the bright external part of the arm of the spiral in the easternmost regions.) The maximum values of the absorption shown on photometric profiles vary from one section to the other. from 0·6 to 1·0 photographic magnitudes. By comparing these with photovisual measurements, made in a similar manner, we find that the maximum of the colour excess also varies between the limits +0·1 and +0·3 magn. By combining various sections and thus reducing accidental errors mainly due to the lack of uniformity in the brightness of NGC 5195, we finally find the maxima given below:

0·8 magn. for total photographic absorption,
0·2 magn. for differential absorption.

Thus the ratio of total photographic absorption to differential absorption (in the international system) is 4·0, a value which is very near to that obtained for the Milky Way, by means of the Stebbins and Whitford curve (3·8, see page 179). This agreement seems to permit us to conclude that "*the absorbing matter of an arm of a spiral has a constitution which is more or less similar to that of the interstellar clouds of our Galaxy*" (Holmberg). Further, it must be noted that the photographic absorption of the arms of the spiral is exactly of the same order of magnitude as the transverse opacity of the Milky Way (in fact for all sections the average opacity from west to east is 0·39 magn.).

Distribution of absorbing matter, true orientation in space and sense of rotation of the spirals

Theoretically the determination of the sense in which the spirals rotate about their axis of revolution is of prime importance. Although to all intents and purposes this problem is beyond the scope of the present work, it is nevertheless essential to indicate briefly how it is connected in practice with the question of the distribution of absorbing matter, to which it lends a further great interest.

On the dynamic theory of Lindblad, the rotation should take place "arms first". For instance, the nebula NGC 5194 (Plate XXIII), seen almost face on, ought to be turning in a retrograde sense (clock-wise). But, in this case, we have no means of showing the rotation because it is too slow to produce an appreciable displacement of the arms on the human time scale.

When the equatorial plane of the spiral is sufficiently inclined to the plane tangential to the celestial sphere, a difference in the velocities of the two extremities of the long axis can be found (V. M. Slipher, 1914). An image of the long axis may be projected on to the slit of a spectrograph. The absorption lines in the end approaching the observer are displaced towards the violet and in the end which recedes from him towards the red. Thus a line corresponding to the whole image of the long axis is inclined so that the approaching side can be distinguished. The angle of inclination of the line also leads to the evaluation of the peripheral velocity in km./sec.

To fix the sense of the rotation, however, it is still necessary to know the true orientation of the nebula in space. As we have already indicated (page 299) the measurement of the apparent axial ratio leads to the angle of inclination i but not to its sign: two planes symmetrical with respect to the tangent plane to the sphere obviously give the same projection on it. It is thus necessary to discover a means of fixing the sign of the inclination, in other words, to distinguish that side of the apparent small axis which is nearest to the observer. The only procedure available is the examination of the marks of absorption.

When the angle i is nearly 90°,[3] it is legitimate to suppose that the dark ring appearing at the periphery is situated in the equatorial plane of the nebula. Thus the end of the apparent short axis nearest to the observer is determined without ambiguity by the curvature of the absorbing band. This is evidently the higher side of the end in the photograph of NGC 4594 reproduced on Plate XXIV (the nebula is seen slightly "from above"); it is the lower end in the case of NGC 4565 (the spiral is seen a little "from below"). The sense of the rotation is thus well determined if we can measure the radial velocity, but the arms are invisible when the nebulae are seen in profile and the solution of the problem thus loses all interest.

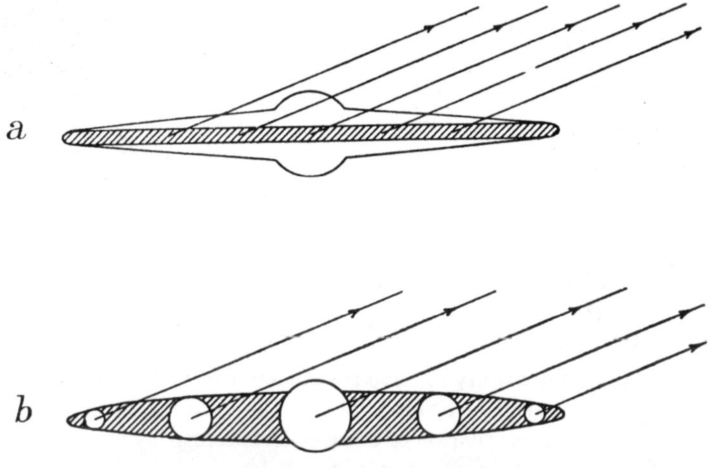

Fig. 49.—Scheme of spirals.
(*a*) Absorbing matter (shaded area) is supposed to be confined in a thin layer near the equatorial plane. If this layer is opaque enough it is visible, since in the direction of the arrows that half of the nebula which is nearest the observer is the most weakened.
(*b*) Absorbing matter (shaded area) is assumed to be outside the nucleus and the whorls (the circles represent sections through these). In this case that half of the nebula which is furthest from the observer is the most weakened by absorption.

The difficulties of interpretation begin in the interesting cases where the inclination is small enough to permit the distinction of the sense of rotation of the spiral. We then know (page 299) that the absorption has as its effect the general diminution of the brightness of one-half of the nebula on one side of the long axis. However, two hypotheses can account for the distribution of absorbing matter:

(a) We can suppose that it is concentrated in a rather thin layer in the neighbourhood of the equatorial plane, as in our Galaxy. Thus Fig. 49 shows clearly that the half of the nebula nearest to the observer ought to be the less bright. This is the point of view adopted by Slipher and, more recently, by Hubble [151]. Combined with the knowledge of radial velocities, this conception forces us to the conclusion that the nebula turn "arms behind", i.e. contrary to Lindblad's theory.

(b) We can suppose, on the other hand, that the absorbing clouds are assembled in the external regions, above and below the equatorial plane where they more or less envelop the bright arms (Fig. 49b). Then the half nearest to the observer is necessarily the most luminous. Lindblad thinks that the first type of distribution (a) predominates in the compact spirals of the "less advanced" types of the Hubble classification (types Sb and Sc). In this case the spirals would turn in the sense predicted by the theory.

Photometric and colorimetric study of numerous nebulae (Andromeda [177, 179], NGC 7331 [178], M 63=NGC 5055 [180] provided Lindblad with arguments in favour of his point of view. In Andromeda, for instance, the north-west side of the nebula, particularly rich in absorption marks, would be the most distant from the observer. By contrast, Holmberg [142] finds arguments in favour of the opposite conception (a) in his photometric study of NGC 3623, as does also G. de Vaucouleurs [355] for the case of NGC 2146.

The statistical and theoretical work of Holmberg

Looking at the question of absorption in the spirals statistically, Holmberg [143] showed that there is a relation between the inclination i of the equatorial plane and the absolute magnitude. To obtain absolute magnitudes, it is first necessary to have the apparent total magnitudes. Holmberg uses those determined at Harvard (Shapley-Ames) after correction for a systematic error. As criterion of distance he successively uses the angular values of the long axis (also corrected for systematic errors) and the radial velocities which are proportional to the distances. The first procedure gives, between the absolute magnitude of the nebula seen in profile ($i=90°$) and seen face on ($i=0°$), an average difference $+0{\cdot}37 \pm 0{\cdot}15$ magn. (108 nebulae); the second gives $+0{\cdot}60 \pm 0{\cdot}32$ magn. (seventy objects). Finally, the cluster of spirals in Virgo leads to an independent determination. As these objects are all at the same distance from the observer, the difference of their absolute magnitudes is equal to the difference of their apparent magnitudes. If we construct separate frequency curves for the magnitudes of elongated nebulae (ratio of axes $\frac{\beta}{a} \leqslant 0{\cdot}4$) and for globular

NGC
5194

NGC
5195

PLATE XXIII.—The external galaxy M 51 in Canes Venatici. The outermost arm (bottom) of the spiral NGC 194 partially covers the irregular galaxy NGC 5195. 193 cm. telescope of the Haute-Provence Observatory.

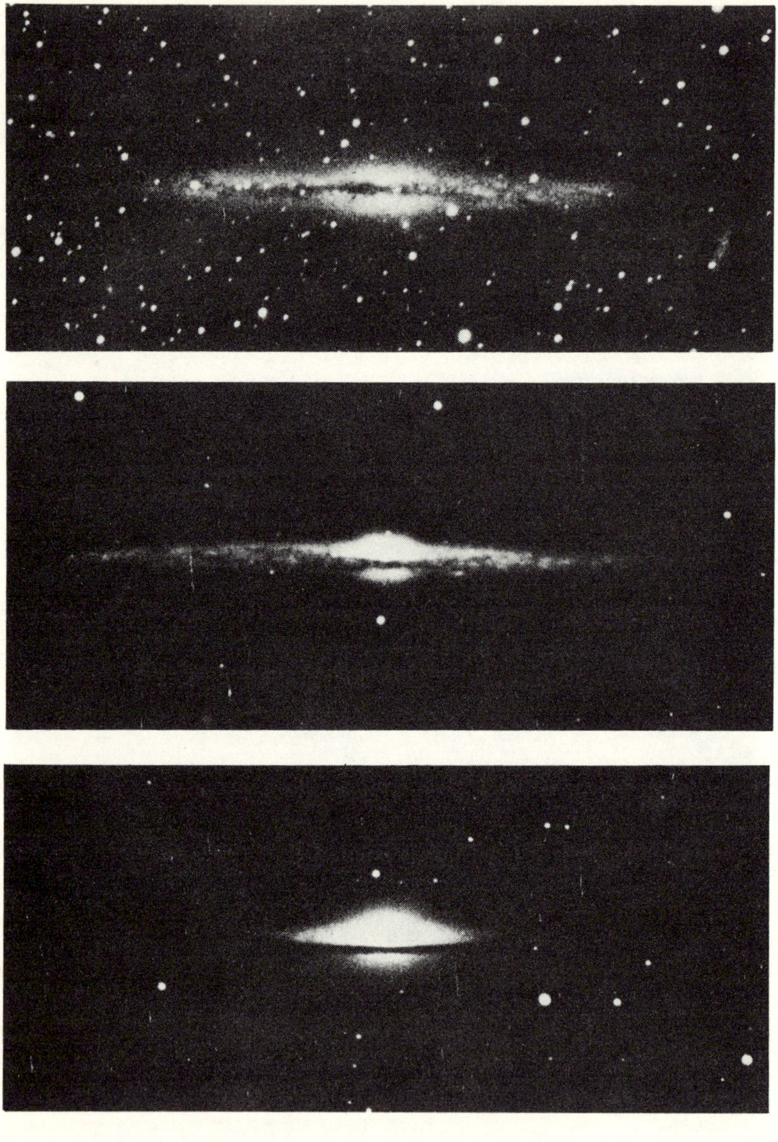

Plate XXIV.—Three spiral nebulae seen in profile, all on the same scale. *Top:* NGC 891 (type Sb). *Centre:* NGC 4565 (Sb). *Foot:* NGC 4595 (Sa). 80 cm. telescope (M. de Kérolyr).

nebulae ($\frac{\beta}{\alpha} \leqslant 0{\cdot}7$), we find between the two curves a systematic discrepancy of approximately $\frac{1}{2}$ magnitude. For $i=90°$ and $i=0°$, the difference would still be $+0{\cdot}60 \pm 0{\cdot}18$ magn. (113 nebulae). Holmberg concluded that the nebulae seen in profile are on the average 0·5 magn. fainter than the nebulae seen face on. The weakening is certainly due to absorption within the nebula itself.

The effect of inclination on colour excess appears less clear and is certainly much smaller. Positive according to photographic and photovisual measurements, it is slightly negative according to the photo-electric measurements of Stebbins and Whitford; in other words, the nebulae are a little bluer when they are seen in profile than when they are seen face on.

Holmberg showed that similar results could be obtained by means of "nebular models" constructed for the purposes of calculation. Fig. 50 represents a section through a meridian of the simplest of them.

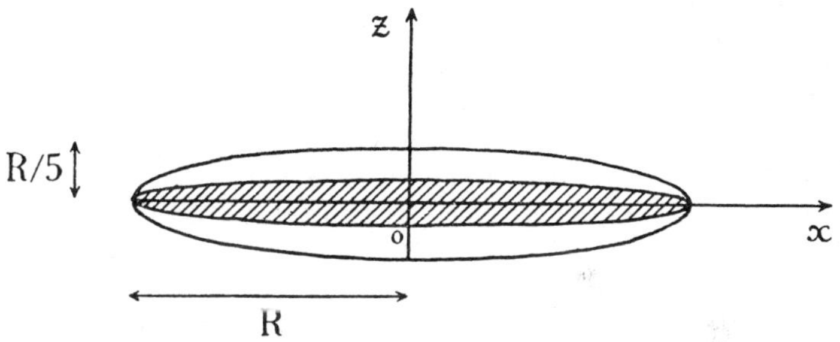

Fig. 50.—Holmberg's model of a spiral.

The "luminous matter" is contained in a flattened ellipsoid of revolution of equatorial radius R and of minor semi-axis $\frac{R}{5}$. Absorbing matter is concentrated in an even flatter ellipsoid with the same major axis, and with minor axis $p\frac{R}{5}$ ($p<1$). Inside the two ellipsoids, the luminous intensity and the co-efficient of absorption are represented by normal distribution functions.

We plot the curves representing, for various values of p, the difference of magnitude Δm between the nebulae seen in profile ($i=90°$, along the Ox axis) and seen face on ($i=0°$, along the Oz axis), as a function of a parameter t. The latter is analogous to the "transverse opacity of the Milky Way" (page 192), which measures in magnitudes, the absorption which a light ray would undergo in crossing the nebula from one pole to the other along the minor axis.[4] The curves show that Δm varies little with t, if this quantity is above 2 magn.; but greatly with p. For t of the order of 2 to 4 magn., Δm changes from $-0{\cdot}5$ magn. to $+1{\cdot}2$ magn., when p increases from 0 to 1. For $p=0{\cdot}4$, we again obtain the difference $\Delta m=+0{\cdot}5$ magn., the value resulting from the statistical study of the spirals. The calculations made for intermediate values of the angle of inclination show that the calculated weakening varies similarly to the observed weakening, as a function of the ratio $\frac{a}{\beta}$ of the apparent axes.

Similarly, Holmberg calculated the difference ΔE between the colour excess of the nebula seen in profile and that of the nebula seen face on[5] and he found that it is always smaller ($|\Delta E| < 0{\cdot}1$ magn.). In certain cases ΔE can become negative, and thus conforms with Stebbins and Whitford's measurements. Thus, for $p=0{\cdot}4$, ΔE is positive or negative according to whether t is smaller or bigger than $2{\cdot}6$ magn.

In spite of its extreme simplicity, Holmberg's scheme with p taken as $0{\cdot}4$ reproduces the results of the statistical study of the spirals very well. The axial ratio of the absorbing ellipsoid would thus be $0{\cdot}4\times0{\cdot}2=0{\cdot}08$.[6]

With this model, Holmberg also constructed theoretical curves of the distribution of brightness over the surface of the nebulae. The photometric profiles calculated for $i=90°$, always taking $p=0.4$, are very similar to those of the spiral NGC 891. Even if the total optical thickness t does not exceed 1 magn., the observer receives less than 10 per cent of the light emitted from the centre. For $i=75°$, the results are already very different. Absorption is noticeable over the entire surface of the nebula and naturally produces a marked dissymmetry, since it weakens that part which is nearest to the observer. With an optical thickness t of the order of 2 magn., the dissymmetry is almost that which Holmberg observed in NGC 623.

In all the cases the effect produced on the colour index is not very marked. At the centre of the dark band of the nebula seen in profile, the colour excess does not exceed $+0{\cdot}2$ to $+0{\cdot}3$ magn. Thus the ratio between photographic absorption and differential absorption, *both taken for the total magnitude of the entire nebula*, can be very much higher than the value which corresponds to the crossing of a simple absorbing screen. If the latter were five for instance, we could obtain for the total effect a ratio of twenty to twenty-five when the optical thickness is very large and the parameter p very small.

Polarization of light in the spirals

Galactic clouds, when they are strongly illuminated by stars, can become bright enough to appear as continuous spectrum nebulae. Thus the light they scatter may be partially polarized, at least in certain directions (page 198). The absorbing clouds of the spirals would probably behave in the same manner if they had the same constitution as those of the Milky Way. In clouds very near the centre of the spiral, the nucleus produces a considerable illumination and conditions are favourable to the investigation of polarization. Öhman (1939–42) [218] effectively managed to show this phenomenon in certain clouds of the nebula in Andromeda, thus proving that they scatter light in the same manner as the Galactic clouds.

These remarkable observations were made by means of the grille polarigraph described in an earlier chapter (page 201).[7] Each section of the nebula isolated by the grille, in the focal plane of the astrograph gives, on the plate, two contiguous images of which one contains vibrations parallel to the bars of the grid, the other perpendicular vibrations. The apparatus, when conveniently orientated with respect to the nucleus of the nebula, permits the evaluation of the ratio of the intensities of the radial and tangential vibrations. The curve of blackening of the plate against illumination having previously been determined, it suffices to compare the microphotometric records of the two juxtaposed images.

Öhman found clear traces of polarization in three different regions of the nebula in Andromeda (M 31). The first is a dark band 5·3′ from the centre, of positional angle 315°, in the north-west region. It shows a "negative" (or tangential) polarization, that is to say with the privileged vibration directed along the radius vector joining the point observed to the centre of the nebulae; the proportion of polarized light reaches 6·2 per cent. The second region is at the same position angle at 2·7′ from the centre; its polarization, still "negative", is 1·3 per cent. With a telescope of longer focus, giving larger images, Öhman found a "positive" (or radial) polarization, with the privileged vibration normal to the radius vector, in a small cloud situated only 15″ from the centre, at a position angle 344°. All these observations were made in blue light; others, made in red light on the first two regions, have not shown clear polarization either, as Öhman supposed, because the proportion of polarized light varies much with the wave-lengths, or else because the intensity of the scattered light is much weaker in the red.

From these observations Öhman endeavoured to get an indication on the actual orientation of the nebulae in space (sign of the angle $i=\pm 19°$). They seemed to him to favour Lindblad's idea: the north-west region of the nebula, dimmed by absorption, is more distant from the observer.[8]

Another interpretation of the phenomenon has recently been suggested by Page [228]. The light received in looking at absorbing clouds could mainly be due to the stars seen through the clouds. Thus there would exist an interstellar polarization similar to that discovered by Hiltner and Hall in the Milky Way (Chapter XV). On this hypothesis, it is not surprising that the polarization should not appear in the red since, in the Milky Way, it seems independent of the wave-length.

Madame Elvius [92] studied the spiral M 63 (NGC 5055) by means of the Öhman polarigraph, and she found, in many regions, proportions of polarized light between 3·5 and 7·4 per cent. Polarization which appeared mainly along a large dark band is obviously related to the absorbing clouds of the nebulae. The privileged vibration is often parallel to the isophotes traced by Lindblad and Delhaye [180].

These observations made in blue light, can still be interpreted in two different ways: polarization of light of the nucleus scattered by the clouds, or polarization of light transmitted by these. Both hypotheses can account for the orientation of the privileged vibration and for the proportion of polarized light, and both seem equally probable. Lindblad and Delhaye have shown that there exists, behind the large dark band, an arm of the spiral more easily visible in red light than in blue light because of the differential absorption of the clouds. It is thus possible that we receive both the transmitted and the polarized light of the clouds. In order to solve this question, Madame Elvius has proposed to study the polarization in M 63 in red light.

Emission lines in the external galaxies

If an observer, placed in another galaxy, could photograph the spectrum of our Milky Way seen by him as if it were a spiral nebula, he would find emission lines in it which in places would be superimposed on the continuous spectrum of all the stars. These would correspond to the largest line-spectrum nebulae and to the H II regions of Strömgren around the hot stars.

Many of the external galaxies also show emission lines in places. Mayall [190], and later Baade and Mayall [14] have investigated the presence of the doublet O II 3727–3729 (not resolved) in spectra of small dispersion, of some hundreds of nebulae obtained at the Lick Observatory and at Mount Wilson. This is the easiest line to detect, not only because of its large intensity, but also by reason of its position at the ultra-violet extremity of spectra obtained with glass prisms, i.e. in a region where the dispersion is larger and the continuous background coming from the stars of the nebulae is already very faint (class G). Only 20 per cent of the elliptical nebulae examined show the doublet of O II but the proportion is larger with spirals and it increases rapidly when we consider nebulae of "more advanced" types. It reaches at least 80 per cent in the classes Sb and particularly in Sc. Thus the line appears in the arms of the spiral and in the bright nodosities, whereas in the types So and Sa, as also in the elliptical nebulae, it is confined to the immediate neighbourhood of the nucleus. After the doublet of O II the lines most frequently observed are those of hydrogen (Hα, Hβ, Hγ) and of O III (5007–4959 A).

In the nearest spirals we can localize exactly, on spectrograms, the small regions which show bright lines. In the Great Nebula in Andromeda (M 31), H. W. Babcock [16] found that their spectra, showing the lines already mentioned, were almost the same as that of the nebula in Orion. The lines are very sharp and Babcock could use them to measure the radial velocity in different points of the nebula. Similarly, twenty-five bright-line objects were found in M 33, which is almost at the same distance as M 31 (Mayall and Aller [191]) and ten objects in M 101 (Seyfert [272]). According to Shain and Mademoiselle Gazé [430] who have photographed them by isolating the Hα line, certain gaseous nebulae contained in M 31, M 33, M 101 and NGC 6822 reach enormous dimensions and their masses could be thousands or tens of thousands of Solar masses. Thus the nebula NGC 604, in M 33, with a density of the order of 100 protons per c.c., ought to have a mass of $2 \cdot 10^5$ Suns—higher than the biggest gaseous nebula of the Milky Way.

At Mount Wilson, in order to detect the gaseous nebulae of M 31, a method similar to that which has been described for the investigation of Galactic emission areas was followed (page 94). The nebula is photographed through two different colour filters on two different emulsions [14]. The first combination uses wave-lengths between 6300 and 6700 A, including the Hα line, the second isolates the region 6900–7500 A, which does not contain strong emission lines. The bright areas appear clearly on the first negatives, but not on the second. Thus, more than 300 areas have been counted in the Great Nebula in Andromeda. Most of them were along the arms of the spiral.[9] Often invisible in blue light, they are probably strongly reddened, which indicates a concentration of grains in the neighbourhood of the arms; this corresponds to the usual association of atoms and of grains as in the Milky Way. The space between the whorls ought to be much more transparent since, in them, we can photograph distant galaxies up to 15′ from the nucleus. Thus, the fact that the globular clusters associated with M 31 (about half of which ought to be found on the "other side" of the nebula) are, in general, more reddened, tends to prove that there is not much smoke between the arms.

Because of their proximity (r=24 kpc.), the Magellanic Clouds, which are irregular galaxies, lend themselves to the investigation of nebular areas showing emission lines. By photographing them either with a prism-objective or directly with a red filter, Henize and Miller [130] have catalogued 150 areas emitting Hα lines in the Great Cloud and forty-three in the Small.

Finally, we know a dozen spirals belonging mainly to the classes Sa and Sb whose *nucleus* shows particularly strong emission lines. Seyfert [273] studied six of them in detail. He found that the bright lines in them were very much broadened, no doubt due to a Doppler effect, of which the amplitude could reach 8500 km./sec. (Balmer lines in NGC 3516 and 7469). Sometimes, we only observe hydrogen lines, and much weaker lines of [O III] (NGC 3516), sometimes a spectrum very rich in forbidden lines similar to that of a planetary nebula at very high excitation, and with lines of [Ne III], [A IV], [Fe VII], etc. This is the case in NGC 1068 where the emission spectrum carries, in the "photographic region", 13 per cent of the total light of the nucleus.

In brief, in the spirals, both bright-line nebulae apparently similar to Galactic nebulae, and also absorbing clouds similar to those of the Milky Way can be found. The presence of dark clouds in the spirals has been known for a long time and has been responsible for orientating those workers who, for a quarter of a century, have endeavoured to show absorption of light in the neighbourhood of the Galactic plane. Despite certain difficulties in interpretation, the study of the distribution of clouds in the spirals can still be useful to our knowledge of the stellar system around us, of which we cannot form a complete picture. But, if our position prevents us from seeing the large outlines of the structure of the Milky Way with one sweep of the eye, it does permit us to study fairly well the quite indiscernible details of even the greatest spirals. And thus the work on diffuse matter dispersed in our Galaxy nevertheless has assisted and will continue to assist our understanding of the phenomena observed in the distant systems.

However, the large family of spirals comprises very different objects, in which diffuse matter is not always distributed in the same way. If some of them, such as the Great Nebula in Andromeda, seem very similar to our Milky Way, there are also very different ones, such as the nebulae studied by Seyfert, where the emission of rarefied gases is much more prominent. On the other hand, diffuse matter is rare or absent in elliptical galaxies.

Diffuse matter and stellar populations

The external regions of the nearest spirals, such as those of Andromeda (Sb) have for long been "resolved into stars", in the sense that we have been able to photograph the brightest stars in them individually. The resolution of stars in elliptical galaxies and of nuclear spirals presents much more difficulty. In 1943, Baade [12] managed to photograph the most luminous stars of the nucleus of Andromeda, up to very near the centre, and those of its two companions: M 32, which is almost spherical, and NGC 205, which is elliptical and weaker (see Plate II). He did this by means of the 2.5 m. telescope at Mount Wilson, using plates sensitive to the red (up to 6700 A), together with a filter which weakened the bright lines of [O I] 6300–6364 A, emitted in the high atmosphere, in order to diminish the background due to the light of the sky. Similarly, Baade [13] resolved the pair of weak elliptical nebulae NGC 147 and NGC 185, situated at almost the same distance as the system of Andromeda.

In red light, the brightest of the stars photographed in the nucleus of Andromeda and in the two companions have a magnitude of +20.0 and their colour index is near to +1.3. Thus their photographic magnitude is almost +21.3; and these stars are much weaker than those photographed in the spirals of the Great Nebula (which partly explains the difficulty in resolving the nucleus). With the modulus of distance adopted by Baade ($m-M=22.4$), they would have as absolute photographic magnitudes $M_{pg}=-1.1$ ($M_{pv}=-2.4$). *This is almost exactly the absolute magnitude of the brightest stars contained in the globular clusters, which also have the same colour index* (see page 30).

Thus, just like the globular clusters, the elliptical galaxies and the nuclei of the spirals contain neither very luminous O and B stars, nor super-giants. Also, the Cepheids of short periods, abundant in the globular clusters, seem just as numerous in the system of Sculptor, studied by Shapley [278], the only elliptical galaxy sufficiently near for us to observe these variables.

Baade concluded that *the population of the elliptical galaxies is purely of the type II* (see page 34). In the advanced spirals, the two populations co-exist, but with different localizations; the type II in the nucleus, the type I in the whorls.[10] Thus, in our Galaxy, we seem to be dealing with a population *I* by reason of the ex-centric situation of the Sun, but the nucleus of the Milky Way ought to be of population II. Thus, on the side of the Galactic centre, we find stars belonging to this category, notably short period Cepheids.[11] Open Galactic clusters which are frequently found in the neighbourhood of the Sun, can be considered as condensations of type I, and the globular clusters, visible above or below the centre of the Milky Way, as condensations of population II. They are also frequent in the elliptical nebulae.

Now, we know already that absorbing clouds, which abound in the arms of the spirals, are absent in the nuclei; they are also absent in elliptical galaxies. Emission areas, which are characteristic of interstellar gases, are also generally much more frequent in arms than in the nuclei, and they are rare in elliptical galaxies. *Thus, interstellar matter appears largely associated with populations of type I.* It is highly probable that absorbing clouds, numerous in the neighbourhood of the Sun, are absent in the nucleus of the Milky Way.

The absence of diffuse matter goes hand in hand with the absence of supergiants. This is an important confirmation of the theory of the formation of stars by condensation of clouds (Spitzer) (see Chapter XVI). If the systems with population II initially contained supergiants, these have been extinct for a long time and owing to the lack of primary matter, it has not been possible for new ones to be formed.

Encounters of galaxies

The galaxies which form the densest clusters, such as those of the Coma Berenices and the Corona Borealis, belong mainly to the class SO.

These SO galaxies do not show any trace of absorbing matter. Their colour index is the same as that of the elliptical nebulae. It is thus very probable that their population belongs to the type II.

On the other hand, since we find many normal spirals in the less dense clusters, in those of Virgo or Hercules for example, Spitzer and Baade (1951) [291] have tried to find out whether the particular composition of the very dense clusters is not due to the more frequent collisions of nebulae.

When two galaxies meet they can cross each other without great damage to the stars of which they are composed. Because of the great mutual distance of the stars and the large relative velocity of the two galaxies (at least some hundreds of kms. per second, often more than 1,000 kms./sec.), the encounters of stars can be neglected.

This event, on the other hand, is catastrophic for interstellar matter. Even if the density of hydrogen is only 0.1 per c.c., each atom does not cross more than 0.1 parsec, on the average, before encountering another atom. Now, the mean relative velocity of the two galaxies would be 2,400 km./sec. in the cluster Coma and the energy communicated by collision to each hydrogen atom corresponds to 7,600 electron volts. The gas is thus rapidly ionized and most of the protons are thrown outside the two galaxies, and so almost all the interstellar matter.

The collisions of the galaxies can thus explain the loss of diffuse matter. We only have to evaluate the frequency of these encounters. Spitzer and Baade estimated that for the last 3.10^9 years (3,000,000,000 years, the presumed age of the universe), each galaxy of the cluster of Coma Berenices has had between 20 and 150 collisions, more than sufficient to account for the abundance of the SO nebulae. In spite of the provisional character of these calculations, it seems very probable that the collision of galaxies can be held responsible for the loss of interstellar matter.

Let us suppose that a galaxy crossing a cluster in a straight line passes at a minimum distance r from the centre. If the other galaxies were immobile, the average number of collisions during this passage would be $\sigma N(r)$, where we denote by σ the collision cross-section of an average galaxy, and by $N(r)$ the total number of galaxies contained in a cylinder of unit section, traversing the cluster at the distance r from the centre. The number of collisions must be multiplied by $\sqrt{2}$ when we taken into account the velocity of the other galaxies. To evaluate σ, we take 4,000 to 10,000 pc. as the average diameter of the galaxies. The value $N(r)$ is drawn from the counts of Page and Wilson in the cluster Coma, 12·5 million parsecs distant. At 10′ from the centre for example, the depth of the cluster is approximately 280 kpc., and with an average velocity of 1,700 km./sec., a value taken from the study of radial velocities, a nebula traverses it in $1{\cdot}7.10^8$ years. Thus, in 3.10^9 years eighteen passages can be made. The total number of collisions of a nebula would thus be

$$18\sqrt{2}\sigma N(r).$$

With $N(r)=1{\cdot}6.10^{-8}$, a value taken from the count of nebulae, we find 22 or 140 collisions according to whether we take 4,000 or 10,000 pc. as distance of collision.

Zwicky's observations

With the 120-cm. Schmidt telescope of Mount Palomar, Zwicky [439, 440] recently discovered that double or multiple galaxies were sometimes linked by large clouds of luminous matter.[12] Some have since been photographed with a telescope of 5-m. aperture. The phenomenon was especially observed in a double galaxy in Aquarius, a double galaxy and a triple galaxy in Hercules. Here are some details of the most fully studied cases: we find in the cluster Coma Virgo, towards 12 h. 30 m. of right ascension and +11°40′ of declination, three galaxies of apparent photographic magnitudes between 15 and 16. Two of them (IC 3481 and its anonymous neighbour) belong to the type SO, the third (IC 3483) to the type Sa. Now, the photographs show, between the three components of the system, a very faint luminous double link whose contrast with the background of the sky is only very small. It describes curved bands of enormous dimensions. From its radial velocity (+7,250 km./sec.), we can evaluate the distance of IC 3483 as 13·2 million parsecs (13·2 Mpc.); as the angular distance from IC 3483 to IC 3481 is 340″, the distance of the two galaxies projected on the plane normal to the line of vision is approximately 22,000 pc. This would be the *minimum* length of the luminous band which links them.

Ought we to see in this "intergalactic luminous matter" an example of exchanges of interstellar matter between neighbouring galaxies? On this hypothesis the luminosity would, no doubt, be due to the excitation of the hydrogen atoms, and the clouds ought to be red because of the preponderance of the Hα line. However, according to the first photographs obtained with a 5-m. telescope, in red light and in blue light, their colour clearly seems blue. Zwicky thinks that there are large streams of stars due to the rupture of the galaxies at the moment of their approach.

In this case, the phenomenon does not seem to be in direct agreement with the loss of interstellar matter envisaged by Spitzer and Baade at the moment of collision of the galaxies. But these observations show at least that the approach and the encounter of galaxies is of relatively frequent occurrence and of great cosmic importance.

NOTES TO CHAPTER XVIII

[1]If α and β are the angular values of the apparent semi-axes ($\beta<\alpha$) for a flat disc we should have

$$\cos i=\frac{\beta}{\alpha}.$$

If we consider the nebula as a flattened ellipsoid of revolution, we have, denoting by f the ratio of the axis of revolution to the equatorial axis ($f<1$),

$$\cos^2 i=\frac{\left(\frac{\beta}{\alpha}\right)^2-f^2}{1-f^2}.$$

[2]Holmberg obtained an opposite result for the case of NGC 891, where the dark band produced a weakening more pronounced in the red than in the blue (differing of the order of 0.8 magn.). But he noted that measurement is probably affected by the chance presence of numerous bright and very white stars in front of the absorbing cloud. Many of them are clearly visible on Plate XXIV.

[3]For i exactly equal to 90°, the problem is indeterminate.

[4]Holmberg represents the luminous intensity of unit volume, at a point (x, y, z) by

$$I(x, y, z)=\frac{1}{f(2\pi)^{3}/_{2}}e^{-\frac{x^2}{2}-\frac{y^2}{2}-\frac{z^2}{2}}$$

and the absorption co-efficient at the same point by

$$k(x, y, z)=\frac{u}{pf(2\pi)^{3/2}}e^{-\frac{x^2}{2}-\frac{y^2}{2}-\frac{z^2}{2p^2f^2}}$$

with $u=\frac{2\pi t}{1.09}$: f is the ratio of the axes of the ellipsoid containing the "luminous matter", and here takes the value $f=0.2$.

One can now evaluate, in magnitudes, the total weakening A of the nebula along Oz ($i=0°$) by a numerical integration from $-\infty$ to $+\infty$.

$$A=-2.5\log\left[\iiint I(x,y,z).10^{-A(x,y,z)}\,dx\,dy\,dz\right] \qquad \text{(XVIII.1)}$$

where $A(x, y, z)$ represents the absorption undergone by the light emitted by unit volume at the point x, y, z. This quantity can be written:

$$A(x, y, z)=t\cdot e^{\frac{-(x^2+y^2)}{2}}\int_{-\infty}^{z}\frac{e^{\frac{-z^2}{2p^2f^2}}}{pf\sqrt{2\pi}}\,dz=t\cdot e^{\frac{-x^2+y^2/_2}{2}}Q(z/pf)$$

The function Q, defined by $Q(a)=\int_{-\infty}^{a}\frac{e^{\frac{-x^2}{2}}}{\sqrt{2\pi}}dx$ is Charlier's probability integral. The required difference in magnitude is $\Delta m=A'-A$

The total weakening A' along the Ox axis is calculated in a similar way by replacing $A(x, y, z)$ in (XVIII.1) by

$$A'(x, y, z)=\frac{t}{pf}\,e^{-\frac{y^2}{z}-\frac{z^2}{2p^2f^2}}Q(x).$$

The required difference in magnitude is $\Delta m=A'-A$.

[5]For this purpose he assumed that, in the case of photovisual measurements, t is four-fifths the value of the photographic measurements, that is to say, that photographic absorption is equal to five times the differential absorption. This ratio seems a little large (see page 179).

[6]Most spirals have a very bright central nucleus not shown in the preceding model. Actually, total absorption and colour excess depend very little on the nuclear concentration. Holmberg, in fact, made his simple scheme more complex by adding to it a spherical nucleus of radius equal to the minor semi-axis ($0.2R$), inside which the distribution of light follows another law, with a concentration more marked towards the centre. The results obtained with three different distribution laws are very similar to those of the original scheme. We always obtain, between $i=90°$ and $i=0°$, a difference $\Delta m=+0.5$ magn. by giving p a value of 0.4 to 0.5.

[7]In the last series of observations Öhman substituted for the rhomb a Wollaston prism which is chosen so as to halve the distance between the bars of the grid placed at the focus of the collimating lens.

[8]In this case, for the first region, the angle of scattering would be of the order of 50°; in the opposite case it would be between 161° and 180°. Öhman seems to prefer the first value for the appearance of so marked a polarization. Actually, we still have no idea of the variation of the proportion of light polarized by the small particles of the cloud as a function of the angle of scattering. In particular we do not know for which values of this angle the polarization can become negative.

[9]This seems to be the case also for M 51, which G. Courtès [61] observed with interference filters which he uses for the study of Galactic areas.

[10]If we try to resolve a spiral of advanced type by, for instance, progressively increasing the exposure time, we should first notice the appearance of a *very small number* of very luminous stars; these would be the O and B or F to M supergiants. A very much longer exposure time is required for photographing a *great number* of fainter stars: the ordinary giants of population I. In the elliptical galaxies or in the nuclei of the spirals there appears at first glance a *large number* of stars of equal luminosity: the K giants of population II, much less luminous than the supergiants, but of greater intensity than the giants of type I.

[11]The high-velocity stars in the Milky Way, studied by Oort in 1926 [219] probably belong to population II, because they do not contain any very luminous O or B stars. In them the proportion of dwarfs to giants is still greater than in the low-velocity stars (population I). Similarly the sub-dwarfs, intermediate between the dwarfs of the principal series and the white dwarfs (page 32) would be high-velocity stars belonging to population II. Some observations indicate that the spectroscopic criteria of absolute magnitude used for the slow stars near the Sun, are no longer valid for the fast stars of population II.

[12]The luminous filaments linking very near galaxies to each other such as NGC 750 and 751 had already been observed long ago but the phenomenon discovered by Zwicky is of another order of magnitude, since the length of the luminous bands can be many times that of the diameter of the galaxies.

CHAPTER XIX

INTERGALACTIC SPACE

INTERGALACTIC space is certainly very transparent since it permits spirals at apparently enormous distances to be photographed. But it is important to find out if there are any manifestations of the existence of dispersed matter in it. "Intergalactic" absorption lines, analogous to interstellar lines, have never been detected in the spectra of the external galaxies. Since a gas spread throughout space would participate in the expansion of the universe which gives rise to a radial velocity proportional to the distance, the lines produced between the nebulae and the observer would be spread over many tens, or even hundreds of angströms and, in consequence, they would be invisible.

Therefore only the effect of a general absorption by scattering in space can be demonstrated. We shall deal successively with the measurements of brightness of the nebulae, with their colour indices, and finally with counts of them. Even if the spatial density of the particles is very small, absorption can become appreciable over very long distances and thus affect the evaluation of the distances.

Intergalactic absorption and brightness of the spirals

Let us recall that the stellar luminosity l of any star can be expressed as

$$l=B\omega, \tag{0.3}$$

B being the average brightness and ω the solid angle subtended by the star. In the case of a source having an appreciable apparent diameter, the two factors B and ω can, in principle, be measured separately. The first is almost independent of distance, and is only affected by absorption, whereas the second depends only on the distance.[1]

The only direct measurements of brightness which we have are the old visual measurements of Wirtz (page 185) [383]. It must be stressed that they refer only to nebulae of very large apparent diameter and which are thus relatively near. By plotting the magnitude per unit solid angle obtained from Wirtz's measurements against the parallaxes previously evaluated, Lundmark [181] found that the brightness of the nebulae decreased as the parallax, as if there existed absorption proportional to distance. Lundmark took as unit of length the distance of the nebula Andromeda which, at that time, was estimated to be 230 kpc., and he evaluated the visual absorption co-efficient as 0·02 magnitudes per "Andromeda", that is to say

$$a_V = \frac{0{\cdot}02}{230} = 8{\cdot}7.10^{-5} \text{ magn./kpc.}$$

This value is approximately 9,000 times smaller than the average photo-visual absorption co-efficient in the neighbourhood of our Galactic plane (~0·78 magn./kpc.).

A later discussion (J. Dufay, 1932) has, however, shown that this phenomenon was illusory. Wirtz compared the focal image of each nebula with an "artificial nebula" whose brightness he varied, but whose diameter remained constant. Thus the two areas, of small brightness, did not generally have the same apparent diameter. Under these conditions experiment shows that, for equal brightnesses, the smallest area would appear the darker. This leads one to attribute a greater brightness to nebulae of large diameter. Further, since the brightness of nebulae is not uniform, the observer involuntarily concentrates his attention on the central region of the large nebulae since this region is the brightest. It is for this reason that the effect is most marked in elliptical nebulae and those of the class Sa which have a nucleus which is much more luminous compared with the peripheral region than is the case in the classes Sb and Sc. These two causes of error, inherent in the method of measurement, act in the same sense. The measurements of Wirtz, therefore, cannot lead to a valid determination of the intergalactic absorption co-efficient.

Though we lack direct measurements we can, however, evaluate the brightness of the nebulae from their total magnitude m and from their apparent diameter α by means of equation

$$m = -\mu 5 \log \alpha + 0{\cdot}26 \tag{0.7}$$

where μ represents the magnitude per unit solid angle (page 22). In the absence of all absorption, we can assume that the average brightness varies only litle from one object to the other, for each category of nebula (μ=constant). In that case m ought to be a linear function of log α, and by plotting m against log α, we should obtain a straight line of slope -5.

Hubble (1926) [148], who used the visual magnitudes of Holeschek, constructed graphs of this kind for various classes of nebulae ($m_V < 13$). The dispersion of the points is large, but we do not observe systematic deviations from the straight line of slope -5. Shapley and Miss Ames [280] obtained a similar result with photographic magnitudes of 2775 nebulae of the cluster Coma Virgo ($11 < m_{pg} < 18$). Thus all this takes place as if intergalactic space were very transparent and the authors conclude that the absorption co-efficient cannot exceed

0·007 magn. per million pc. or megaparsecs (Mpc.).

Actually, the Harvard photographic magnitudes of the nebulae are very inaccurate (as are also the visual magnitudes of Holeschek) and subject to systematic errors (Bigay 1951 [37]). In rediscussing all the known facts, Eigenson (1939 [88]) could only conclude that we could not find any trace of absorption up to nebulae of the thirteenth magnitude and that, in consequence, the absorption co-efficient cannot exceed some hundredths of magnitudes per megaparsec. The question can, without doubt, be re-examined by means of the correct photographic magnitudes determined by Bigay [37]. For this, however, we must also have very accurate and homogeneous values of the apparent diameters, which are badly defined and always difficult to evaluate.

In the meantime, we must be content with rough estimates. We may assume without difficulty that intergalactic absorption, if it did exist, could not reduce the brightness of the nebulae of the cluster in Boötes, whose apparent distance is approximately 70 megaparsecs, to 1/10 (or perhaps 1/5) of its original value. The absorption, in magnitude, is thus lower than

$$\Delta m = 2{\cdot}5 \text{ magn., or even } 1{\cdot}75 \text{ magn.}$$

By means of the relations

$$\log\left(\frac{r'}{r}\right) = \frac{\Delta m}{5} \text{ and } \Delta m = ar,$$

we can easily evaluate r, and then a. According to whether we take

$$\Delta m = 2{\cdot}5 \text{ magn.} \left(\frac{B}{B_0} = 0{\cdot}1\right) \text{ or } \Delta m = 1{\cdot}75 \text{ magn.} \left(\frac{B}{B_0} = 0{\cdot}2\right)$$

we find $r = 22{\cdot}5$ Mpc. or $r = 31{\cdot}2$ Mpc.

and $a = 0{\cdot}11$ magn./Mpc. or $a = 0{\cdot}056$ magn./Mpc.

The latter values which are 10,000 and 20,000 times smaller than the average photographic absorption co-efficient in the Milky Way, can be considered as upper limits (in fact, probably too high) of the intergalactic absorption co-efficient.

The reddening of the elliptical galaxies

An intergalactic absorption due to solid grains or to atoms or molecules ought to be accompanied by appreciable reddening.

Stebbins and Whitford [306] measured the photo-electric colour indices of twenty-six elliptic nebulae whose photographic magnitudes range between 9·5 and 18·2 and whose radial velocities range between 1,240 and 38,900 km./sec. Throughout this interval the colour index increases with the radial velocity v, and its variation is well represented by the straight line

$$C = +0{\cdot}84 + 1{\cdot}33.10^{-5}v,$$

when v is expressed in km./sec. Further, since the radial velocity of a galaxy is proportional to its distance, the colour index increases linearly with it. For the most distant nebulae belonging to the cluster in Boötes, the reddening attains +0·52 magn. Here we have an incontestable phenomenon established by precise measurements. But its interpretation is not simple.

The shifting of the spectral lines moves the entire spectrum towards the red. A radiation, emitted with wave-length λ arrives at the observer with wave-length $\lambda + d\lambda = \lambda\left(1 + \frac{v}{c}\right)$, where c is the velocity of light. The energy curve is thus deformed and the colour index is increased. To evaluate this effect, Stebbins and Whitford traced the energy curve of a nearby elliptical nebula (M 32=NGC 224 near the Great Nebula in Andromeda M 31), by means of photo-electric measurements using six colour filters. They then multiplied all the wave-lengths by the factor $1 + \frac{v}{c} = 1{\cdot}13$, in order to arrive at the energy curve of the nebula in Boötes (v=38,900 km./sec.). Finally, they read off from the two curves, the ordinates corresponding to the effective wave-lengths 4320 and 5290 A of the two filters used in the study of elliptical nebulae. The variation of colour index due to the displacement towards the red is +0·22 magn.; it is thus much lower than the observed reddening. The residue

$$0{\cdot}52 - 0{\cdot}22 = +0{\cdot}30 \text{ magn.}$$

could be attributed to a differential absorption proportional to the distance.

If we suppose that the scattering particles are the same as in the Milky Way, the total photographic absorption can be evaluated by means of the absorption curve of Stebbins and Whitford (page 174). Repeating what we have done previously (page 179), we find

$$\chi = \frac{a_{4260}}{(a_{4320} - a_{5290})} \simeq 4{\cdot}7.$$

Whence

$$\Delta m_{pg} = 4{\cdot}7 \times 0{\cdot}30 = 1{\cdot}41 \text{ magn.}$$

The apparent distance of the cluster in Boötes is approximately r= 70 Mpc. Taking absorption into account, its true distance would be given by

$$5 \log r = 5 \log r' - 1{\cdot}41,$$

whence $r=36{\cdot}7$ Mpc., and the co-efficient absorption would be

$$a_{pg}=\frac{1{\cdot}41}{r}=0{\cdot}038 \text{ magn./Mpc.}^2$$

It is approximately 30,000 times smaller than in the neighbourhood of the equatorial plane of our Milky Way.

Intergalactic absorption and count of nebulae

When no account is taken of absorption, the spatial density of the stars in the Milky Way seems to decrease in all directions the further we get from the Sun. It is precisely this appearance which originally led us to suspect the existence of interstellar absorption (page 143). The same ought to be the case in intergalactic space, even if the absorption there is much smaller, because of the enormous distances involved.

Now, on the hypothesis of perfectly transparent space, the counts of nebulae by apparent magnitudes have led Hubble to the conclusion that the average number of galaxies per unit volume could be considered as independent of the distance. This is shown by the relation

$$\log N_m = Const. + 0{\cdot}6\, m, \tag{X.3}$$

verified by the counts of Hubble [150], where N_m represents the number of galaxies brighter than the magnitude m contained in unit solid angle (see page 193).

On the hypothesis of intergalactic absorption, the true distances r are less than the apparent distances r'. Thus, if the apparent spatial density of the nebulae $D'(r')$ seems constant, the true density $D(r)$ ought to increase more and more with the distance in all directions. This conclusion seems quite absurd. It is just as probable that intergalactic absorption does not exist, or at least that it is so small that it has not appreciably affected the counts of nebulae made so far. Let us then see if the absorption co-efficient deduced from the measurements of Stebbins and Whitford is compatible with Hubble's observations.

The curves of Fig. 51 represent, for various values of the absorption co-efficient, the variation of the ratio $\frac{D'(r')}{D(r)}$ as a function of the true distance r (in dotted lines), and as a function of the apparent distance r' (in straight lines).[3] If the distribution of the galaxies in space were truly uniform ($D(r)$ is a constant), these curves would give the relative variations of the apparent density $D'(r')$.

Let us for the moment consider the curves which correspond to $a=0{\cdot}04$ magn./Mpc. $D'(r')$ decreases very quickly with increase of r' and falls to one-tenth of its initial value at an apparent distance of 60 Mpc., a value much smaller than that involved in Hubble's observations. An absorption co-efficient of this order of magnitude is thus quite incompatible with the counts of the galaxies and we are forced to conclude that *the residual reddening of the elliptical nebulae is not due to intergalactic absorption.*

The ageing of elliptical galaxies

The most probable explanation of the phenomenon has been suggested by Stebbins and Whitford themselves [306]. It is based on the ageing of the elliptical galaxies.

The light which we receive from the nebula M 32 in Andromeda, which is at a distance of 230 kpc., has been emitted for $3{\cdot}26\times2{\cdot}3\times10^5\simeq 7{\cdot}5.10^5$ years (750,000 years), but that which comes to us from the cluster

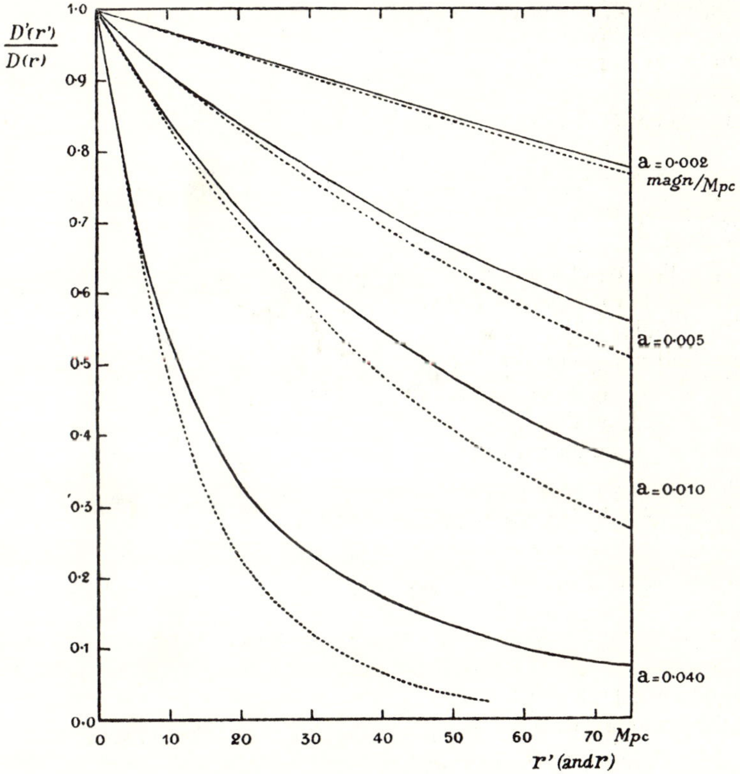

Fig. 51.—Variations of the ratio $\frac{D'(r')}{D(r)}$ as a function of the true distance r (in dotted lines), and the apparent distance r' (in thick lines). $\frac{D'(r')}{D(r)}$ is the ratio between the apparent density and the true spatial density of the extragalactic nebulae, for different values of the intergalactic absorption co-efficient.

in Boötes, whose distance is 70 Mpc. (if we neglect all absorption), has been emitted for $3{\cdot}26 \times 7 \times 10^7 \simeq 2{\cdot}3.10^8$ years (230,000,000 years). Thus, at one and the same time, we observe the nearest galaxies as they are now (or at least, within 1,000,000 years) and the most distant galaxies as they were some 200,000,000 years ago. It is quite possible that they should have evolved in an appreciable manner during this time, which is perhaps one-twentieth or one-tenth of the age attributed to the universe.

Radiation of stars of the main sequence could not have been changed in an appreciable fashion (see page 285), but if initially there existed in these galaxies supergiants whose life is shorter, they ought to have become more and more faint. Perhaps the same is the case with certain giants. If they were essentially red stars, their disappearance would have left those galaxies more blue in which very luminous stars have not been able to re-form. Thus the distant elliptical galaxies would be redder than the nearest because, at the time when the radiation which we receive today was actually emitted, they still contained red stars of great luminosity. Their extinction has naturally diminished the intensity of the blue radiation also, but in a smaller proportion, so that the absolute magnitudes of the distant galaxies must be corrected. This time the correction would tend to increase the photometric distances slightly.[4]

The interpretation which we have just given is naturally only valid for elliptical galaxies, the only ones studied by Stebbins and Whitford in their first work. We already know that elliptical galaxies do not contain clouds of particles capable of re-forming supergiants. On the other hand, the situation is quite different in the spirals where the condensations of clouds can, to the point of exhaustion, again give rise to very luminous stars.

More recently Whitford (1949) studied both ellipsoids and spirals in the cluster of the Corona Borealis whose apparent distance is approximately 40 Mpc. He found that the colour excesses of the spirals were almost exactly those predicted from the shift of the whole spectrum towards the red, whereas the elliptical nebulae showed a supplementary colour excess of 0·15 magn. These observations thus confirm both the explanation of the residual reddening of the elliptical galaxies and the current ideas on the birth of the supergiants.[5]

Transparency of intergalactic space

Neither measurements of the brightness nor of the colour index of the distant nebulae have been able to prove definitely the existence of any kind of absorption in intergalactic space. On the contrary—counts of the nebulae give evidence for a perfect transparency. But, even apart from the uncertainty of the observations, the interpretation of the results raises difficulties.

The magnitudes of the nebulae ought to be corrected to account for the shift of the spectrum towards the red. Today it appears probable that this shift is really due to the expansion of the universe. It is, therefore, necessary to take account, amongst other things, of the fact that the number of photons received per unit time is reduced when the source recedes from the observer. This new correction is equal to the first if we use bolometric magnitudes.

In his first calculations [150], Hubble had made only the simple correction. A new discussion, made in collaboration with Tolman [152] has not led to a clear decision between the two points of view, but it was thought that the double correction was a little too high. Since then the authors have remarked that it would be imprudent to neglect the possibility of variations in the luminous intensity of the galaxies during 200,000,000 years. Later on, Schatzman showed [261] that one could completely reconcile the observations of Hubble with the theory of the expanding universe by supposing that the luminous intensity of the galaxies decreases with time. The order of magnitude of the new correction appears plausible, and could correspond to the progressive extinction of the giants.[6]

In the presence of these difficulties, it seems impossible to say with what degree of approximation the counts of Hubble verify the hypothesis of a roughly uniform spatial distribution of the galaxies. Again we can only try to fix the upper limit of the absorption co-efficient, this time from the counts of the nebulae.

Let us return to Fig. 51 and let us now consider the curves for $a=$ 0·01, 0·005 and 0·002 magn./Mpc. The first two still lead to so rapid a decrease in the apparent spatial density that they seem incompatible with the counts. For $r'=60$ Mpc., $D'(r')$ would, in fact, be reduced to the fraction 0·42 or the fraction 0.62 of its original value. By contrast, the third curve at the same apparent distance only shows a decrease in density of 14 per cent. Numerically, this value is perhaps not incompatible with Hubble's results, because of the uncertainty of the results themselves and of their interpretation. It thus seems reasonable to take, as the upper limit of the average photographic absorption co-efficient, in intergalactic space

$$a_{pg}=0{\cdot}002 \text{ magn./Mpc.}$$

(2 magnitudes per 1,000,000,000 parsecs), a value which is much lower than the rough limit indicated by the brightness of the spirals. Fig. 51 shows that, up to at least a hundred megaparsecs, the influence of such an absorption on the evaluation of the distances would be practically negligible.

In all our remarks we have adopted the usual scale of distance of external galaxies. But, in a recent communication, Baade [395] announced that the distances of all the spirals ought to be approximately doubled. The evaluation of the distances is essentially based on the period of the Cepheids, which is associated with their absolute magnitude according to the law of Miss Leavitt; the law is particularly well established for the so-called "cluster" variables of period less than one day. Now, up to the present, in order to fix the distances of the external galaxies, reference was made to ordinary Cepheids, which were brighter and of period above one day, for which the graph of the period-luminosity curve is less certain. It is the recent work on the cluster variables of lower brightness in the external galaxies which has led to the large correction indicated by Baade. If we now apply the correction to the study of intergalactic absorption, the upper limit of the absorption co-efficient becomes

$$a_{pg}=0.001 \text{ magn./Mpc.},$$

which is 1 magnitude per 1,000,000,000 parsecs, a value which is 1,000,000 times smaller than the average absorption co-efficient in the neighbourhood of the equatorial plane of the Milky Way.

If such an absorption were produced by solid particles identical to those of the clouds of the Milky Way, the specific mass of intergalactic space would be of the order of

$$10^{-6}\times 10^{-26}=10^{-32} \text{ gm./c.c.},$$

but if these grains were accompanied by the atoms of a gas, in the same proportion as in the Galactic clouds, the density would then be approximately

$$10^{-6}\times 3.10^{24}=3.10^{-30} \text{ gm./c.c.}$$

It would thus be of the same order of magnitude as the density which, according to Hubble, would be obtained by distributing the matter of all the galaxies uniformly throughout all space ($\sim 10^{-30}$ gm./c.c.). Finally, if the particles responsible for the absorption were atoms or molecules only, the density of space would attain a very much larger and highly improbable value.[7]

It is hardly necessary to underline the entirely arbitrary character of these hypotheses on the properties of an absorbing medium whose very existence is in doubt.

Many astrophysicists consider intergalactic space to be entirely empty. On the other hand, Zwicky [440] thinks that the disintegration of galaxies has scattered in space—and mainly in the neighbourhood of the clusters of galaxies—groups of stars, individual stars, dust and gases. These he has tried to demonstrate for some years by the powerful means at his disposal at Mount Palomar. We have seen above (page 313) that he has discovered immense luminous clouds which link neighbouring galaxies. He has also found, near the pole of the Milky Way, a great number of variable and very faint blue stars. The brightness of some of them does not exceed the twentieth photographic magnitude. Some could represent the first members of the intergalactic population. On this basis, Zwicky estimated that the average density of the universe is rather larger than is generally imagined ($\backsim 10^{-26}$ gm./c.c., instead of 10^{-30} to 10^{-28}). Even if these points of view were acceptable, it would not affect the validity of the preceding calculations: today nothing would justify the affirmation of the existence of absorbing particles in intergalactic space.[8]

NOTES TO CHAPTER XIX

[1]It depends exclusively on the distance, only in the case of a star of uniform brightness, with well-defined contours. The brightness of elliptical or spiral nebulae decreases progressively towards the edges, and the weakening due to the absorption can change the apparent diameter measured.

[2]Stebbins and Whitford take $\varkappa=4.0$, whence $a=0.030$ magn./Mpc., but only the order of magnitude is of importance here.

[3]The calculation is quite analogous to that of Schalén in the case of stars (page 152). According to equation (VII.6) we have

$$\frac{D'(r')}{D(r)}=e^{-3kr/2}\left(1+\frac{kr}{2}\right)^{-1},$$

with $k=\frac{a}{1.086}$ Mpc.$^{-1}$ when a is expressed in magn./Mpc. the curves $\frac{D'(r')}{D(r)}$ as a function of r are easily constructed. As for r' it is, of course, evaluated from

$$\log\left(\frac{r'}{r}\right)=\frac{ar}{5}.$$

[4]The correction corresponding to the residual colour excess of the cluster in Boötes would lead to its apparent distance being multiplied by 1.17.

[5]G. de Vaucouleurs [353] proposed a different explanation for the residual colour excess of the ellipsoids. In a preliminary paper, Stebbins and Whitford [305], instead of constructing the energy curve of a nearby nebula, simply considered it as a black body at 5,500°K. In this case, the colour excess due to the shift towards the red would only be 0.08 magn. for the cluster in Boötes. G. de Vaucouleurs observed that, because of the accumulation of absorption lines and bands, the energy curve of the Solar spectrum (type dG2) would greatly differ from that of a black body at 6,000° in the region of the small wave-lengths. The shift towards the red would further produce the large ultra-violet depression observed in the Solar spectrum in the region of the transmission band of a blue filter. By applying the Solar energy curve of E. Pettit to nearby ellipsoids, Vaucouleurs calculated the colour excess for the cluster in Boötes as +0.47 magn. which is very near the observed magnitude (+0.52 magn.).
Stebbins and Whitford, however, have stated that although the energy curve of M 32, for instance, can, near its maximum, be considered as that of a dG3 star, it would differ from it by a slower decrease towards the large and the small wave-lengths. This is obviously due to a mixture of spectra of stars of different classes; the K and M stars increase the red part of the spectrum, the A and F stars increase the violet. Thus, the energy curve of an ellipsoid certainly does not show a depression comparable to that of the Solar spectrum at the beginning of the ultra-violet. Thus the only relevant point arising from Vaucouleurs's work is the necessity of plotting the energy curve of the elliptical galaxies with greater precision, so that the colour excess due to the shift to the red can be calculated more accurately.

[6]The very general calculation of Schatzman does not distinguish between elliptical and spiral galaxies.

[7]We have seen that to account for an absorption of 1.1 magn./kpc. there would have to be nearly 300 electrons per c.c. and 700 times as many hydrogen molecules ($2{\cdot}1.10^{5}$ per c.c.). Even an absorption 10^{6} times smaller would still require 0.4 molecules of H_2 per c.c., and the density would then be $0{\cdot}7.10^{-24}$ gm./c.c., i.e. the same order of magnitude as in the Milky Way.

[8]The theory of Spitzer and Baade (page 312) predicts the expulsion of interstellar matter when galaxies collide. But the high temperature generated by these encounters would produce the complete vaporization of solid particles, and the extreme rarefication of the expelled gases makes it impossible to consider a later reconstitution of grains.

BIBLIOGRAPHY AND INDEX

BIBLIOGRAPHY

Collective Works and Monographs

Bart. J. BOK. *The distribution of Stars in Space*, 1937 [40].

HENRI MINEUR. *L'espace interstellaire*. Presses Universitaires, France, 1947.

P. SWINGS. *Les spectres des nébuleuses gazeuses.* Paris, Hermann, 1935.

B. VORONTZOV-VELYAMINOV. *Gaseous nebulae and novae.* Second Edition, Moscow and Leningrad, 1948 [360] (in Russian, with a large bibliography).

K. WURM. *Die planetarischen Nebel*, Berlin, Akademie-Verlag, 1951.

H. ZANSTRA. *Théorie de l'emission de la lumière par les nébuleuses gazeuses.* Paris, Hermann, 1936.

J. L. GREENSTEIN. *Interstellar Matter* (J. A. HYNEK, Astrophysics, A Topical Symposium, N.Y. McGraw Hill, 1951).

L. H. ALLER. *Astrophysics. Nuclear Transformations, Stellar Interiors and Nebulae.* Chapter 5: "The Planetary Nebulae", Chapter 6: "The Interstellar Medium " (N.Y. Ronald Press, 1954).

Accounts on Interstellar Matter

L'absorption de la lumière dans l'espace interstellaire. Paris July 1937; *Annales d'Astrophysique*, 1, 1938.

Symposium on Interstellar Matter, Cambridge (Mass.), December 1946; Harvard Monographs No. 7, 1948.

Problems of Cosmical Aerodynamics, Paris, August 1949; Central Air Documents Office, 1951.

Summaries

C. S. BEALS. *Interstellar Matter.* Monthly Notices of R.A.S., 102, 1942, pp. 97–106 [28].

C. S. BEALS. *The Material in Interstellar Space.* Popular Astronomy, 52, 1944, pp. 47–66 [29].

S. ROSSELAND. *Absorption of Light in Space.* Monthly Notices of R.A.S., 98, 1939, pp. 301–10 [144].

C. SCHALÉN. *Neuere Untersuchungen über die interstellare Materie.* Naturwissenschaften, 1949, pp. 33–41; *Uppsala Astr. Obs. Medd.* 2, No. 6 [260].

F. H. SEARES. *The Dust in Space*. Publ. Astr. Soc. Pacific 52, 1940, pp. 80–115 [267].

Special Reference

Abbreviations used for the periodicals quoted:

A. J.=Astronomical Journal.
An. Ap.=Annales d'astrophysique.
Ap. J.=Astrophysical Journal.
B.A.N.=Bulletin of the Astronomical Institute of the Netherlands.
C.R.=Comptes rendus de l'Academie des Sciences de Paris.
A.J.U.S.S.R.=Astronomical Journal of the U.S.S.R.
M.N.=Monthly Notices of the Royal Astronomical Society.
P.A.S.P.=Publications of the Astronomical Society of the Pacific.
Zs.f.Ap.=Zeitschrift für Astrophysik.

[1] ADAMS (W. S.). *Ap. J.*, 93, 1941, p. 11.
[2] ADAMS (W. S.). *Ap. J.*, 97, 1943, pp. 105; 109, 1949, p. 354.
[3] ALLER (L. H.). *Ap. J.*, 93, 1941, pp. 236; 113, 1951, p. 125.
[4] ALLER (L. H.). *Ap. J.*, 108, 1948, p. 462.
[5] ALLER (L. H.). *Ap. J.*, 113, 1951, p. 120.
[6] ALTER (G.). *Accounts and observ.* Czechoslovak Astro. Soc., Prague, No. 10, 1949.
[7] AMBARZUMIAN (V.). *Bull. Abastumani Ap. Obs.*, 4, 1939, p. 17. *Trans. Int. Astr. Union*, 7, 1950, p. 452.
[8] AMBARZUMIAN (V.) and GORDELADSE. *Bull. Abastumani Obs.*, 2, 1937, p. 37.
[9] ANDRILLAT (H.). *C.R.*, 231, 1950, pp. 1432: 234, 1952, p. 62.
[10] BAADE (W.). *Ap. J.*, 96, 1942, p. 188.
[11] BAADE (W.). *Ap. J.*, 97, 1943, p. 119.
[12] BAADE (W.). *Ap. J.*, 100, 1944, p. 137.
[13] BAADE (W.). *Ap. J.*, 100, 1944, p. 147.
[14] BAADE (W.) and MAYALL (N. U.). *Problems Cosmical Aerodyn.*, 1951, p. 165.
[15] BAADE (W.) and MINKOWSKI (R.). *Ap. J.*, 86, 1937, p. 119 and p. 123.
[16] BABCOCK (H. W.). *Lick Obs. Bull.*, 19, 1939, p. 41.
[17] BAKER (R. H.). *Harvard Circular*, No. 426, 1939.
[18] BAKER (R. H.) and NANTKES (E.), *Ap. J.*, 99, 1944, p. 125.
[19] BARBIER (D.). *An. Ap.*, 7, 1944, p. 80.
[20] BARBIER (D.). *An. Ap.*, 8, 1945, p. 35.
[21] BARBIER (D.). *An. Ap.*, 11, 1948, p. 193.
[22] BARNARD (E. E.). *Ap. J.*, 49, 1919, pp. 1 and 360; *Atlas of the Selected Regions of the Milky Way*, Carnegie Institution of Washington, 1927.
[23] BATCHELOR (G. K.). *Proc. Royal Soc.* London, A. 201, 1949, p. 405; *Problems Cosmical Aerodyn.*, 1951, p. 149.

[24] Bates (D. R.) and Spitzer (L. Jr.). *Ap. J.*, 113, 1951, p. 441.
[25] Beals (C. S.). *M.N.* 93, 1933, pp. 585; 94, 1934, p. 663; *Publ. Dominion Ap. Obs. Victoria*, 6, 1937, p. 333.
[26] Beals (C. S.). *M.N.*, 96, 1936, p. 661.
[27] Beals (C. S.)., *P.A.S.P.*, 49, 1937, p. 115; *Ap. J.*, 87, 1938, p. 568.
[28] Beals (C. S.). *M.N.*, 102, 1942, p. 97.
[29] Beals (C. S.). *Popular Astron.*, 52, 1944, p. 47.
[30] Beals (C. S.) and Blanchet (G. H.). *M.N.*, 98, 1938, p. 398.
[31] Becker (W.). *Veröf.* Univers. Sternwarte Berlin-Babelsberg, 10, No. 3, 1933; *Zs.f.Ap.*, 5, 1932, p. 101.
[32] Becker (W.). *Zs.f.Ap.*, 17, 1939, p. 285.
[33] Bennett (A. L.). *Ap. J.*, 85, 1937, p. 257.
[34] Berg (M.). *Bull. Obs. Pulkovo*, 15, No. 2, 1936, p. 1.
[35] Berman (L.). *Lick Obs. Bull.*, 15, 1930, p. 86.
[36] Berman (L.). *Lick Obs. Bull.*, 18, 1937, p. 57.
[37] Bigay (J. H.). *Journ. Observateurs*, 34, 1951, p. 89; *An. Ap.*, 14, 1951, p. 319.
[38] Blumer (H.). *Zs. f. Physik*, 32, 1925, p. 119; 38, 1926, p. 304; 39, 1926, p. 195.
[39] Bok (B. J.). *Harvard Coll. Obs. Circular*, No. 371, 1931.
[40] Bok (B. J.). *The Distribution of Stars in Space.* Univers. Chicago Press, 1937.
[41] Bok (B. J.) *Harvard Monographs*, No. 7, 1948, p. 53.
[42] Bok (B. J.) and Reilly (E.). *Ap. J.*, 105, 1947, p. 255.
[43] Bok (B. J.), Olmsted (M.) and Boutelle (B. D.). *Ap. J.*, 100, 1949, p. 21.
[44] Bok (B. J.) and Randall-Arons (J. M.). *Ap. J.*, 101, 1945, p. 280.
[45] Bok (B. J.) and Wright (F. W.). *Ap. J.*, 101, 1945, p. 300.
[46] Bolton (J. G.). *Nature*, 162, 1948, p. 141.
Bolton (J. G.) and Stanley (G. J.). *Australian Journal Sc. Research*, A, 2, 1949, p. 139.
[47] Bowen (I. S.). *P.A.S.P.*, 39, 1927, p. 295; *Ap. J.*, 67, 1928, p. 1.
[48] Bowen (I. S.). *Ap. J.*, 81, 1935, p. 1.
[49] Bowen (I. S.). *Rev. Modern Physics*, 8, No. 2, 1936.
[50] Bowen (I. S.) and Wyse (A. B.). *P.A.S.P.*, 50, 1938, p. 348. *Lick Obs. Bull.*, 19, 1939, p. 1.
[51] Calvert (R. L.). *Ap. J.*, 114, 1951, p. 123.
[52] Campbell (W. W.) and Moore (J. H.). *Public. Lick Obs.*, 13, 1918, p. 75.
[53] Carpenter (E. F.). *Publ. Amer. Astr. Soc.*, 7, 1931, p. 25; *P.A.S.P.*, 45, 1933, p. 227.

[54] Cederblad (S.). *Medd. Lunds Astr. Obs.*, Series II, No. 119, 1946.
[55] Cernuschi (F.). *Ap. J.*, 105, 1947, p. 241.
[56] Chandrasekhar (S.). *Ap. J.*, 103, 1946, p. 350.
[57] Chandrasekhar (S.). *Ap. J.*, 110, 1949, p. 110.
[58] Cillie (G.). *M.N.*, 92, 1932, p. 820; 96, 1936, p. 771.
[59] Collins (O. C.). *Ap. J.*, 86, 1937, p. 529.
[60] Courtes (G.). *C.R.*, 232, 1951, p. 795 and p. 1283.
[61] Courtes (G.). *C.R.*, 234, 1952, p. 506.
[62] Croze (F.) and Mihul (C.). *C.R.* ,185, 1927, p. 702.
[63] Curtiss (H. D.). *Publ. Lick Obs.*, 13, 1918, p. 45.
[64] Danjon (A.) and Couderc (P.). *An. Ap.*, 13, 1950, p. 211.
[65] Davis (L. Jr.) and Greenstein (J. L.). *Phys. Rev.*, 75, 1949, p. 1605.
[66] Davis (L. Jr.) and Greenstein (J. L.). *Ap. J.*, 114, 1951, p. 206.
[67] Debye (P.). *Ann. d. Physik*, 30, 1909, p. 57.
[68] Divan (L.). *C.R.*, 233, 1951, p. 284.
[69] Douglas (A. E.) and Herzberg (G.). *Ap. J.*, 94, 1941, p. 381.
[70] Dufay (J.). *C.R.*, 196, 1933, p. 101; *Journ. Observateurs*, 16, 1933, p. 77.
[71] Dufay (J.). *C.R.*, 201, 1935, p. 1323.
[72] Dufay (J.). *An. Ap.*, 1, 1938, p. 195.
[73] Dufay (J.). *An. Ap.*, 12, 1949, p. 306.
[74] Dufay (J.). *C.R.*, 228, 1949, p. 1277.
[75] Dufay (J.) and Liau (S. P.). *C.R.*, 196, 1933, p. 1372; *An. Ap.*, 1, 1938, p. 219.
[76] Dufay (J.) and Smoukovich (D.). *C.R.*, 208, 1939, p. 1481.
[77] Dufay (J.) and Wang Shih-Ky, *Journ. Observateurs*, 18, 1935, p. 193.
[78] Duke (D.). *Ap. J.*, 113, 1951, p. 100.
[79] Dunham (Th. Jr.). *P.A.S.P.*, 49, 1937, p. 26.
[80] Dunham (Th. Jr.). *Nature*, 139, 1937, p. 246.
[81] Dunham (Th. Jr.). *Proc. Amer. Phil. Soc.*, 81, 1939, p. 277.
[82] Dunham (Th. Jr.). *Publ. Amer. Astr. Soc.*, 10, 1940, p. 113.
[83] Dunham (Th. Jr.) and Adams (W. S.). *P.A.S.P.*, 49, 1937, p. 5.
[84] Dunham (Th. Jr.). *P.A.S.P.*, 53, 1941, p. 341.
[85] Duyvendack (J. J. L.). *P.A.S.P.*, 54, 1942, p. 91.
[86] Eddington (A. S.). *The Internal Constitution of the Stars*, Cambridge University Press, 1926.
[87] Eigenson (M. S.). *Poulkovo Obs. Circular*, No. 26–7, 1939, p. 24.
[88] Eigenson (M. S.). *Poulkovo Obs. Circular*, No. 26–7, 1939, p. 28.
[89] Elvey (C. T.). *Ap. J.*, 74, 1931, p. 298.
[90] Elvey (C. T.) and Mehlin (T. G.). *Ap. J.*, 75, 1932, p. 354.
[91] Elvey (C. T.) and Roach (F. E.). *Ap. J.*, 85, 1937, p. 213.
[92] Elvius (A.). *Stockholm Obs. Ann.*, 17, 1951, No. 4.

[93] Elvius (T.). *Stockholm Obs. Ann.*, 16, 1951, No. 5.
[94] Evans (J. W.). *Ap. J.*, 93, 1941, p. 275.
[95] Ewen (H. I.) and Purcell (E. M.). *Nature*, 168, 1951, p. 356; *A. J.*, 56, 1951, p. 125.
[96] Fabry (Ch.). *Journ. Physique*, 5th Series, 6, 1916, p. 207; *Ap. J.*, 45, 1917, p. 269.
[97] Fabry (Ch.), Buisson (H.) and Bourget (H.). *Journ. Phys.*, 5th Series, 4, 1914, p. 357; *Ap. J.*, 40, 1914, p. 241.
[98] Fermi (E.). *Phys. Rev.*, 75, 1949, p. 1169.
[99] Gans (R.). *Ann. d. Physik*, 37, 1912, p. 881.
[100] Gerasimovic (B. P.). *Harvard Obs. Circular*, No. 339, 1929.
[101] Gerasimovic (B. P.). *Zs. f. Ap.*, 4, 1932, p. 265.
[102] Gliese (W.) and Walter (K.). *Zs. f. Ap.*, 29, 1951, p. 94.
[103] Götz (F. W. P.). *Astr. Nachrichten*, 255, 1925, p. 63.
[104] Greaves (W. H.), Davidson (C.) and Martin (E.). *M.N.*, 88, 1928, p. 703; 89, 1928, p. 125.
[105] Greenstein (J. L.). *Harvard Obs. Circular*, No. 422, 1937.
[106] Greenstein (J. L.). *Harvard Ann*, 105, 1937, p. 359.
[107] Greenstein (J. L.). *Ap. J.*, 87, 1938, p. 151.
[108] Greenstein (J. L.). *Ap. J.*, 87, 1938, p. 581.
[109] Greenstein (J. L.). *Ap. J.*, 104, 1946, p. 403.
[110] Greenstein (J. L.). *Ap. J.*, 104, 1946, p. 414.
[111] Greenstein (J. L.). *Harvard Obs. Monographs*, No. 7, 1948, p. 19.
[112] Greenstein (J. L.) and Aller (L. H.). *Ap. J.*, 111, 1950, p. 328.
[113] Greenstein (J. L.) and Henyey (I. G.). *Ap. J.*, 89, 1939, p. 647.
[114] Greenstein (J. L.) and Henyey (I. G.). *Ap. J.*, 93, 1941, p. 327.
[115] Greenstein (J. L.) and Henyey (I. G.). *Ap. J.*, 96, 1942, p. 78.
[116] Greenstein (J. L.) and Page (T. L.). *Ap. J.*, 114, 1951, p. 106.
[117] Greenstein (J. L.) and Struve (O.). *Ap. J.*, 90, 1939, p. 625.
[118] Guintini (P.). *An. Ap.*, 5, 1942, p. 114.
[119] Hall (J. S.). *Ap. J.*, 79, 1934, p. 145.
[120] Hall (J. S.). *Ap. J.*, 84, 1936, p. 372.
[121] Hall (J. S.). *Ap. J.*, 85, 1937, p. 145.
[122] Hall (J. S.). *Science*, 109, 1949, p. 166.
[123] Hall (J. S.). *Ap. J.*, 56, 1951, p. 40.
[124] Hall (J. S.) and Mikesell (A. H.). *Publ. U.S. Naval Obs.*, Series II, 17, part 1, 1950.
[125] Hanbury Brown (R.) and Hazard (C.). *M.N.*, 111, 1951, p. 357.
[126] Hartmann (J.). *Ap. J.*, 19, 1904, p. 268.
[127] Hartwig (G.). *Zs. f. Ap.*, 17, 1939, p. 191.
[128] Heerschen (D. S.). *Ap. J.*, 114, 1951, p. 132.
[129] Heger (M. L.). *Lick Obs. Bull.*, 10, 1919, p. 59; 10, 1921, p. 141.

[130] HENIZE (K. G.) and MILLER (F. D.). *Publ. Obs. Michigan*, 10, 1951, p. 75.
[131] HENYEY (I. G.). *Ap. J.*, 84, 1936, p. 609.
[132] HENYEY (I. G.). *Ap. J.*, 85, 1937, p. 107.
[133] HENYEY (I. G.) and GREENSTEIN (J. L.). *An. Ap.* 3, 1940, p. 117.
[134] HENYEY (I. G.) and KEENAN (P. C.). *Ap. J.*, 91, 1940, p. 625.
[135] HERBIG (G.). *P.A.S.P.*, 58, 1946, p. 163.
[136] HERBIG (G.). *Ap. J.*, 113, 1951, p. 697.
[137] HILTNER (W. A.). *Ap. J.*, 106, 1947, p. 231.
[138] HILTNER (W. A.). *Science*, 109, 1949, p. 165. *Nature*, 163, 1949, p. 283.
[139] HILTNER (W. A.). *Ap. J.*, 109, 1949, p. 471.
[140] HILTNER (W. A.). *Phys. Rev.*, 78, 1950, p. 170.
[141] HILTNER (W. A.). *Ap. J.*, 114, 1951, p. 241.
[142] HOLMBERG (E.). *Medd. Lunds Astr. Obs.*, Series II, No. 114, 1945.
[143] HOLMBERG (E.). *Medd. Lunds Astr. Obs.*, Series II, No. 120, 1947.
[144] HOLMBERG (E.). *Medd. Lunds Astr. Obs.*, Series I, No. 170, 1950.
[145] HOLMBERG (E.). *Medd. Lunds Astr. Obs.*, Series II, No. 128, 1950.
[146] HUBBLE (E.). *Ap. J.*, 56, 1922, p. 162.
[147] HUBBLE (E.). *Ap. J.*, 56, 1922, p. 400.
[148] HUBBLE (E.). *Ap. J.*, 64, 1926, p. 321.
[149] HUBBLE (E.). *Ap. J.*, 71, 1930, p. 231.
[150] HUBBLE (E.). *Ap. J.*, 79, 1934, p. 8.
[151] HUBBLE (E.). *Ap. J.*, 97, 1943, p. 112.
[152] HUBBLE (E.) and TOLMAN (R. C.). *Ap. J.*, 82, 1935, p. 302.
[153] HUNTER (A.) and MARTIN (E. G.). *M.N.*, 100, 1940, p. 669.
[154] JANSKY (K. G.). *Proc. Inst. Radio Eng.*, 20, 1932, p. 1920.
[155] JANSKY (K. G.). *Proc. Inst. Radio Eng.*, 25, 1937, p. 1517.
[156] JANSSEN (E. M.). *Ap. J.*, 103, 1946, p. 380.
[157] JONES (H. S.). *M.N.*, 75, 1914, p. 4.
[158] JOY (A.). *Ap. J.*, 89, 1939, p. 356.
[159] JOY (A.). *Ap. J.*, 102, 1945, p. 168.
[160] KALINIAK (A. A.), KRASSOVSKY (W. I.) and NIKONOV (W. B.). *Docklady Acad. Sc.* U.S.S.R., 66, 1949, p. 25.
[161] KAPTEYN (J. C.). *A. J.*, 24, 1904, p. 115.
[162] KAPTEYN (J. C.). *Ap. J.*, 29, 1909, p. 46.
[163] KAPTEYN (J. C.). *Ap. J.*, 55, 1922, p. 302.
[164] KARMAN (Th. von). *Problems cosm. aerodyn.*, 1951, p. 129.
[165] KEENAN (P. C.). *Ap. J.*, 84, 1936, p. 600.
[166] KHABIBULLIN (S. T.). *A.J.U.S.S.R.*, 27, 1950, p. 105.
[167] KHOLOPOV (P. N.). *A.J.U.S.S.R.*, 27, 1950, p. 110.
[168] KIENLE (H.). *M.N.*, 88, 1928, p. 700.
[169] KIPPER. *A.J.U.S.S.R.*, 27, 1951, p. 321.

[170] Klüber (H. von). *Zs. f. Ap.*, 6, 1933, p. 259; 13, 1937, p. 174.
[171] Kopff (A.). *Publ. Königstuhl Heidelberg*, I, 1902, p. 177.
[172] Kramers (H. A.) and Ter Haar (D.). *B.A.N.*, 10, 1946, p. 137.
[173] Kukarkin (B. W.). *A.J.U.S.S.R.*, 17, 1940, p. 13.
[174] Lehman-Balanowskaja (I. W.). *Bull. Obs. Poulkovo*, 14, 1935, p. 1. (No. 118).
[175] Lehmann (G.). *Cahiers de Physique*, No. 10, 1942, p. 1.
[176] Lindblad (B.). *Nature*, 135, 1935, p. 133.
[177] Lindblad (B.). *An. Ap.*, 1, 1938, p. 173.
[178] Lindblad (B.). *Stockholm Obs. Ann.*, 14, No. 3, 1942.
[179] Lindblad (B.) and Brahde (R.). *Ap. J.*, 104, 1946, p. 211.
[180] Lindblad (B.) and Delhaye (J.). *Stockholm Obs. Ann.*, 15, No. 9, 1949.
[181] Lundmark (K.). *M.N.*, 85, 1925, p. 865.
[182] Lyot (B.) and Dollfus (A.). *C.R.*, 228, 1949, p. 1773.
[183] Malmquist (K. G.). *Stockholm Obs. Ann.*, 13, No. 4, 1939; *Uppsala Astr. Obs. Ann.*, 1, No. 7, 1943 and No. 8, 1944.
[184] Markaryan. *Contr. Burakan Obs.*, No. 1, 1946.
[185] Markowitz (W.). *A. J.*, 56, 1951, p. 134.
[186] Martel (M. T.). *An. Ap.*, 12, 1949, p. 275.
[187] Martel (M. T.). *An. Ap.*, 13, 1950, p. 102.
[188] Martel (M. T.). *C.R.*, 232, 1951, p. 2183.
[189] Mayall (N. U.). *P.A.S.P.*, 49, 1937, p. 101.
[190] Mayall (N. U.). *Lick Obs. Bull.*, 19, 1939, p. 33.
[191] Mayall (N. U.) and Aller (L. H.). *Ap. J.*, 95, 1942, p. 5.
[192] Mayall (N. U.) and Oort (J. H.). *P.A.S.P.*, 54, 1942, p. 95.
[193] McKellar (A.). *P.A.S.P.*, 52, 1940, p. 307; *Publ. Dominion Obs.* Victoria, 7, 1941, p. 15.
[194] Melnikov (O. A.). *Poulkovo Obs. Circ.*, No. 21, 1937, p. 3.
[195] Menzel (D. H.). *Trans. Intern. Astr. Union*, 7, 1950, p. 468.
[196] Menzel (D. H.), Aller (L. H.), Baker (J. G.), Hebb (M. H.), Goldberg (L.) and Shortley (G. H.). "Physical Processes in Gaseous Nebulae"; *Ap. J.*, I, 85, 1937, p. 330; II, 86, 1937, p. 70; III, 88, 1938, p. 52; IV, 88, 1938, p. 313; V, 88, 1938, p. 422; VI, 89, 1939, p. 587; VII, 90, 1939, p. 271; VIII, 90, 1939, p. 601; IX, 91, 1940, p. 307; X, 92, 1940, p. 408; XI, 93, 1941, p. 178; XII, 93, 1941, p. 195; XIII, 93, 1941, p. 230; XIV, 93, 1941, p. 236; XV, 93, 1941, p. 244; XVI, 94, 1941, p. 30; XVII, 94, 1941, p. 436; XVIII, 102, 1945, p. 239.
[197] Merrill (P. W.). *P.A.S.P.*, 46, 1934, p. 206; *Ap. J.*, 83, 1936, p. 126.
[198] Merrill (P. W.). *P.A.S.P.*, 58, 1946, p. 354.
[199] Merrill (P. W.) and Humason (M. L.). *P.A.S.P.*, 50, 1938, p. 212.

[200] Merrill (P. W.), Sanford (R. F.), Wilson (O. C.) and Burwell (C. G.). *Ap. J.*, 86, 1937, p. 274.
[201] Merrill (P. W.) and Wilson (O. C.). *Ap. J.*, 87, 1938, p. 9.
[202] Mie (G.). *Ann. d. Physik*, 25, 1908, p. 377.
[203] Mineur (H.). *An. Ap.*, 1, 1938, p. 97.
[204] Minkowski (R.). *Ap. J.*, 96, 1942, p. 199.
[205] Minkowski (R.). *Ap. J.*, 97, 1943, p. 128.
[206] Minkowski (R.). *Ap. J.*, 107, 1948, p. 106.
[207] Minkowski (R.). *P.A.S.P.*, 61, 1949, p. 151.
[208] Minkowski (R.). *Publ. Obs. Univ. Michigan*, 10, 1951, p. 25.
[209] Morgan (W. W.). *Ap. J.*, 90, 1939, p. 632.
[210] Morgan (W. W.) and Sharpless (S.). *Ap. J.*, 103, 1946, p. 249.
[211] Muller (C. A.) and Oort (J. H.). *Nature*, 168, 1951, p. 357.
[212] Müller (H.). *Astr. Nachr.*, 269, 1939, p. 57.
[213] Müller (H.) and Hufnagel (L.). *Zs. f. Ap.*, 9, 1935, p. 231.
[214] Muller (R.). *Zs. f. Ap.*, 3, 1931, pp. 261 and 369; 4, 1932, p. 365.
[215] Nantkes (E.) and Baker (R. H.). *Ap. J.*, 107, 1948, p. 113.
[216] Nassau (J. J.) and Mac Rae (P. A.), *Ap. J.*, 110, 1949, p. 40.
[217] Nicolet (M.). *An. Ap.*, 14, 1951, p. 249.
[218] Ohman (Y.). *P.A.S.P.*, 54, 1942, p. 72; *Stockholm Obs. Ann.*, 14, No. 4, 1942.
[219] Oort (J. H.). *Groningen Publ.*, No. 40, 1926.
[220] Oort (J. H.). *B.A.N.*, 6, 1932, p. 248.
[221] Oort (J. H.). *B.A.N.*, 8, 1938, p. 233.
[222] Oort (J. H.). *An. Ap.*, 1, 1938, p. 71.
[223] Oort (J. H.). *M.N.*, 106, 1946, p. 159.
[224] Oort (J. H.). *Problems Cosm. Aerodyn.*, 1951, p. 118.
[225] Oort (J. H.) and Van De Hulst (H. C.). *B.A.N.*, 10, 1946, p. 187.
[226] Page (T. L.). *M.N.*, 96, 1936, p. 627; *Ap. J.*, 96, 1942, p. 78.
[227] Page (T. L.). *Ap. J.*, 108, 1948, p. 157.
[228] Page (T. L.). *P.A.S.P.*, 63, 1951, p. 142.
[229] Page (T. L.) and Greenstein (J. L.). *Ap. J.*, 114, 1951, p. 98.
[230] Pahlen (E. Von der). *Astr. Nachr.*, 238, 1930, p. 269.
[231] Pannekoek (A.). *Proc. Kon. Akad. Amsterdam.* 23, 1920, p. 707.
[232] Pannekoek (A.). *Publ. Astr. Inst. Univ. Amsterdam*, No. 7, 1942.
[233] Parenago (P. P.). *A.J.U.S.S.R.*, 17, 1940, No. 4, p. 1; 22, 1945, No. 3, p. 130.
[234] Parenago (P. P.). *A.J.U.S.S.R.*, 23, 1946, p. 69.
[235] Pasternack (S.). *Ap. J.*, 92, 1940, p. 129.
[236] Pawsey (J. L.). *Nature*, 168, 1951, p. 358.
[237] Piddington (J. H.). *M.N.*, III, 9151, p. 45.
[238] Plaskett (H. H.). *Publ. Dominion Ap. Obs. Victoria*, 4, 1928, p. 187.

[239] Plaskett (J. S.) and Pearce (J. A.). *Publ. Dominion Ap. Obs.* 5, 1935, p. 167.
[240] Prodell (A. G.) and Kusch (P.). *Phys. Rev.*, 79, 1950, p. 1009.
[241] Reber (G.). *Ap. J.*, 91, 1940, p. 621; 100, 1944, p. 279.
[242] Redman (R. O.). *M.N.*, 96, 1936, p. 588.
[243] Redman (R. O.) and Shirley (E. G.). *M.N.*, 98, 1938, p. 613.
[244] Rosseland (S.). *M.N.*, 98, 1939, p. 301.
[245] Rudnick (J.). *Ap. J.*, 83, 1936, p. 394.
[246] Ryle (M.), Smith (F. G.) and Elsmore (B.). *M.N.*, 110, 1950, p. 508.
[247] Sanford (R. F.). *Ap. J.*, 86, 1937, p. 136.
[248] Sanford (R. F.). *Ap. J.*, 110, 1949, p. 117.
[249] Sanford (R. F.), Merrill (P. W.) and Wilson (O. C.). *P.A.S.P.*, 50, 1938, p. 58; 51, 1939, p. 238 (Sanford).
[250] Schalén (C.). *Astr. Nachr.*, 236, 1929, p. 249.
[251] Schalén (C.). *Medd. Astr. Obs. Uppsala*, No. 37, 1928; No. 55, 1932.
[252] Schalén (C.). *Arkiv. f. Mat. Astr. o. Fys.*, 22 A, No. 15, 1930, p. 1; 22 B, No. 8, 1931, p. 1.
[253] Schalén (C.). *Medd. Astr. Obs. Uppsala*, No. 58, 1934; No. 61, 1935.
[254] Schalén (C.). *Medd. Astr. Obs. Uppsala*, No. 64, 1936.
[255] Schalén (C.). *Uppsala Astr. Obs. Ann.*, 1, No. 2, 1939.
[256] Schalén (C.). *An. Ap.*, 1, 1938, p. 60.
[257] Schalén (C.). *Zs. f. Ap.*, 17, 1939, p. 260.
[258] Schalén (C.). *Uppsala Astr. Obs. Ann.*, 1, No. 9, 1945.
[259] Schalén (C.). *Uppsala Astr. Obs. Ann.*, 2, No. 5, 1948.
[260] Schalén (C.). *Medd. Astr. Obs. Uppsala*, 2, No. 6, 1949.
[261] Schatzman (E.). *An. Ap.*, 10, 1947, p. 14.
[262] Schatzman (E.). *An. Ap.*, 12, 1949, p. 161.
[263] Schatzman (E.). *An. Ap.*, 13, 1950, p. 227 and 367.
[264] Schlüter (A.) and Biermann (L.). *Zs. f. Naturforschung* 5a, 1950, p. 237.
[265] Schoenberg (E.) and Jung (B.). *Astr. Nachr.*, 247, 1932, p. 413.
[266] Schoenberg (E.) and Jung (B.). *Astr. Nachr.*, 253, 1934, p. 261; *Mitt. Sternw. Breslau*, 4, 1937.
[267] Seares (F. H.). *P.A.S.P.*, 52, 1940, p. 80.
[268] Seares (F. H.) and Joyner (M.). *Ap. J.*, 98, 1943, p. 302.
[269] Seaton (J.). *M.N.*, III, 1951, p. 368.
[270] Seeliger (H. von). *Sitz. Ber. d. K. Bayer. Ak. d. Wiss.*, 25, 1909 and 1911, p. 460.
[271] Sekido (Y.), Masuda (T.), Yoshida (S.) and Wada (M.). *Phys. Rev.*, 83, 1951, p. 658.
[272] Seyfert (C. K.). *Ap. J.*, 91, 1940, p. 261.
[273] Seyfert (C. K.). *Ap. J.*, 97, 1943, p. 28.

[274] SHAIN (G. A.) and GAZÉ (V. F.). *Isvestia Obs. Crimea*, 6, 1951, p. 3.

[275] SHAIN (P. F.). *Poulkovo Obs. Circ.*, No. 11, 1936, p. 3.

[276] SHAIN (P. F.). *Bull. Abastumani Ap. Obs.*, 7, 1943, p. 189.

[277] SHAPLEY (H.). *Contr. Mount Wilson Obs.*, No. 116, 1915.

[278] SHAPLEY (H.). *Proc. Nat. Acad. Washington*, 25, 1939, p. 565.

[279] SHAPLEY (H.). *Publ. Obs. Univ. Michigan*, 10, 1951, p. 79.

[280] SHAPLEY (H.) and AMES (A.). *Harvard Obs. Bull.*, No. 864, 1929, p. 1.

[281] SHAPLEY (H.) and WILSON (H. H.). *Harvard Circ.*, 275, 1925.

[282] SHERMAN (F.). *Ap. J.*, 90, 1939, p. 630.

[283] SLOCUM (L. T.). *Lick Obs. Bull.*, 15, 1931, p. 123.

[284] SMITH (C. E.). *Lick. Obs. Bull.*, 18, 1937, p. 39.

[285] SMITH, ELSKE (V. P.) and VAN DE HULST (H. C.). *A. J.*, 56, 1951, p. 141.

[286] SPITZER (L. Jr.). *Ap. J.*, 93, 1941, p. 369; 94, 1942, p. 232.

[287] SPITZER (L. Jr.). *Ap. J.*, 95, 1942, p. 329.

[288] SPITZER (L. Jr.). *Ap. J.*, 107, 1948, p. 6; 109, 1949, p. 337.

[289] SPITZER (L. Jr.). *Harvard Monographs*, No. 7, 1948, p. 86; *Physics Today*, Sept. 1948, p. 7.

[290] SPITZER (L. Jr.). *Ap. J.*, 108, 1948, p. 276.

[291] SPITZER (L. Jr.) and BAADE (W.). *Ap. J.*, 113, 1951, p. 413.

[292] SPITZER (L. Jr.), EPSTEIN (I.) and LI-HEN. *An. Ap.*, 13, 1950, p. 147.

[293] SPITZER (L. Jr.) and GREENSTEIN (J. L.). *Ap. J.*, 114, 1951, p. 407.

[294] SPITZER (L. Jr.) and SAVEDOFF (M. P.). *Ap. J.*, 111, 1950, p. 593.

[295] SPITZER (L. Jr.) and TUKEY (J. W.). *Science*, 109, 1949, p. 461. *Ap. J.*, 114, 1951, p. 187.

[296] STEBBINS (J.) and HUFFER (C. M.). *Publ. Washburn Obs.*, 15, 1934, p. 217.

[297] STEBBINS (J.), HUFFER (C. M.) and WHITFORD (A. E.). *Ap. J.*, 91, 1940, p. 20.

[298] STEBBINS (J.), HUFFER (C. M.) and WHITFORD (A. E.). *Ap. J.*, 90, 1939, p. 459; 92, 1940, p. 193.

[299] STEBBINS (J.), HUFFER (C. M.) and WHITFORD (A. E.), *Ap. J.*, 94, 1941, p. 215.

[300] STEBBINS (J.) and WHITFORD (A. E.). *Ap. J.*, 84, 1936, p. 132.

[301] STEBBINS (J.) and WHITFORD (A. E.). *Ap. J.*, 86, 1937, p. 268.

[302] STEBBINS (J.). and WHITFORD (A. E.) *Ap. J.*, 98, 1943, p. 20.

[303] STEBBINS (J.). and WHITFORD (A. E.) *Ap. J.*, 102, 1945, p. 318.

[304] STEBBINS (J.). and WHITFORD (A. E.) *Ap. J.*, 106, 1947, p. 235.

[305] STEBBINS (J.). and WHITFORD (A. E. *Ap. J.*, 78, 1948, p. 204.

[306] STEBBINS (J.). and WHITFORD (A. E.) *Ap. J.*, 108, 1948, p. 413.

[307] STICKER (R.). *Veröf.* Bonn. Heft 30, 1937.

[308] STODDARD (L. G.). *Ap. J.*, 102, 1945, p. 267.

[309] STRATTON (J. A.) and HOUGHTON (H. G.). *Phys. Rev.*, 38, 1931, p. 159.

[310] STROHMEIER (W.). *Zs. f. Ap.*, 17, 1939, p. 83.

[311] STROHMEIER (W.). *Zs. f. Ap.*, 27, 1950, p. 49.

[312] STOY (R. H.). *M.N.*, 93, 1933, p. 588.

[312a] STOY (R. H.). *P.A.S.P.*, 51, 1938, p. 233.

[313] STRÖMGREN (B.). *Ap. J.*, 89, 1939, p. 526.

[314] STRÖMGREN (B.). *Ap. J.*, 108, 1948, p. 242.

[315] STRÖMGREN (B.). *Särtyck ur Astronomiska Sällskajet Tycho Brahe, Arsbok*, 1949.

[316] STRÖMGREN (B.). *Problems Cosm Aerodyn.*, 1951, p. 7.

[317] STRUVE (O.). *Ap. J.*, 65, 1927, p. 163.

[318] STRUVE (O.). *Ap. J.*, 67, 1928, p. 353.

[319] STRUVE (O.). *M.N.*, 89, 1929, p. 567.

[320] STRUVE (O.). *Ap. J.*, 77, 1933, p. 153.

[321] STRUVE (O.). *Ap. J.*, 86, 1937, p. 614.

[322] STRUVE (O.). *An Ap.*, 1, 1938, p. 143.

[323] STRUVE (O.) and ELVEY (C. T.). *Ap. J.*, 83, 1936, p. 162.

[324] STRUVE (O.) and ELVEY (C. T.). *Ap. J.*, 88, 1938, p. 364; 89, 1939, p. 119 and 517. STRUVE (O.), ELVEY (C. T.) and LINKE (W.). *Ap. J.*, 90, 1939, p. 301.

[325] STRUVE (O.), ELVEY (C. T.) and KEENAN (P. C.). *Ap. J.*, 77, 1933, p. 274.

[326] STRUVE (O.), ELVEY (C. T.) and ROACH (F. E.). *Ap. J.*, 84, 1936, p. 219.

[327] STRUVE (O.), KEENAN (P. C.) and HYNEK (J. A.). *Ap. J.*, 79, 1934, p. 1.

[328] STRUVE (O.), VAN BIESBROEK (G.) and ELVEY (C. T.). *Ap. J.*, 87, 1938, p. 559.

[329] SWINGS (P.). *An. Ap.*, 1, 1938, p. 39.

[330] SWINGS (P.) and DESIRANT (M.). *Ciel et Terre*, 1939, No.5. SWINGS (P.) and ÖHMAN (Y.). *The Observatory*, 62, 1939, p. 150.

[331] SWINGS (P.) and JOSE (P. D.). *P.A.S.P.*, 61, 1949, p. 181.

[332] SWINGS (P.) and ROSENFELD (L.). *Ap. J.*, 86, 1937, p. 483.

[333] TCHENG MAO-LIN and DUFAY (J.). *An. Ap.*, 7, 1944, p. 143.

[334] TER HAAR (D.). *Ap. J.*, 100, 1944, p. 288.

[335] TRUMPLER (R. J.) *Lick. Obs. Bull.*, 14, 1930, p. 154.

[336] TRUMPLER (R. J.). *P.A.S.P.*, 42, 1930, p. 267.

[337] TRUMPLER (R. J.). *Ap. J.*, 91, 1940, p. 186.

[338] TOWNES (C. H.). *Ap. J.*, 105, 1947, p. 235.

[339] *U.R.S.I. Bruits radioelectriques solaires et galactiques. Rapport special*, No. 1, Brussels, 1950.

[340] VANÄS (E.). *Uppsala Astr. Obs. Ann.*, 1, No. 1, 1939; *Medd. Astr. Obs. Uppsala*, No. 81, 1941.
[341] VAN DE HULST (H. C.). *Nederl. Tijdschrift v. Naturkunde*, 11, 1945, p. 201.
[342] VAN DE HULST (H. C.). *Astr. Res. Obs. Utrecht*, 11, part 1, 1946.
[343] VAN DE HULST (H. C.). *Astr. Res. Obs. Utrecht*, 11, part 2, 1949.
[344] VAN DE HULST (H. C.). *The atmospheres of the Earth and Planets*, Chicago, 1948. "Scattering in the atmospheres", p. 49.
[345] VAN DE HULST (H. C.). *Ap. J.*, 112, 1950, p. 1.
[346] VAN DE HULST (H. C.). *Problems Cosm. Aerodyn.*, 1951, p. 45.
[347] VAN DE HULST (H. C.). *A. J.*, 56, 1951, p. 144.
[348] VAN DE KAMP (P.). *A. J.*, 40, 1930, p. 145.
[349] VAN DE KAMP (P.). *A. J.*, 42, 1932, p. 97.
[350] VAN RHIJN (P. J.). *Ap. J.*, 43, 1916, p. 36.
[351] VAN RHIJN (P. J.). *Groningen Publ.*, No. 47, 1936, p. 9.
[352] VAN RHIJN (P. J.). *Groningen Publ.*, No. 51, 1946, p. 23.
[353] VAUCOULEURS (G. de). *C.R.*, 227, 1948, p. 466.
[354] VAUCOULEURS (G. de). *An. Ap.*, 11, 1948, p. 247.
[355] VAUCOULEURS (G. de). *An. Ap.*, 13, 1950, p. 362.
[356] VELGHE (A.). *Ann. Royal Belgian Obs.*, 3rd Series, 5, 1950, No. 3.
[357] VISSOTZKY (A. N.) and WILLIAMS (E. T. R.). *Ap. J.*, 77, 1933, p. 301.
[358] VORONTZOV-VELYAMINOV (B.). *Astr. Nachr.*, 243, 1931, p. 165.
[359] VORONTZOV-VELYAMINOV (B.). *Zs. f. Ap.*, 8, 1934, p. 195; *The Observatory*, 62, 1939, p. 213. *A.J.U.S.S.R.*, 27, 1950, No. 5.
[360] VORONTZOV-VELYAMINOV (B.). *Gaseous Nebulae and novae*, vol. 1, p. 588, 2nd Edition, Moscow and Leningrad, 1948.
[361] VORONTZOV-VELYAMINOV (B.) and PARENAGO (P.). *A.J.U.S.S.R.*, 8, 1932, p. 296.
[362] WALLENQUIST (A. K.). *Medd. astr. Obs. Uppsala*, No. 71, 1937.
[363] WALTER (K.). *Zs. f. Ap.*, 20, 1941, p. 256.
[364] WANG-SHIH-KY. *C.R.*, 201, 1935, p. 1326.
[365] WANG-SHIH-KY. *Publ. Obs. Lyon*, 1, No. 19, 1936.
[366] WARWICK (C.). *Proc. Nat. Acad. Washington*, 36, 1950, p. 415.
[367] WEAVER (H. F.). *Ap. J.*, 110, 1949, p. 190.
[368] WEIZSÄCKER (C. F. von). *Problems Cosm. Aerodyn.*, 1951, p. 200.
[369] WERNBERG (G.). *Uppsala Astr. Obs. Ann.*, 1, No. 4, 1941.
[370] WESTERHOUT (G.) and OORT (J. H.). *B.A.N.*, 11, 1951, p. 323.
[371] WESTGATE (C.). *Ap. J.*, 78, 1933, p. 65.
[372] WHIPPLE (F. L.). *Ap. J.*, 104, 1946, p. 1.
[373] WHIPPLE (F. L.). *Harvard Monographs*, No. 7, 1948, p. 109.
[374] WHIPPLE (F. L.) and GREENSTEIN (J. L.). *Proc. Nat. Acad.* Washington, 23, 1937, p. 177.

[375] Whitford (A. E.). *Ap. J.*, 107, 1948, p. 102.
[376] Whitney (W. T.) and Weston (E. B.). *Ap. J.*, 107, 1948, p. 371.
[377] Wilkens (H.). *Publ. Obs. Astr. La Plata*, 23.
[378] Williams (E. G.). *Ap. J.*, 79, 1934, p. 280.
[379] Williams (E. T. R.). *Ap. J.*, 75, 1932, p. 386.
[380] Wilson (O. C.). *P.A.S.P.*, 58, 1946, p. 210.
[381] Wilson (R. E.). *Ap. J.*, 92, 1940, p. 170.
[382] Wilson (O. C.) and Merrill (P. W.). *Ap. J.*, 86, 1937, p. 44.
[383] Wirtz (C.). *Medd. Lunds Astr. Obs.* Series II, No. 20, 1923.
[384] Wolf (M.). *Astr. Nachr.*, 219, 1923, p. 109.
[385] Wright (W. H.). *Publ. Lick Obs.*, 13, 1918, p. 193.
[386] Wyse (A. B.). *Ap. J.*, 95, 1942, p. 356.
[387] Zanstra (H.). *Publ. Dominion Ap. Obs. Victoria*, 4, 1931, p. 209.
[388] Zanstra (H.). *Zs. f. Ap.*, 2, 1931, p. 1.
[389] Zanstra (H.). *Zs. f. Ap.*, 2, 1931, p. 329; *M.N.*, 93, 1932, p. 131.
[390] Zanstra (H.). *The Observatory*, 59, 1936, p. 314.
[391] Zug (R. S.). *Lick Obs. Bull.*, 16, 1933, p. 119; 18, 1937; p. 89.
[392] Zwicky (F.). *Ap. J.*, 93, 1941, p. 411.

[393] Atanasijevic (I.). *C.R.*, 235, 1952, p. 130.
[394] Baade (W.). *P.A.S.P.*, 58, 1946, p. 249.
[395] Baade (W.). *Communication Congres U.A.I.*, Rome, Sept. 1952.
[396] Binnendijk (L.). *Ap. J.*, 115, 1952, p. 428.
[397] Blaauw (A.). *B.A.N.*, 11, 1952, p. 458.
[398] Bok (B. J.) and Van Wijk (U.). *A. J.*, 75, 1952, p. 213.
[399] Chamberlain (J. W.). *A. J.*, 57, 1952, p. 158.
[400] Colacevich (A.). *Rend. Accad. Nat. dei Lincei*, Series VI, 17, 1933, p. 1065.
[401] Courtès (G.). *L'Astronomie*. 66, 1952, p. 221.
[402] Divan (L.). *An. Ap.*, 15, 1952, p. 237; *C.R.*, 235, 1952, p. 126.
[403] Dufay (J.). *Congres de l'U.A.I.*, Rome, Sept. 1952; *Sky and Telescope*, 12, 1952, p. 41.
[404] Dufay (J.), Bigay (J. H.) and Berthier (P.). *C.R.*, 235, 1952, p. 120; *Vistas in Astronomy*.
[405] Dufay (J.), Bigay (J. H.), Berthier (P.) and Texereau (J.). *C.R.*, 235, 1952, p. 1284.
[406] Gazé (V. F.) and Shain. *Izvestia Obs. Crimea*, 9, 1952, p. 52.
[407] Gourzadian (G.). *A.J.U.S.S.R.*, 29, 1952, p. 111.
[408] Güttler (A.). *Zs. f. Ap.*, 31, 1952, p. 1.
[409] Gyllenberg (W.). *Lund Obs. Medd.*, Series II, 6, No. 52, 1929.
[410] Hoerner (S. von). *Zs. f. Ap.*, 30, 1951, p. 17.
[411] Jentzch (C.) and Unsold (A.). *Zs. f. Ap.*, 25, 194[illegible] [illegible] 37[illegible].

[412] KAHN (F. D.). *M.N.*, 112, 1952, p. 518.
[413] KERR (F. J.) *Sky and Telescope*, 12, 1953, p. 59.
[414] LYTTKENS (E.). *Uppsala Astr. Obs. Ann.*, 2, No. 9, 1951.
[415] MARTEL (M. T.). *C.R.*, 237, 1953, p. 787.
[416] MCCUSKEY (S. W.). *Ap. J.*, 109, 1949, p. 139.
[417] MCCUSKEY (S. W.). *Ap. J.*, 109, 1949, p. 414.
[418] MCCUSKEY (S. W.). *Ap. J.*, 113, 1951, p. 672.
[419] MCCUSKEY (S. W.). *Ap. J.*, 115, 1952, p. 479.
[420] MCCUSKEY (S. W.) and SEYFERT (C. K.). *Ap. J.*, 106, 1947, p. 1.
[421] MELNIKOV (O. A.). *A.J.U.S.S.R.*, 24, 1947, p. 73.
[422] MINKOWSKI (R.). *Ann. Ap.*, 9, 1946, p. 97.
[423] OORT (J. H.). *Commun. to Intern. Astr. U.*, Rome, Sept. 1952.
[424] PARENAGO (P. P.). *Adv. in Astr. Sci.*, 4, 1948, Moscow.
[425] RAMSEY (W. H.). *Private Communication.*
[426] SCHALEN (C.). *Uppsala Astr. Obs. Ann.*, 3, No. 5, 1952.
[427] SCHOENBERG (E.). *Zs. f. Naturforschung*, 4, No. 3, 1949.
[428] SCHOENBERG (E.) and TOLMAR (Gy.). *Abhand d. Bayer. Akad. d. Wis.*, 56, 1949.
[429] SHAIN (G. A.) and GAZÉ (V. F.). *Izvestia Obs. Crimea*, 8, 1952, p. 80.
[430] SHAIN (G. A.) and GAZÉ (V. F.) *Symposium on the Evolution of the Stars. Congress of the U.A.I.*, Rome, September 1952.
[431] SHAIN (G. A.). and GAZÉ (V. F). *Atlas of Diffuse Galactic Nebulae*, published by the Ac. Sc. of the U.S.S.R. Observatory of the Crimea, 1952.
[432] SHARPLESS (S.) and OSTERBROK (D.). *Ap. J.*, 115, 1952, p. 89.
[433] *Sky and Telescope*, 11, 1952, p. 138.
[434] STRUVE (O.). *Sky and Telescope*, 10, 1951, p. 215.
[435] SWINGS (P.) and SWENSSON (J. W.). *An. Ap.*, 15, 1952, p. 290.
[436] TEXEREAU (J.). *L'Astronomie*, 66, 1952, p. 349.
[437] WALRAVEN (Th.) and FOKKER (A. D.). *B.A.N.*, 11, 1952, p. 441.
[438] WESTON (E. B.). *A. J.*, 57, 1952, p. 28.
[439] ZWICKY (F.). *Communication Congres U.A.I.*, Rome, Sept. 1952.
[440] ZWICKY (F.). *P.A.S.P.*, 64, 1952, p. 242.

Supplementary list of recent books, etc.

[441] ANDRILLAT (H.). *Journ. Phys.*, 14, 1953, p. 105; *C.R.*, 238, 1954, p. 1781; *Thesis*, Paris, 1954.
[442] GARSTANG (R. H.). *M.N.*, 111, 1951, p. 115; *Ap. J.*, 115, 1952, p. 506.
[443] LILLER (W.) and ALLER (L. H.). *Ap. J.*, 120, 1954, p. 48.
[444] SEATON (M. J.). *Phil. Trans., A.*, 245, 1953, p. 469; *Proc. Royal Soc., A*, 218, 1953, p. 400.

[445] SEATON (M. J.). *M.N.*, 114, 1954, p. 154.
[446] ALLER (L. H.). *Ap. J.*, 120, 1954, p. 401.
[447] ALLER (L. H.). *The Atmosphere of the Sun and Stars*, Ronald Press Co., N.Y., 1953, p. 327.
[448] BARBIER (D.) and ANDRILLAT (H.). *C.R.*, 238, 1954, p. 1099.
[449] MINKOVSKI (R.). *P.A.S.P.*, 65, 1953, p. 161.
[450] SEATON (M. J.). *An. Ap.*, 17, 1954, p. 196.
[451] FESSENKOV (G. A.) and ROJKOVSKY (D. A.). *Atlas of Gaseous Nebulae*, published by the Academy of Science of Kazakistan, 1953.
[452] DUFAY (J.). *L'Astronomie*, 68, 1954, p. 361.
[453] DUFAY (M.). *Journ. Phys.*, 15, 1954, p. 50.
[454] *U.R.S.I. Les sources discrètes d'emission radioélectrique extra-terrestre. Rapport spécial*, No. 3. Brussels, 1954.
[455] VAN DE HULST (H. C.), MÜLLER (C. A.) and OORT (J. H.), *B.A.N.* 12, 1954, p. 117.
[456] KWEE (K. K.), MÜLLER (C. A.) and WESTERHOUT (G.). *B.A.N.* 12, 1954, p. 211.
[457] MÜNCH (G.). *P.A.S.P.*, 65, 1953, p. 179.
[458] BURBIDGE (E. M.) and BURBIDGE (G. R.). *P.A.S.P.*, 65, 1953, p. 292.
[459] VAN RHIJN (P. J.). *Gronigen Publ.* No. 53, 1949.
[460] DUFAY (J.), BERTHIER (P.) and MORIGNAT (B.). *C.R.*, 239, 1954, p. 478.
[461] HILTNER (W. A.). *Ap. J.*, 120, 1954, p. 454.
[462] CAYREL and SCHATZMAN (E.). *Colloque d'Astrophysique de Liége* (1954).

INDEX OF NAMES

INDEX OF SUBJECTS

SOME DOVER SCIENCE BOOKS

CHANCE, LUCK AND STATISTICS: THE SCIENCE OF CHANCE,
Horace C. Levinson

Theory of probability and science of statistics in simple, non-technical language. Part I deals with theory of probability, covering odd superstitions in regard to "luck," the meaning of betting odds, the law of mathematical expectation, gambling, and applications in poker, roulette, lotteries, dice, bridge, and other games of chance. Part II discusses the misuse of statistics, the concept of statistical probabilities, normal and skew frequency distributions, and statistics applied to various fields—birth rates, stock speculation, insurance rates, advertising, etc. "Presented in an easy humorous style which I consider the best kind of expository writing," Prof. A. C. Cohen, Industry Quality Control. Enlarged revised edition. Formerly titled *The Science of Chance.* Preface and two new appendices by the author. Index. xiv + 365pp. 5⅜ x 8. Paperbound $2.00

BASIC ELECTRONICS,
prepared by the U.S. Navy Training Publications Center

A thorough and comprehensive manual on the fundamentals of electronics. Written clearly, it is equally useful for self-study or course work for those with a knowledge of the principles of basic electricity. Partial contents: Operating Principles of the Electron Tube; Introduction to Transistors; Power Supplies for Electronic Equipment; Tuned Circuits; Electron-Tube Amplifiers; Audio Power Amplifiers; Oscillators; Transmitters; Transmission Lines; Antennas and Propagation; Introduction to Computers; and related topics. Appendix. Index. Hundreds of illustrations and diagrams. vi + 471pp. 6½ x 9¼.
Paperbound $2.75

BASIC THEORY AND APPLICATION OF TRANSISTORS,
prepared by the U.S. Department of the Army

An introductory manual prepared for an army training program. One of the finest available surveys of theory and application of transistor design and operation. Minimal knowledge of physics and theory of electron tubes required. Suitable for textbook use, course supplement, or home study. Chapters: Introduction; fundamental theory of transistors; transistor amplifier fundamentals; parameters, equivalent circuits, and characteristic curves; bias stabilization; transistor analysis and comparison using characteristic curves and charts; audio amplifiers; tuned amplifiers; wide-band amplifiers; oscillators; pulse and switching circuits; modulation, mixing, and demodulation; and additional semiconductor devices. Unabridged, corrected edition. 240 schematic drawings, photographs, wiring diagrams, etc. 2 Appendices. Glossary. Index. 263pp. 6½ x 9¼. Paperbound $1.25

GUIDE TO THE LITERATURE OF MATHEMATICS AND PHYSICS,
N. G. Parke III

Over 5000 entries included under approximately 120 major subject headings of selected most important books, monographs, periodicals, articles in English, plus important works in German, French, Italian, Spanish, Russian (many recently available works). Covers every branch of physics, math, related engineering. Includes author, title, edition, publisher, place, date, number of volumes, number of pages. A 40-page introduction on the basic problems of research and study provides useful information on the organization and use of libraries, the psychology of learning, etc. This reference work will save you hours of time. 2nd revised edition. Indices of authors, subjects, 464pp. 5⅜ x 8.
Paperbound $2.75

THE RISE OF THE NEW PHYSICS (formerly THE DECLINE OF MECHANISM), *A. d'Abro*
This authoritative and comprehensive 2-volume exposition is unique in scientific publishing. Written for intelligent readers not familiar with higher mathematics, it is the only thorough explanation in non-technical language of modern mathematical-physical theory. Combining both history and exposition, it ranges from classical Newtonian concepts up through the electronic theories of Dirac and Heisenberg, the statistical mechanics of Fermi, and Einstein's relativity theories. "A must for anyone doing serious study in the physical sciences," *J. of Franklin Inst.* 97 illustrations. 991pp. 2 volumes.
Vol. 1 Paperbound $2.25, Vol. 2 Paperbound $2.25, The set $4.50

THE STRANGE STORY OF THE QUANTUM, AN ACCOUNT FOR THE GENERAL READER OF THE GROWTH OF IDEAS UNDERLYING OUR PRESENT ATOMIC KNOWLEDGE, *B. Hoffmann*
Presents lucidly and expertly, with barest amount of mathematics, the problems and theories which led to modern quantum physics. Dr. Hoffmann begins with the closing years of the 19th century, when certain trifling discrepancies were noticed, and with illuminating analogies and examples takes you through the brilliant concepts of Planck, Einstein, Pauli, de Broglie, Bohr, Schroedinger, Heisenberg, Dirac, Sommerfeld, Feynman, etc. This edition includes a new, long postscript carrying the story through 1958. "Of the books attempting an account of the history and contents of our modern atomic physics which have come to my attention, this is the best," H. Margenau, Yale University, in *American Journal of Physics.* 32 tables and line illustrations. Index. 275pp. 5⅜ x 8.
Paperbound $1.75

GREAT IDEAS AND THEORIES OF MODERN COSMOLOGY, *Jagjit Singh*
The theories of Jeans, Eddington, Milne, Kant, Bondi, Gold, Newton, Einstein, Gamow, Hoyle, Dirac, Kuiper, Hubble, Weizsäcker and many others on such cosmological questions as the origin of the universe, space and time, planet formation, "continuous creation," the birth, life, and death of the stars, the origin of the galaxies, etc. By the author of the popular *Great Ideas of Modern Mathematics.* A gifted popularizer of science, he makes the most difficult abstractions crystal-clear even to the most non-mathematical reader. Index. xii + 276pp. 5⅜ x 8½.
Paperbound $2.00

GREAT IDEAS OF MODERN MATHEMATICS: THEIR NATURE AND USE, *Jagjit Singh*
Reader with only high school math will understand main mathematical ideas of modern physics, astronomy, genetics, psychology, evolution, etc., better than many who use them as tools, but comprehend little of their basic structure. Author uses his wide knowledge of non-mathematical fields in brilliant exposition of differential equations, matrices, group theory, logic, statistics, problems of mathematical foundations, imaginary numbers, vectors, etc. Original publications, 2 appendices. 2 indexes. 65 illustr. 322pp. 5⅜ x 8. Paperbound $2.00

THE MATHEMATICS OF GREAT AMATEURS, *Julian L. Coolidge*
Great discoveries made by poets, theologians, philosophers, artists and other non-mathematicians: Omar Khayyam, Leonardo da Vinci, Albrecht Dürer, John Napier, Pascal, Diderot, Bolzano, etc. Surprising accounts of what can result from a non-professional preoccupation with the oldest of sciences. 56 figures. viii + 211pp. 5⅜ x 8½.
Paperbound $1.50

FIVE VOLUME "THEORY OF FUNCTIONS" SET BY KONRAD KNOPP

This five-volume set, prepared by Konrad Knopp, provides a complete and readily followed account of theory of functions. Proofs are given concisely, yet without sacrifice of completeness or rigor. These volumes are used as texts by such universities as M.I.T., University of Chicago, N. Y. City College, and many others. "Excellent introduction . . . remarkably readable, concise, clear, rigorous," *Journal of the American Statistical Association.*

ELEMENTS OF THE THEORY OF FUNCTIONS,
Konrad Knopp
This book provides the student with background for further volumes in this set, or texts on a similar level. Partial contents: foundations, system of complex numbers and the Gaussian plane of numbers, Riemann sphere of numbers, mapping by linear functions, normal forms, the logarithm, the cyclometric functions and binomial series. "Not only for the young student, but also for the student who knows all about what is in it," *Mathematical Journal.* Bibliography. Index. 140pp. 5⅜ x 8. Paperbound $1.50

THEORY OF FUNCTIONS, PART I,
Konrad Knopp
With volume II, this book provides coverage of basic concepts and theorems. Partial contents: numbers and points, functions of a complex variable, integral of a continuous function, Cauchy's integral theorem, Cauchy's integral formulae, series with variable terms, expansion of analytic functions in power series, analytic continuation and complete definition of analytic functions, entire transcendental functions, Laurent expansion, types of singularities. Bibliography. Index. vii + 146pp. 5⅜ x 8. Paperbound $1.35

THEORY OF FUNCTIONS, PART II,
Konrad Knopp
Application and further development of general theory, special topics. Single valued functions. Entire, Weierstrass, Meromorphic functions. Riemann surfaces. Algebraic functions. Analytical configuration, Riemann surface. Bibliography. Index. x + 150pp. 5⅜ x 8. Paperbound $1.35

PROBLEM BOOK IN THE THEORY OF FUNCTIONS, VOLUME 1.
Konrad Knopp
Problems in elementary theory, for use with Knopp's *Theory of Functions,* or any other text, arranged according to increasing difficulty. Fundamental concepts, sequences of numbers and infinite series, complex variable, integral theorems, development in series, conformal mapping. 182 problems. Answers. viii + 126pp. 5⅜ x 8. Paperbound $1.35

PROBLEM BOOK IN THE THEORY OF FUNCTIONS, VOLUME 2,
Konrad Knopp
Advanced theory of functions, to be used either with Knopp's *Theory of Functions,* or any other comparable text. Singularities, entire & meromorphic functions, periodic, analytic, continuation, multiple-valued functions, Riemann surfaces, conformal mapping. Includes a section of additional elementary problems. "The difficult task of selecting from the immense material of the modern theory of functions the problems just within the reach of the beginner is here masterfully accomplished," *Am. Math. Soc.* Answers. 138pp. 5⅜ x 8. Paperbound $1.50

A HISTORY OF PHYSICS: IN ITS ELEMENTARY BRANCHES (THROUGH 1925), INCLUDING THE EVOLUTION OF PHYSICAL LABORATORIES,
Florian Cajori
Revised and enlarged edition. The only first-rate brief history of physics. Still the best entry for a student or teacher into the antecedents of modern theories of physics. A clear, non-mathematical, handy reference work which traces in critical fashion the developments of ideas, theories, techniques, and apparatus from the Greeks to the 1920's. Within each period he analyzes the basic topics of mechanics, light, electricity and magnetism, sound, atomic theory and structure of matter, radioactivity, etc. A chapter on modern research: Curie, Kelvin, Planck's quantum theory, thermodynamics, Fitzgerald and Lorentz, special and general relativity, J. J. Thomson's model of an atom, Bohr's discoveries and later results, wave mechanics, and many other matters. Much bibliographic detail in footnotes. Index. 16 figures. xv + 424pp. 5⅜ x 8.
Paperbound $2.50

OPTICKS, *Sir Isaac Newton*
In its discussions of light, reflection, color, refraction, the wave theory and the corpuscular theory of light, this work is packed with scores of insights and discoveries. In its precise and practical discussion of construction of optical apparatus, contemporary understandings of phenomena it is truly fascinating to modern physicists, astronomers, mathematicians. Foreword by Albert Einstein. Preface by I. B. Cohen of Harvard University. 7 pages of portraits, facsimile pages, letters, etc. cxvi + 414pp. 5⅜ x 8. Paperbound $2.25

DIALOGUES CONCERNING TWO NEW SCIENCES,
Galileo Galilei
This classic of experimental science, mechanics, engineering, is as enjoyable as it is important. A great historical document giving insights into one of the world's most original thinkers, it is based on 30 years' experimentation. It offers a lively exposition of dynamics, elasticity, sound, ballistics, strength of materials, the scientific method. "Superior to everything else of mine," Galileo. Trans. by H. Crew, A. Salvio. 126 diagrams. Index xxi + 288pp. 5⅜ x 8.
Paperbound $1.75

TREATISE ON ELECTRICITY AND MAGNETISM,
James Clerk Maxwell
For more than 80 years a seemingly inexhaustible source of leads for physicists, mathematicians, engineers. Total of 1082pp. on such topics as Measurement of Quantities, Electrostatics, Elementary Mathematical Theory of Electricity, Electrical Work and Energy in a System of Conductors, General Theorems, Theory of Electrical Images, Electrolysis, Conduction, Polarization, Dielectrics, Resistance, etc. "The greatest mathematical physicist since Newton," Sir James Jeans. 3rd edition. 107 figures, 21 plates, 1082pp. 5⅜ x 8.
Vol. 1 Paperbound $2.50, Vol. 2 Paperbound $2.50
The set $5.00

Prices subject to change without notice.

Available at your book dealer or write for free catalogue to Dept. Adsci, Dover Publications, Inc., 180 Varick St., N.Y., N.Y. 10014. Dover publishes more than 150 books each year on science, elementary and advanced mathematics, biology, music, art, literary history, social sciences and other areas.